MATHEMATICAL FINANCE

MATHEMATICAL FINANCE

M. J. Alhabeeb

Professor of Economics and Finance
Isenberg School of Management
University of Massachusetts Amherst
Amherst, Massachusetts

WILEY

A JOHN WILEY & SONS, INC., PUBLICATION

Published by John Wiley & Sons, Inc., Hoboken, New Jersey
Published simultaneously in Canada

For general information on our other products and services or for technical support, please contact our Customer Care Department within the United States at (800) 762-2974, outside the United States at (317) 572-3993 or fax (317) 572-4002.

Wiley also publishes its books in a variety of electronic formats. Some content that appears in print may not be available in electronic formats. For more information about Wiley products, visit our web site at www.wiley.com.

Library of Congress Cataloging-in-Publication Data:

Alhabeeb, M. J., 1954–
    Mathematical finance / M. J. Alhabeeb.
        p. cm.
    Includes bibliographical references and index.
    ISBN 978-0-470-64184-2 (cloth)
    1. Finance–Mathematical models. 2. Investments–Mathematics. 3. Business mathematics. I. Title.
    HG106.A44 2011
    332.01′5195–dc22

                                                                    2011007717

Printed in the United States of America

10 9 8 7 6 5 4 3 2 1

Mathematics is a wonderful, mad subject, full of imagination, fantasy, and creativity that is not limited by the petty details of the physical world, but only by the strength of our inner light.

Gregory Chaitin, *"Less Proof, More Truth,"* New Scientist, July 28, 2007

* * *

Money is a terrible master, but an excellent servant.

P. T. Barnum

The best investment I have ever made has been the investment in my children, which has rewarded me with the highest perpetual rate of return.

I dedicate this book to my children, MJ Junior and Reema, and to the memory of my parents, who invested in me through their toil and tears.

CONTENTS

PREFACE

Teaching college finance for decades made me see firsthand how most of our students face math difficulties and how they view their lack of fluency in the quantitative analysis as a big obstacle in their learning journey. Their common struggle often manifests itself as a hindering factor in their academic progress, and a gap that may become more difficult to fill later as they advance in their professional careers. My students have often expressed their discouragement and frustration at standard mathematical textbooks, which usually focus on technical math and treat finance and financial subjects on the side. I wanted this book to reverse that approach and place the focus on finance and financial problems—but in their computational sense and mathematical language. In other words, I wanted to redirect the emphasis toward problem solving of major financial issues, using mathematical methods as tools. I am hoping that in doing so, I will satisfy one of our students' major learning needs, help facilitate their academic march to higher levels, and equip them with the fundamental skills that will continue to be helpful beyond graduation. The book is written as a direct reflection of a long classroom experience in which the accumulation of knowledge of the material has been enhanced by the diversity of students' learning styles and their various educational requirements. Specifically and deliberately, my primary attention has been directed to the role of mathematical formulas to illustrate and clarify the underlying fundamental mathematical logic of problem solving. It is to provide students with an additional opportunity to solidify their understanding of financial problems and to be able to analyze and interpret the solutions. Secondary to the mathematical formulas are the traditional table values, which are also utilized in this text.

As calculators and computers have become more sophisticated and widely and readily available to every student, the need to take a step back and reem-phasize the utilization of the traditional mathematical methods seemed evident and real. It is essentially the need to help strengthen the basic comprehension of how theories work, which is an essential intellectual aspect of our learning that should not be lost to advances in computers, no matter how important the latter are in our lives. For this reason, the traditional approach in this book is meant to balance out the prevalent use of increasingly sophisticated digital pro-grams and methods of problem solving. Included in these technological advances are the standard financial calculators that every student can have. Despite their tremendous help from the students' perspective, these calculators have certainly reduced the systematic process of learning the basic methods of solving complex problems. Using advanced calculators has increasingly meant learning how to punch the right keys without the need to know the underlying problem structure, or to comprehend the scientific logic behind the calculated operations. To this end, the book intentionally skips the use of financial calculators and computers to cut down on dependency on such methods as easy but blind ways of problem solving, and to shift back to the use of formulas and tables. However, unlike typical books in mathematics of finance, there is less emphasis here on derivations and proofs, although a few are included for their strong relevance in certain situations.

This book is intended primarily for upper-level undergraduate and first-year graduate students in business and public administration as well as in economics and related majors. It would be very helpful for those who have to prepare for professional examinations in actuarial programs as well as in the CFP, ChFC, CPA, CLU, PFS, and AFC professional tests. It can also serve as an invaluable reference for researchers and market analysts everywhere. The material is essential to a general understanding of the field of finance and all the related financial and business areas.

It has been felt over the years that neither the sheer volume of subjects and issues covered in a one-semester finance course, nor the nature of existing texts would normally allow an instructor to exercise the desired rigor beyond general conceptual exposition and basic analysis. Basically, there is no or little chance to go deep into the mathematical treatment of the issues and details of problem solving in every finance topic. This book offers the opportunity to be rigorous enough either in a separate course dedicated to a mathematical approach to finance, or in selected areas within existing regular finance courses, in order to utilize the most profound computational applications in finance. Furthermore, and to the best of my knowledge, this book is the first to address mathematical finance in its inclusive sense, covering major topics in corporate finance, entrepreneurial and managerial finance, and personal finance. It is a culmination of teaching many finance courses over more than two decades. The book includes adequate, thorough, and balanced material covering the entire range of financial applications. It contains numerous solved examples, all of which are word problems constructed from real-life issues to emphasize the applicational nature of the material, as opposed to lingering on the technical details of finding and proving that are typical in mathematical approaches. Because of the theoretical nature of the material, the language and approach can very well be universal. I strove to write it in language that is both direct and simple and to present it in an easy and friendly manner to make the mathematical material less threatening and more reassuring, for smooth handling of the major mathematical issues at hand. The exposition method chosen lays out the fundamental theoretical concepts first, followed by step-by-step solved examples to transmit the conceptual structure into a quantitative format. Because of my belief that deep knowledge and solid skills in the mathematical applications cannot be gained only by reading, and without working out problems and practicing actual detailed solutions, each unit ends with a summary of the concepts, a list of formulas, and a good number of exercises.

The material is divided into eight units. Unit I is a mathematical introduction to refresh readers' memory of the most fundamental mathematical concepts relevant to the financial topics discussed throughout the remainder of the book. This unit includes three chapters, covering numbers, exponents, and logarithms; mathematical progressions; and statistical measures. Unit II covers the mathematics of the time value of money, discussed in four chapters: simple interest; bank discount; compound interest; and annuities. Unit III is on the mathematics of debt and leasing, that is addressed in three chapters: credit and loans; mortgage debt; and leasing. Unit IV covers the mathematics of capital budgeting and depreciation in two chapters: capital budgeting and depreciation and depletion. Unit V includes two chapters: break-even analysis and leverage. Unit VI is on the mathematics of investment, discussed in five chapters: stocks; bonds; mutual funds; options; and cost of capital and ratio analysis. Unit VII includes two chapters: measuring return and risk and the capital asset pricing model. The final unit, on the mathematics of insurance, includes three chapters: life annuities; life insurance; and property and casualty insurance. The appendix includes mathematical tables that are needed to work some of the examples and exercises.

As I come to the end of work on this book, it is my pleasure to acknowledge the assistance of many people. These friends contributed constructively in bringing this book to existence, although I remain solely responsible for any flaw or shortcoming that may remain. My heartfelt appreciation goes first to my students in several universities over the years. Their quest for learning, questions, concerns, struggles, and worries, their corrections of my slip-ups in the classroom, and even their deadly mistakes on exams were the inspiration without which this book could not have been written. My many thanks go to my colleague and friend Professor Joe Moffitt for his moral support; and to my friend Sev Yates for her sincere support, and continuous encouragement. Special thanks go to my Wiley editor, Susanne Steitz-Filler, who oversaw this project from beginning to end. She did her job in a highly professional manner and with remarkable capability. Many thanks go to Jackie Palmieri and Rosalyn Farkas also at Wiley, and to Romaine Heldt Project Manager at Laserwords, for their deligent job and dedication. I am also indebted to the professional assistance of Peg Cialek, my department's senior secretary, who tirelessly typed the entire manuscript with competence, patience, and sharp eyes for mathematical notation. My special thanks are due to my former graduate assistant, Don Hedeman, for his beautiful job in rendering all the diagrams and graphs. Special thanks also go to my former undergraduate student Heather Sullivan, who is now a proud senior financial analyst in Boston. Heather gave me her class notes which were much more neat and organized than my own, and served as a blueprint for the manuscript of this book. My deep appreciations are for my friend, the highly talented artist Anna Kubaszewska for digitally painting the beautiful image on the cover, according to my fussy requirements. Last but not least, I thank all the anonymous reviewers of the manuscript for their helpful and constructive comments and suggestions.

M. J. ALHABEEB

Belchertown, Massachusetts

UNIT I

Mathematical Introduction

1 Numbers, Exponents, and Logarithms

1.1. NUMBERS

It is essential to understand numbers and their fractions and decimals in finance. Interest rates, time, and financial ratios are commonly expressed by fractions and decimals, and that is where most of the common mistakes are made, especially when it comes to working out computational problems. The common numbers we normally see and use, such as 3, −5, and 0, are called **whole numbers** or **integers** (see Figure 1.1). Along with their partials or fractions, they form what are called **rational numbers**. Conversely, **irrational numbers** are those numbers that cannot terminate or repeat, such as $\sqrt{3}$ and e. Both rational and irrational numbers are called **real numbers** and stand opposite to unreal or **imaginary numbers**, such as the square root of a negative 1 ($\sqrt{-1}$) and its multiples, such as $4\sqrt{-1}$.

1.2. FRACTIONS

A **fraction** is a part of a whole, expressed as a numerator (the number on top of a division line) divided by a denominator (the number on the bottom). A fraction such as $\frac{3}{8}$ represents, for example, three slices of a pizza that has been cut into eight slices. Consequently, there is no fraction with a zero denominator, since we cannot take parts out of nothing. Also, a fraction with a zero numerator makes no sense. However, in a common fraction the numerator is smaller than the denominator. Such a fraction is called a **proper fraction**, such as $\frac{2}{5}$ or $\frac{13}{20}$. When the numerator is larger than the denominator, the fraction would be called an **improper fraction**, such as $\frac{10}{8}$ or $\frac{5}{4}$, referring to a number of parts that can make up more than the whole, as in 10 slices, which would make up a pizza and a quarter: $\frac{10}{8} = \frac{8}{8} + \frac{2}{8} = 1\frac{1}{4}$. The form $1\frac{1}{4}$ is called a **mixed number**, as it is a combination of a whole number (1) and a fraction ($\frac{1}{4}$). If the numerator and denominator are equal, they would form a whole 1, and that is not a fraction. If either the numerator or denominator of a fraction or both contain other fractions,

Mathematical Finance, First Edition. M. J. Alhabeeb.
© 2012 John Wiley & Sons, Inc. Published 2012 by John Wiley & Sons, Inc.

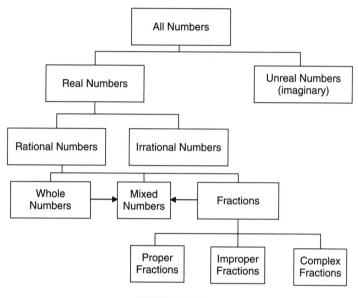

FIGURE 1.1

the whole fraction would be called a **complex fraction**, such as

$$\frac{\frac{5}{8}}{9}, \quad \frac{2}{\frac{3}{7}}, \quad \frac{\frac{1}{3}}{\frac{5}{8}}$$

One of the most significant characteristics of fractions is that the value of a fraction would not change if we multiply or divide both terms by the same number to produce higher- or lower-term fractions.

Example 1.2.1   Convert $\frac{2}{3}$ into a higher-term fraction and $\frac{9}{15}$ into a lower-term fraction.

We can augment the first fraction into a higher-term fraction by multiplying both the numerator and denominator by 5:

$$\frac{2}{3} = \frac{2 \times 5}{3 \times 5} = \frac{10}{15}$$

and we can reduce the second fraction into a lower-term fraction by dividing both the numerator and denominator by 3:

$$\frac{9}{15} = \frac{9 \div 3}{15 \div 3} = \frac{3}{5}$$

A combination of a whole number and a fraction (a mixed number), can also be written as an improper fraction.

Example 1.2.2   Convert the mixed number $2\frac{5}{6}$ into a fraction.

$$2\frac{5}{6} = \frac{(6 \times 2) + 5}{6} = \frac{17}{6}$$

Example 1.2.3   Convert the fraction $\frac{14}{3}$ into a mixed number.

$$\frac{14}{3} = 14 \div 3 = 4 + \frac{2}{3} = 4\frac{2}{3}$$

If two or more fractions have the same denominator, they can be added or subtracted by adding or subtracting their numerators only.

$$\frac{7}{11} - \frac{5}{11} = \frac{2}{11}$$

$$\frac{9}{5} + \frac{3}{5} = \frac{12}{5}$$

But if they have different denominators, a common denominator must be found.

$$\frac{3}{5} + 2\frac{2}{3} = \frac{3}{5} + \frac{8}{3} = \frac{9 + 40}{15} = \frac{49}{15} = 3\frac{4}{15}$$

$$\frac{5}{4} - \frac{1}{3} = \frac{15 - 4}{12} = \frac{11}{12}$$

When multiplying fractions, both numerators and denominators are multiplied:

$$\frac{2}{7} \times \frac{3}{5} = \frac{6}{35}$$

and when dividing fractions, the process is turned into multiplication, the **dividend** (the first fraction) being multiplied by the reciprocal (reverse) of the **divisor** (the second fraction):

$$\frac{5}{8} \div \frac{2}{3} = \frac{5}{8} \times \frac{3}{2} = \frac{15}{16}$$

If the fractions have the same denominators, only the numerators are divided; the denominators cancel each other.

$$\frac{6}{7} \div \frac{4}{7} = \frac{6}{4} = \frac{3}{2}$$

1.3. DECIMALS

A **decimal** is a quotient resulting from dividing the numerator of a fraction by its denominator. It is therefore a fraction expressed by the use of a decimal point.

The number of figures to the right of the decimal point is called the number of **decimal places**.

$$\frac{5}{21} = 5 \div 21 = .2381$$

In this case there are many figures to the right of the decimal point, but it is rounded to only four decimal places. We can also convert a decimal back to a common fraction:

$$.125 = \frac{125}{1,000}$$

and we can reduce it by dividing both terms by 125:

$$\frac{125 \div 125}{1,000 \div 125} = \frac{1}{8}$$

Another example:

$$3.4 = 3\frac{4}{10} = 3\frac{2}{5} = \frac{17}{5}$$

One of the most common mistakes made by students occurs when expressing interest rates in decimal format.

Example 1.3.1   What is the monthly rate of $8\frac{1}{4}\%$ annual interest?

$$8\frac{1}{4}\% = .0825$$
$$.0825 \div 12 = .006875$$

Example 1.3.2   What is the weekly rate of $7\frac{1}{3}\%$ annual interest?

$$7\frac{1}{3}\% = .0733$$
$$.0733 \div 52 = .00141$$

1.4. REPETENDS

A **repetend** is a *nonterminate* decimal in which a certain figure is repeated indefinitely after the decimal point. For example, the decimal form of the fraction $\frac{2}{3}$ is a repetend because dividing 2 by 3 results in a .6 followed by nonending decimal places of 6's.

$$\frac{2}{3} = .666666\ldots$$

1.5. PERCENTAGES

A **percentage** is a fraction whose denominator is 100. Therefore, all percentages express all the hundredths possible, including multiples of hundredths that exceed the 100th part, as in 250%. A percentage is expressed by a decimal that has two decimal places, such as .86, or by using a percent sign, as in 86%.

$$34\% = \frac{34}{100} = .34$$

$$9\tfrac{1}{4}\% = \frac{9\tfrac{1}{4}}{100} = \frac{\tfrac{37}{4}}{100} = \frac{37}{4} \times \frac{1}{100} = \frac{37}{400} = .0925$$

It is crucial in finance to be able to convert an interest rate from a common fraction to a decimal percentage.

In adding and subtracting decimals, the figures should be arranged vertically in columns and the decimal points should be lined up.

$$
\begin{array}{r}
.015 \\
+.00167 \\
\hline
.01667
\end{array}
\qquad
\begin{array}{r}
9.398 \\
-2.1 \\
\hline
7.298
\end{array}
$$

In multiplying decimals, the figures should be multiplied first without their decimal points. Then the combined decimal places of the multiplied figures should be applied to the answer.

$$.52 \times .0039 \times .117 = 52 \times 39 \times 117$$
$$= 237276$$
$$= .000237276$$

Since we have nine combined decimal places, the decimal point of the answer should be placed three places to the left of the answer; that is, there are nine places to the right of the decimal point.

In dividing decimals, we utilize the following procedure:

1. Move the decimal point to the right end of the divisor.
2. Move the decimal point of the dividend to the right by the same number of decimal places as in the divisor.
3. Place the decimal point of the quotient in the same position as in the dividend.
4. Divide the changed figures.

$$14.976 \div 2.4$$

2.4 (the divisor) would be 24

and 14.976 (the dividend) would be 149.76

$$\begin{array}{r} 6.24 \\ 24\overline{)149.76} \\ \underline{144} \\ 57 \\ \underline{48} \\ 96 \\ \underline{96} \\ 0 \end{array}$$

1.6. BASE AMOUNT, PERCENTAGE RATE, AND PERCENTAGE AMOUNT

The use of percentages is one of the most common applications in finance. To better understand percentages we note three variables:

- The base (B), which is the entire amount.
- The percentage amount (P), which is the partial amount resulting from applying the percentage.
- The percentage rate (R) to the base.

$$P = B \cdot R$$

Example 1.6.1   If a person pays a 28% tax rate on his extra income of $12,850, he would pay $3,598 in taxes.
$B = 12,850$, $R = .28$, and P is the amount of taxes, which would be

$$P = 12,850(.28) = 3,598$$

We can also obtain B and R in terms of the other variables:

$$B = \frac{P}{R}$$

Example 1.6.2   If someone paid $12 as a 15% tips in a restaurant, how much did the dinner cost?
The restaurant bill would be B, the tips are P, and the rate (R) is 15%.

$$B = \frac{P}{R}$$
$$= \frac{12}{.15} = \$80$$

The percentage rate (R) can also be obtained by

$$R = \frac{P}{B}$$

Example 1.6.3   If a person has earned $196.90 interest on saved amount of $3,580, what would the interest rate on saving have been?

$$R = \frac{P}{B}$$

$$= \frac{196.90}{3,580}$$

$$= .055 = 5.5\%$$

1.7. RATIOS

A **ratio** is a form of relative comparison between two values. Mathematically, a ratio is expressed as a quotient of two numbers, where the top number is called the **first term** and the bottom is called the **second term**. For example, if the measurements of a room are 36 feet long and 12 feet wide, the ratio of the length to the width would be $\frac{36}{12} = \frac{3}{1}$ (or $3:1$), which represents the fact that the length is three times the width of the room, thus establishing the relative comparison between the two dimensions. This relation can also be expressed in reverse, to show how the width relates to the length, where $\frac{12}{36} = \frac{1}{3}$. We can say that the width of the room is one-third of its length. Ratios have many applications in finance, especially in financial ratios, investment, discount, interest, taxes, and insurance premiums. The ratio of A to B is A/B or $A : B$. It is possible that the ratio is expressed only by A, which means that the value of B has to be 1, as in the example of room measurements above. When the length ratio is 3 and the width is 1, we know that the relationship of the length to the width is 3 times, or $3 : 1$. It is also possible that the ratio is expressed as a sequence between more than two variables, such as $A : B : C : D$, which represents sequential ratios of A to B, B to C, and C to D.

Example 1.7.1   A building has the following measurements:

Length (L): 600 feet

Width (W): 200 feet

Height (H): 100 feet

Perimeter (P): 1,600 feet

Then a sequential ratio of $L : W : H : P$ can be obtained as $3 : 2 : \frac{1}{16}$ or $3 : 2 : .0625$.

$$\frac{L}{W} = \frac{600}{200} = 3$$

$$\frac{W}{H} = \frac{200}{100} = 2$$

$$\frac{H}{P} = \frac{100}{1,600} = \frac{1}{16} = .0625$$

1.8. PROPORTIONS

A **proportion** defines the equality between two ratios, such as $A : B = C : D$, which reads "A is to B as C is to D." The outer terms, A and D, are called the **extremes**, and the inner terms, B and C, are called the **means**. Writing the proportion in an equation of ratios leads to another equation, obtained by cross-multiplication of the terms where the product of the extremes would equal the product of the means.

$$\frac{A}{B} = \frac{C}{D} \quad \text{and} \quad AD = BC$$

Also, it leads to equating the ratio of the first extreme to the second mean, A/C, with the ratio of the first mean to the second extreme, B/D.

$$\frac{A}{C} = \frac{B}{D}$$

Example 1.8.1

$$\frac{3}{7} = \frac{.75}{1.75} \equiv \frac{3}{.75} = \frac{7}{1.75}$$

Also,

$$3(1.75) = 7(.75)$$

$$5.25 = 5.25$$

1.9. ALIQUOTS

An **aliquot** is any divisor by which a dividend is divided, leaving a whole number quotient without a remainder. For example, if we divide a 100 by 2 or 5 or 10, we get 50, 20, and 10, respectively. Also, if we divide 100 by $11\frac{1}{9}$ or $6\frac{1}{4}$, we get 9 and 16, respectively. Therefore, all these numbers, such as 2, 5, 10, $11\frac{1}{9}$, and $6\frac{1}{4}$, are aliquots of 100. Aliquots have a special practicality in finance. Table 1.1 shows the common aliquots of 100 by which all quotients are whole numbers with no remainder. Table 1 in the Appendix shows more aliquots.

TABLE 1.1

Quotient $(Q) = \frac{100}{A}$	Aliquot (A) (Divisor)	Quotient $(Q) = \frac{100}{A}$	Aliquot (A) (Divisor)
2	50	12	$8\frac{1}{3}$
3	$33\frac{1}{3}$	13	$7\frac{9}{13}$
4	25	14	$7\frac{1}{7}$
5	20	15	$6\frac{2}{3}$
6	$16\frac{2}{3}$	16	$6\frac{1}{4}$
7	$14\frac{2}{7}$	20	5
8	$12\frac{1}{2}$	25	4
9	$11\frac{1}{9}$	30	$3\frac{1}{3}$
10	10	40	$2\frac{1}{2}$
11	$9\frac{1}{11}$	50	2

1.10. EXPONENTS

Exponential and logarithmic functions are widely used in finance and economics, especially in interest compounding, investment, and asset depreciation and appreciation, as well as in many issues of economic growth. An **exponent** is a number representing the power to a certain base. It refers to the number of times the base is multiplied by itself. For example, in X^3, 3 is the exponent of the base X, and it means that X is multiplied by itself three times: $X \cdot X \cdot X = X^3$. Also, 2^4 means that the base 2 is multiplied four times, as the exponent is 4.

$$2 \times 2 \times 2 \times 2 = 16$$

Common forms of exponents in finance are

$$(1 + r)^n = (1 + r)(1 + r) \ldots (1 + r) \text{ to } n \text{ times}$$

and

$$(1 - d)^k = (1 - d)(1 - d) \ldots (1 - d) \text{ to } k \text{ times}$$

1.11. LAWS OF EXPONENTS

Assuming that X is a base and that a and b are exponents, where $a \neq 0$ and $b \neq 0$, we can summarize the **laws of exponents** as follows:

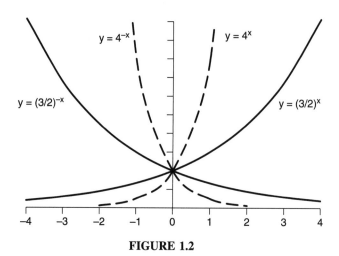

FIGURE 1.2

$X^a \cdot X^b = X^{a+b}$	example: $3^2 \cdot 3^3 = 3^{2+3} = 243$
$\dfrac{X^a}{X^b} = X^{a-b}$	example: $\dfrac{3^4}{3^2} = 3^{4-2} = 3^2 = 9$
$(X^a)^b = X^{ab}$	example: $(3^2)^3 = 3^{2\cdot3} = 3^6 = 729$
$(XY)^a = X^a Y^a$	example: $(2 \times 3)^2 = 2^2 \times 3^2 = 4 \times 9 = 36$
$\left(\dfrac{X}{Y}\right)^a = \dfrac{X^a}{Y^a}$	example: $\left(\dfrac{2}{3}\right)^2 = \dfrac{2^2}{3^2} = \dfrac{4}{9}$
$\dfrac{1}{X^a} = X^{-a}$	example: $\dfrac{1}{3^2} = 3^{-2} = \dfrac{1}{9}$
$\sqrt[a]{X} = X^{1/a}$	example: $\sqrt[2]{9} = 9^{1/2} = 3$
$\sqrt[a]{X^b} = X^{b/a}$	example: $\sqrt[2]{9^4} = 9^{4/2} = 9^2 = 81$
$X^0 = 1 (X \neq 0)$	example: $3^0 = 1$

1.12. EXPONENTIAL FUNCTION

In an **exponential function** a constant (a) has a variable exponent (x) where a has to be larger than zero but does not equal 1. It usually describes a constant rate of growth and takes the form of an equation to depict how a dependent variable such as Y would grow according to certain changes in an independent variable like x. That is,

$$Y = a^x \quad \text{where} \quad a > 0 \quad \text{but} \quad a \neq 1$$

TABLE 1.2

X	Y value: $Y = \left(\dfrac{3}{2}\right)^x$	Y value: $Y = \left(\dfrac{3}{2}\right)^{-x}$
1	$\left(\dfrac{3}{2}\right)^1 = \dfrac{3}{2}$	$\left(\dfrac{3}{2}\right)^{-1} = \dfrac{2}{3}$
2	$\left(\dfrac{3}{2}\right)^2 = \dfrac{9}{4}$	$\left(\dfrac{3}{2}\right)^{-2} = \dfrac{4}{9}$
3	$\left(\dfrac{3}{2}\right)^3 = \dfrac{27}{8}$	$\left(\dfrac{3}{2}\right)^{-3} = \dfrac{8}{27}$
−1	$\left(\dfrac{3}{2}\right)^{-1} = \dfrac{2}{3}$	$\left(\dfrac{3}{2}\right)^{-(-1)} = \dfrac{3}{2}$
−2	$\left(\dfrac{3}{2}\right)^{-2} = \dfrac{4}{9}$	$\left(\dfrac{3}{2}\right)^{-(-2)} = \dfrac{9}{4}$
−3	$\left(\dfrac{3}{2}\right)^{-3} = \dfrac{8}{27}$	$\left(\dfrac{3}{2}\right)^{-(-3)} = \dfrac{27}{8}$

X	Y value: $Y = 4^x$	Y value: $Y = 4^{-x}$
1	$4^1 = 4$	$4^{-1} = \frac{1}{4}$
2	$4^2 = 16$	$4^{-2} = \frac{1}{16}$
3	$4^3 = 64$	$4^{-3} = \frac{1}{64}$
−1	$4^{-1} = \frac{1}{4}$	$4^{-(-1)} = 4$
−2	$4^{-2} = \frac{1}{16}$	$4^{-(-2)} = 16$
−3	$4^{-3} = \frac{1}{64}$	$4^{-(-3)} = 64$

Table 1.2 and Figure 1.2 show two exponential functions under three assumed values of x on each side of the x-axis.

$$Y = \left(\frac{3}{2}\right)^x$$

$$Y = \left(\frac{3}{2}\right)^{-x}$$

$$Y = 4^x$$

$$Y = 4^{-x}$$

1.13. NATURAL EXPONENTIAL FUNCTION

The **natural exponential function** is a function where the base (a) is set to equal the natural base (e).

$$Y = a^x \rightarrow Y = e^x$$

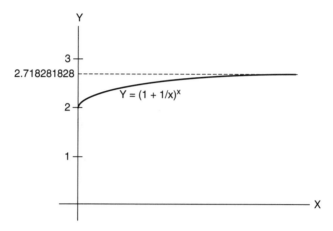

FIGURE 1.3

where

$$e = \lim_{X \to \infty} \left(1 + \frac{1}{X}\right)^x$$
$$= 2.718281828$$

(see Figure 1.3). This function describes continuous growth at constant rates, such as the growth of population, as well as the negative rates of capital growth, which refer to depreciation of assets.

1.14. LAWS OF NATURAL EXPONENTS

The following are the **laws of the natural exponents**:

$$e^0 = 1$$

$$e^1 = e = 2.718281828$$

$$e^a \cdot e^b = e^{a+b}$$

$$(e^a)^b = e^{ab}$$

$$\frac{e^a}{e^b} = e^{a+b}$$

1.15. SCIENTIFIC NOTATION

Numbers can be written in scientific notation for convenience. The general form of **scientific notation** is

$$a(10)^x$$

where a is a number between 1 and 10 and x is an integer. The procedure for writing a number in scientific notation can be summarized as follows:

1. Assume that there is a decimal point even if there is not.
2. Move the decimal point to the right of the first nonzero digit (first from the left) of the given number.
3. Give x (the power of 10) a value equal to the number of places the decimal point was moved.
4. Give x a positive sign if the decimal point was moved to the left and a negative sign if the decimal point was moved to the right.

Example 1.15.1 Write 425 in scientific notation.

$$425 = 425.0$$

Move the decimal point to the right of 4, which would be a movement of two places to the left. The 2 would become a power of 10, a positive power:

$$425 = 425.0 = 4.25(10)^2$$

Example 1.15.2 Write .000359 in scientific notation.

Since 3 is the first nonzero digit, we move the decimal point to the right of 3, which would be a movement of four places to the right. The 4 would become a power of 10, a negative power because we moved the decimal point to the right.

$$.000359 = 3.59(10)^{-4}$$

1.16. LOGARITHMS

A logarithm is a special exponent. It is either a common logarithm or a natural logarithm. A **common logarithm** is an exponent to which a base of 10 must be raised to yield a certain number. For example, if that certain number is 15, then 10 would have to be raised to the power of 1.176 to yield 15. It is written as $\log_{10} 15 = 1.176$ and read "the common logarithm of 15 is 1.176."

$$(10)^{1.176} = 15$$

A **natural logarithm** is an exponent to which a natural base of e must be raised in order to yield a certain number. For example, if that certain number is 50, then e (which is equal to 2.71828) has to be raised to the power of 3.912 to yield 50.

$$(2.71828)^{3.912} \equiv \log_e 50 = \ln 50 = 3.912$$

It is read "the natural log of 50 is 3.912."

1.17. LAWS OF LOGARITHMS

The following are the most common **laws of logarithms**:

$$\log X + \log Y = \log XY$$
$$\log X - \log Y = \log \frac{X}{Y}$$
$$\log X^a = a \log X$$
$$\log \sqrt[a]{X} = \log X^{1/a} = \frac{1}{a} \log X = \frac{\log X}{a}$$
$$\log 1 = 0$$
$$\log 10 = 1$$

1.18. CHARACTERISTIC, MANTISSA, AND ANTILOGARITHM

It is clear that $\log 100$ is 2, and $\log 1{,}000$ is 3 because $100 = 10^2$ and $1{,}000 = 10^3$. This fact leads to the conclusion that the log of any number between 100 and 1,000 has to be equal to something between 2 and 3. In other words, it would be 2 plus a fraction (or decimal). For example, $\log 346$ is 2.5391, and $\log 973$ is 2.9881. To find the log of any number such as these, where the power of 10 is not an exact number, we can go by the following:

1. Express the number in scientific notation, where

$$N = a \cdot 10^X \qquad 1 < a < 10$$

   Scientific notation makes it possible to divide the number into two parts, multiplied by each other. The first part is represented by a, and the second part is 10^x.

2. Use the common logarithm table (Table 2 of the Appendix) to obtain the log of the number, which would be the addition of the logs of the two parts. The log of a would be called the **mantissa**, and the power of 10, x, would be called the **characteristic**.

Example 1.18.1   Find log 346.

1.
$$346 = 3.46(10)^2$$
$$\log 346 = \log 3.46 + \log 10^2$$
$$= \log 3.46 + 2\log 10$$
$$= \log 3.46 + 2 \quad \text{since} \quad \log 10 = 1$$

2. log 3.46 is the mantissa, and 2 is the characteristic. To find log 3.46, we look it up in the table of mantissas (Table 2 in the Appendix). We would ignore the decimal point in 3.46 and consider it as 34 and 6. In the table we look up row 34 and column 6 to read the mantissa, which would be 5391. Since all mantissas are assumed to be preceded by a decimal point, we read the mantissa as .5391.

$$\log 346 = \log 3.46 + 2$$
$$= .5391 + 2$$
$$= 2.5391$$

Example 1.18.2   Find the log of .00935.

1.
$$.00935 = 9.35(10)^{-3}$$
$$\log .00935 = \log 9.35 + (-3)$$

2. Looking up row 93 and column 5 in the table of mantissas reveals that the mantissa is .9708.

$$\log .00935 = \log 9.35 + (-3)$$
$$= .9708 - 3$$
$$= -2.0292$$

An **antilogarithm** is the inverse of a logarithm. In Example 1.18.2 we found that log .00935 was −2.0292. Therefore, the antilogarithm of −2.0292 would be .00935. To find the antilog using the table, we reverse the procedure of finding the log.

Example 1.18.3   Find the antilog of 2.6170.

From the method described above we can conclude that 2.6170 represents a character of 2 and a mantissa of .6170. We look up the mantissa on the table and track it left to 41 and up to 4. We conclude that the antilog must be 414. To verify, we set it up in reverse to see if the log is correct.

$$\log 414 = 2.6170$$

1.19. LOGARITHMIC FUNCTION

A **logarithmic function** is a function of the form $Y = \log_a X$. In logarithmic terms this means that the value of the dependent variable Y is the power to which a would be raised in order to be equal to the value of the independent variable X. These three variables take the following values:

$$a > 0 \quad \text{but} \quad a \neq 1$$
$$X > 0$$
$$-\infty < Y < \infty$$

The logarithmic function can also be defined as the reverse of the exponential function, such that

$$Y = \log_a X \text{ can be written as :}$$
$$X = a^y$$

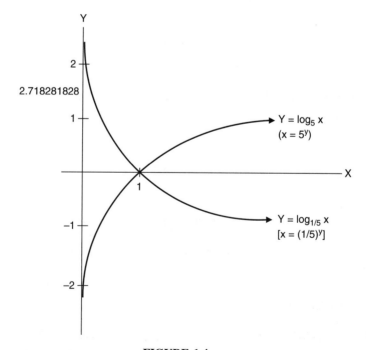

FIGURE 1.4

TABLE 1.3

Y	$X = 5^y$	$X = (1/5)^y$
1	5	$\dfrac{1}{5}$
2	25	$\dfrac{1}{25}$
0	1	1
-1	$\dfrac{1}{5}$	5
-2	$\dfrac{1}{25}$	25

Although the base a can be any positive number but not 1, the most common and practical values are $a = 10$, which defines the common logarithmic function, and $a = e$, which is equal to 2.71828 and defines the natural or **napierian logarithmic function**. Figure 1.4 and Table 1.3, show two examples of logarithmic functions:

$$\log_5 X = Y \quad \text{or} \quad X = 5^y$$

and

$$\log_{1/5} X = Y \quad \text{or} \quad X = \left(\frac{1}{5}\right)^y$$

2 Mathematical Progressions

2.1. ARITHMETIC PROGRESSION

An **arithmetic progression** is a sequence of terms where each term exceeds the preceding term by a fixed difference called the **common difference**. For example, 5, 7, 9, ... is an arithmetic progression with a common difference of 2. Assume that an arithmetic progression starts with a term (a_1) and progresses by a common difference of d. Therefore, the progression would be written

$$a_1, a_1 + d, a_1 + d + d, a_1 + d + d + d, \ldots, a_n - d, a_n$$

$$a_1, a_1 + d, a_1 + 2d, a_1 + 3d, \ldots, a_n - 3d, a_n - 2d, a_n - d, a_n \qquad (1)$$

If the third term is $a_1 + 2d$, and the fourth term is $a_1 + 3d$, the nth term has to be

$$\boxed{a_n = a_1 + (n-1)d}$$

where a_1 is the first term, n is the total number of terms, and d is the common difference. If we reverse sequence (1) above, we get

$$a_n, a_n - d, a_n - 2d, a_n - 3d \ldots, a_1 + 3d, a_1 + 2d, a_1 + d, a_1 \qquad (2)$$

The summations of the progression in (1) and (2) are

$$S_n = a_1 + (a_1 + d) + (a_1 + 2d) + (a_1 + 3d) + \cdots + (a_n - 3d)$$
$$+ (a_n - 2d) + (a_n - d) + a_n \qquad (3)$$

$$S_n = a_n + (a_n - d) + (a_n - 2d) + (a_n - 3d) + \cdots + (a_1 + 3d)$$
$$+ (a_1 + 2d) + (a_1 + d) + a_1 \qquad (4)$$

Adding (3) to (4), we obtain

$$2S_n = (a_1 + a_n) + (a_1 + a_n) + \cdots + (a_n + a_1) + (a_n + a_1)$$
$$2S_n = n(a_1 + a_n)$$

Mathematical Finance, First Edition. M. J. Alhabeeb.
© 2012 John Wiley & Sons, Inc. Published 2012 by John Wiley & Sons, Inc.

Therefore, the S_n value would be

$$S_n = \frac{n}{2}(a_1 + a_n)$$

which is the general formula for the sum of the progression.

Example 2.1.1   What would be the final and 12th terms of the arithmetic progression of 15 terms that starts 17, 12, 7?

$$a_1 = 17 \qquad d = -5$$
$$a_n = a_1 + (n - 1)d$$
$$a_{15} = 17 + (15 - 1) - 5 = -53 \qquad \text{the final term}$$

The 12th term would be $a_1 + 11d = 17 + 11(-5) = -38$ and the entire progression would be

1st	2nd	3rd	4th	5th	6th	7th	8th	9th	10th	11th	12th	13th	14th	15th
17	12	7	2	−3	−8	−13	−18	−23	−28	−33	−38	−43	−48	−53

Example 2.1.2   What would be the common difference in an arithmetic progression of 10 terms starting with 5 and ending with 32?

$$a_n = a_1 + (n - 1)d$$
$$32 = 5 + (10 - 1)d$$
$$32 = 5 + 9d$$
$$27 = 9d$$
$$3 = d$$

and the entire progression would be

1st	2nd	3rd	4th	5th	6th	7th	8th	9th	10th
5	8	11	14	17	20	23	26	29	32

Example 2.1.3   How many terms are in the arithmetic progression 23, 30, 37, down to 72?

$$a_1 = 23 \quad a_n = 72 \qquad d = 7$$
$$a_n = a_1 + (n - 1)d$$
$$72 = 23 + (n - 1)7$$

$$72 - 23 = 7n - 7$$

$$72 - 23 + 7 = 7n$$

$$\frac{56}{7} = n$$

$$8 = n$$

and the entire progression would be

1st	2nd	3rd	4th	5th	6th	7th	8th
23	30	37	44	51	58	65	72

Example 2.1.4   A loan of $3,000 carrying 9% interest is to be paid off $150 a month. The monthly balances and the interest on them would form two arithmetic progressions. Find the total interest paid.

Since the principal balance is going to be reduced gradually by a fixed $150 a month, the principal balance progression will be

$$3,000, 2,850, 2,700, \ldots, 150$$

Since the monthly interest $= .09/12 = .0075$, each term of the balance will be subject to the monthly interest:

$$3,000 \times .0075 = 22.50$$

$$2,850 \times .0075 = 21.37$$

$$2,700 \times .0075 = 20.25$$

Therefore, the interest progression is

$$22.50, 21.37, 20.25, \ldots, 1.125$$

The number of payments

$$n = \frac{3,000}{150} = 20$$

The total interest paid is the summation of all interests, or S_n:

$$S_n = \frac{n}{2}(a_1 + a_n)$$

$$= \frac{20}{2}(22.50 + 1.12)$$

$$= 236.20$$

Example 2.1.5   What would be the principal balance at the end of the first year in Example 2.1.4? What would the interest be for the 17th payment?

The principal balance at the 12th term would be

$$a_{12} = a_1 - 11d$$
$$= 3,000 - 11(150)$$
$$= 1,350$$

The principal balance at the 17th payment would be

$$a_{17} = a_1 - 16d$$
$$= 3,000 - 16(150)$$
$$= 600$$

Interest on the 17th payment would be

$$600 \times .0075 = 4.50$$

or

$$a_{17} = a_1 - 16d$$
$$= 22.50 - 16(1.125)$$
$$= 4.50$$

2.2. GEOMETRIC PROGRESSION

A **geometric progression** is a sequence of terms in which each term can be obtained by multiplying the preceding term by a constant called a **common ratio**. A common ratio is obtained by dividing each term by the term immediately preceding it. For example, 2, 4, 8, 16, ... is a geometric progression with a common ratio of 2, since

$$\frac{16}{8} = 2, \qquad \frac{8}{4} = 2, \qquad \frac{4}{2} = 2$$

and

$$16 = 8 \times 2, \qquad 8 = 4 \times 2, \qquad 4 = 2 \times 2$$

Assume that a geometric progression begins with the term a and progresses by a common ratio r. Therefore, the progression would be written

$$a_1, a_1 r, a_1 \cdot r \cdot r, a_1 \cdot r \cdot r \cdot r, \ldots, a_1 \cdot r^{n-2}, a_1 \cdot r^{n-1}, a_1 r^n$$
$$a_1, a_1 r, a_1 r^2, a_1 r^3, \ldots, a_1 r^{n-2}, a_1 r^{n-1}, a_1 r^n$$

and the nth term has to be

$$a_n = a_1 r^{n-1}$$

where a_1 is the first term, r is the common ratio, and n is the total number of terms. The summation of the progression S_n would be

$$S_n = a_1, a_1 r, a_1 r^2, a_1 r^3 + \cdots + a_1 r^{n-2} + a_1 r^{n-1} \tag{1}$$

Multiplying by r, we obtain

$$r S_n = a_1 + a_1 r^2 + a_1 r^3 + a_1 r^4 + \cdots + a_1 r^{n-1} + a_1 r^n \tag{2}$$

Subtracting (2) from (1) yields

$$S_n - r S_n = a_1 - a_1 r^n$$
$$S_n (1 - r) = a_1 (1 - r^n)$$

$$S_n = \frac{a_1 (1 - r^n)}{1 - r}$$

Example 2.2.1   Find the sixth term of the geometric progression 9, -3, 1, $\ldots$.

The common ratio r is

$$\frac{-3}{9} = -\frac{1}{3} \quad \text{or} \quad \frac{1}{-3} = -\frac{1}{3}$$

The sixth term (a_6) is

$$a_n = a_1 r^{n-1}$$

$$a_6 = 9 \left(-\frac{1}{3} \right)^{6-1}$$

$$= -\frac{1}{27}$$

and the entire progression would be

1st	2nd	3rd	4th	5th	6th
9	-3	1	$-\frac{1}{3}$	$\frac{1}{9}$	$1\frac{1}{27}$

Example 2.2.2

(a) Find the sum of the geometric progression that begins with 4, has the common ratio $\frac{5}{16}$, and contains five terms.

$$S_n = a_1 \cdot \frac{1 - r^n}{1 - r}$$

$$S_5 = \frac{4\left[1 - \left(\frac{5}{16}\right)^5\right]}{1 - \frac{5}{16}}$$

$$S_{10} = 5.80$$

(b) Write all five terms of the progression.

The entire progression would be

1st	2nd	3rd	4th	5th
4	$\dfrac{5}{4}$	$\dfrac{25}{64}$	$\dfrac{125}{1,024}$	$\dfrac{625}{16,384}$

Example 2.2.3

(a) How many terms are there in the geometric progression that begins with 4, has a common ratio of 2, and has a sum of 252?

$$S_n = a_1 \frac{1 - r^n}{1 - r}$$

$$252 = 4 \cdot \frac{1 - 2^n}{1 - 2}$$

$$64 = 2^n$$

$$\log 64 = n \log 2$$

$$\frac{\log 64}{\log 2} = n$$

$$\frac{1.806179974}{3010299957} = n$$

$$6 = n$$

(b) Write all the terms of the progression.

The entire progression would be

1st	2nd	3rd	4th	5th	6th
4	8	16	32	64	128

Example 2.2.4 A printing machine is purchased for $15,000. Its annual depreciation rate is 20%. What will its value be at the end of 10 years?

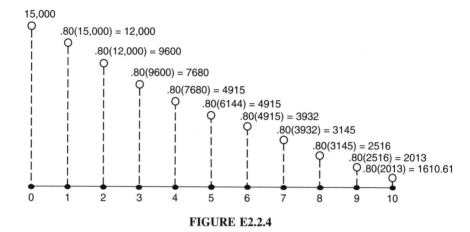

FIGURE E2.2.4

This is a geometric progression problem. The first term (a_1) is the purchase price, $15,000 (see Figure E2.2.4). Since the depreciation rate is 20%, the value of the machine in any year would be 80% of its value the preceding year. Therefore, the common ratio r is 80%. Since we calculate 10 values after the initial value, n would be 11.

$$a_n = a_1 r^{n-1}$$
$$a_{11} = 15,000(.80)^{11-1}$$
$$= 1,610.61$$

and the entire progression would be

1st	2nd	3rd	4th	5th	6th	7th	8th	9th	10th	11th
15,000	12,000	9,600	7,680	6,144	4,915	3,932	3,145	2,516	2,013	1,610.61

2.3. RECURSIVE PROGRESSION

In a **recursive progression** each term is the sum of the two terms preceding it rather than being determined by a common difference or a common ratio. For example, the third term is obtained by adding the second term to the first, the fourth term is obtained by adding the third term to the second, and so on. In this case, a recursive progression requires knowing a minimum of the first and second terms. The nth term of the recursive progression would be written

$$\boxed{a_n = a_{n-1} + a_{n-2}}$$

Example 2.3.1   Finish the following recursive progression down to the seventh term: 5, 7, ...

$$3\text{rd term} = 5 + 7 = 12$$

$$4\text{th term} = 12 + 7 = 19$$

$$5\text{th term} = 19 + 12 = 31$$

$$6\text{th term} = 31 + 19 = 50$$

$$7\text{th term} = 50 + 31 = 81$$

So the entire progression is

1st	2nd	3rd	4th	5th	6th	7th
5	7	12	19	31	50	81

Example 2.3.2   What is the 10th term of the recursive progression that has $a_1 = 2$ and $a_2 = 6$?

$$a_n = a_{n-1} + a_{n-2}$$

$$a_3 = a_{3-1} + a_{3-2}$$

$$= a_2 + a_1$$

$$= 6 + 2$$

$$= 8 \tag{1}$$

$$a_4 = a_{4-1} + a_{4-2}$$

$$= a_3 + a_2$$

$$= 8 + 6$$

$$= 14 \tag{2}$$

$$a_5 = a_{5-1} + a_{5-2}$$

$$= a_4 + a_3$$

$$= 14 + 8$$

$$= 22 \tag{3}$$

$$a_6 = a_{6-1} + a_{6-2}$$

$$= a_5 + a_4$$

$$= 22 + 14$$

$$= 36 \tag{4}$$

$$a_7 = a_{7-1} + a_{7-2}$$

$$= a_6 + a_5$$

$$= 36 + 22$$

$$= 58 \tag{5}$$

$$a_8 = a_{8-1} + a_{8-2}$$
$$= a_7 + a_6$$
$$= 58 + 36$$
$$= 94 \qquad\qquad (6)$$
$$a_9 = a_{9-1} + a_{9-2}$$
$$= a_8 + a_7$$
$$= 94 + 58$$
$$= 152 \qquad\qquad (7)$$
$$a_{10} = a_{10-1} + a_{10-2}$$
$$= a_9 + a_8$$
$$= 152 + 94$$
$$= 246 \qquad\qquad (8)$$

and the entire progression is

1st	2nd	3rd	4th	5th	6th	7th	8th	9th	10th
2	6	8	14	22	36	58	94	152	246

2.4. INFINITE GEOMETRIC PROGRESSION

If we consider the sum of n terms S_n in the geometric progression,

$$S_n = a_1 \frac{1 - r^n}{1 - r}$$
$$= \frac{a_1}{1 - r} - \frac{a_1 r^n}{1 - r}$$

when the common ratio (r) value lies between -1 and $1 (-1 < r < 1)$, we can conclude that as n increases without bound, the term r^n would approach zero and S_n would approach $a_1/(1 - r)$. Therefore, that sum would become the sum of an **infinite geometric progression**, and it is expressed as

$$S_n = \frac{a_1}{1 - r} \qquad \text{when} \quad -1 < r < 1$$

Example 2.4.1   Find the sum of the infinite geometric progression $1, \frac{1}{2}, \frac{1}{4}, \frac{1}{8}, \frac{1}{16}$.

$$S_n = \frac{a_1}{1 - r}$$

$$= \frac{1}{1 - \frac{1}{2}}$$

$$= \frac{1}{\frac{1}{2}}$$

$$= 2$$

which means: As the number of terms in this progression increases to infinity, the sum of all terms reaches only 2. We express this fact by

$$\lim_{n \to \infty} S_n = 2$$

2.5. GROWTH AND DECAY CURVES

Growth and decay curves are graphic representations of an exponential function of the form

$$\boxed{Y = ab^x}$$

where a is positive $(a > 0)$. It is the value of b that determines whether the curve is a growth or a decay curve:

- For a growth curve, b value has to be larger than 1: $(b > 1)$.
- For a decay curve, b value has to be larger than zero but less than 1: $(0 < b < 1)$.

Growth curves are helpful in tracking down and predicting trends in economic or biological growth, such as the growth of bacteria or the spread of an epidemic. Decay curves can be used to model, for example, economic decline or infant mortality.

Example 2.5.1   Suppose that a spill of crude oil in the ocean is estimated by the exponential function $Y = 2(5)^x$, where $x = 0$ is the day that an underwater pipe broke; negative x represents the days before, and positive x, the days after, the incident (see Figure E2.5.1). The amount of spill will grow according to

$$Y = 2(5)^x$$

Let's try to obtain the values of X and Y in Table E2.5.1, which stand for the number of days and the size of spill measured by hundreds of barrels. So the spill will grow exponentially from 200 barrels in the first day of the accident to 625,000 barrels in the fifth day after the accident.

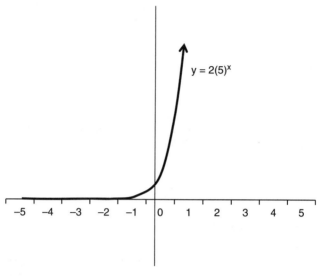

FIGURE E2.5.1

TABLE E2.5.1

X (days)	Y (100 barrels)
−5	.0006
−4	.0032
−3	.016
−2	.08
−1	.4
0	2
1	10
2	50
3	250
4	1,250
5	6,250

Example 2.5.2 If we assume that all attempts to repair the broken pipe in Example 2.5.1 failed and that the breakage conditions stayed the same and did not worsen, how many barrels of oil will spill in 10 days?

The answer is to replace X with 10 days and solve for Y:

$$Y = 2(5)^x$$
$$= 2(5)^{10}$$
$$= 19,531,250$$

and since it is measured by a unit of 100 barrels, the total spill will be

$$Y = 19,531,250 \times 100 = 1,953,125,000$$

Example 2.5.3   Suppose that the annual insurance premium for a small business to transport its product to distributers can be estimated by the function

$$Y = 750(1.25)^x$$

(see Figure E2.5.3), where

$$X = 0 \quad \text{is the premium in 2011}$$
$$X = 1 \quad \text{is the premium in 2012}$$
$$X = -1 \quad \text{is the premium in 2010, and so on}$$

What would the premium be in 2011 and in 2015, and how much was it in 2008?

Premium in 2011:

$$Y = 750(1.25)^0$$
$$= 750$$

Premium in 2015:

$$Y = 750(1.25)^4$$
$$= 1,831.05$$

Premium in 2008:

$$Y = 750(1.25)^{-3}$$
$$= 384$$

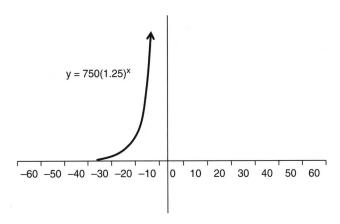

FIGURE E2.5.3

The decay curve is no less important for the estimation of cases that bear exponential development. For example, infant death in a country can be estimated by the function

$$Y = 10^5(.89)^x$$

where

$X = 0$	is a reference to any year
$X = 1$	is a reference to a year after
$X = -1$	is a reference to a year before

We can see the curve as shown in Figure 2.1 and we can, for example, calculate the mortality rate in 2011 and in 2014 and look at how much it was in 2001.

In 2011:

$$Y = 10^5(.89)^x$$
$$= 10^5(.89)^0$$
$$= 100,000$$

In 2014:

$$Y = 10^5(.89)^3$$
$$= 70,497$$

In 2001:

$$Y = 10^5(.89)^{-10}$$
$$= 320,700$$

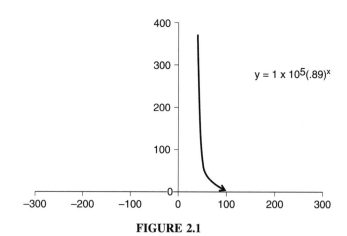

FIGURE 2.1

So mortality is decreasing as time goes forward, and that is what a decreasing exponential function expresses.

Example 2.5.4 Suppose that due to certain medical advances and increasing health awareness, the seasonal rate of catching a certain contagious flu strain is represented by the exponential function

$$Y = 30(.6)^x$$

(see Figure E2.5.4), where

$X =$ 0 refers to the season preceding the advances
        and new preventive measures

$X =$ 1 is a year later

$X = -1$ is a year earlier

What would be the spreading rate of that flu when $X = 0$, and what was it five years earlier and five years later, given that the cases are measured by a 100-case unit?

At the beginning of the flu epidemic,

$$Y = 30(.6)^0$$

$$= 30$$

and since each unit is a 100-case unit, the total number of cases would be

$$y = 30 \times 100 = 3{,}000 \text{ cases}$$

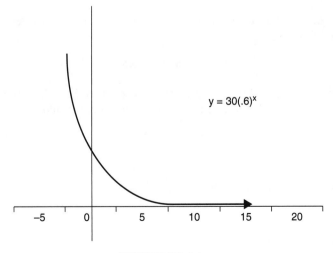

FIGURE E2.5.4

Five years earlier,

$$Y = 30(.6)^{-5}$$
$$= 386$$
$$= 386 \times 100 = 38{,}600 \text{ cases}$$

Five years later,

$$Y = 30(.6)^{5}$$
$$= 2.33$$
$$= 2.33 \times 100 = 233 \text{ cases}$$

2.6. GROWTH AND DECAY FUNCTIONS WITH A NATURAL LOGARITHMIC BASE

With more and more experience, mathematicians found that a natural logarithmic base of $e = 2.71828$ produces better approximations to such exponential functions as growth and decay. It is therefore more appropriate to express the function $(Y = ab^x)$ by

$$\boxed{Y = ae^x}$$

and

$$\boxed{Y = ae^{-x}}$$

where the value of e^x or e^{-x} can be obtained by tables or calculations. Table 3 in the Appendix displays the values of e^x and e^{-x} from 0 to 9.9.

Example 2.6.1 The annual profit for an industrial firm is expressed by the exponential function

$$Y = 3{,}000 + 7{,}500e^{27x}$$

where x is how long the firm has been selling its product. So if the firm introduced the product to the market last year, $x = 1$ and the profit would be

$$Y = 3{,}000 + 7{,}500e^{27(1)}$$
$$= 12{,}824$$

and in the fifth year the profit would be

$$Y = 3{,}000 + 7{,}500e^{27(5)}$$
$$= 31{,}930$$

3 Statistical Measures

3.1. BASIC COMBINATORIAL RULES AND CONCEPTS

Basic combinatorics constitute the fundamental mathematical rules and concepts of counting and ordering. Their understanding plays an important role in comprehending and accepting more advanced concepts of probability and mathematical expectations.

The N_i Rule

If an event i can occur in N_i possible ways throughout n events, the number of ways in which the sequence of n events can be written as:

$$N_1, \ N_2, \ N_3, \ldots, \ N_n$$

Example 3.1.1   Suppose that of the six managerial positions (P_i) announced by a company, each can be filled by two managers. We can order the positions as follows, where $n = 6$:

$$P_i = P_1, \ P_2, \ P_3, \ P_4, \ P_5, \ P_6$$

and since there are two managers for each (a and b), we can rewrite the order as

$$P_i^a = P_1^a, \ P_2^a, \ P_3^a, \ P_4^a, \ P_5^a, \ P_6^a$$
$$P_i^b = P_1^b, \ P_2^b, \ P_3^b, \ P_4^b, \ P_5^b, \ P_6^b$$

Mathematical Finance, First Edition. M. J. Alhabeeb.
© 2012 John Wiley & Sons, Inc. Published 2012 by John Wiley & Sons, Inc.

The n-Factorial ($n!$) Rule

The number $n!$, where n is any positive integer, is the number of ways in which n objects can be ordered. Mathematically, it is calculated by multiplying the number n by all numbers below it in descending order, descending one unit at a time until the last unit, 1:

$$n! = n(n - 1)(n - 2)(n - 3) \qquad \text{given that} \quad 0! = 1 \qquad (1)$$

Example 3.1.2 Suppose that you receive five letters on the same day. If the letters were sent on different days, the chance that you would be reading them in the order in which they were sent would be calculated by 5!:

$$5! = (5)(5 - 1)(5 - 2)(5 - 3)(5 - 4)$$
$$= (5)(4)(3)(2)(1)$$
$$= 120$$

The mn Rule

If there are m and n elements, and each is ordered as

$$A_1, \ A_2, \ A_3, \dots, A_m$$
$$B_1, \ B_2, \ B_3, \dots, B_n$$

we can form a total N that is equal to mn which contains one element from each group:

$$N = mn$$

Example 3.1.3 Suppose that three companies come to a business school to choose one graduating student from each of four departments: finance, marketing, accounting, and management. Let the companies be m, the departments be n, and the total graduating students chosen be N (see Figure E3.1.3):

$$N = mn$$
$$= 3(4)$$
$$= 12$$

Example 3.1.4 Let's assume that six trains go back and forth between two stations in New York City. In how many ways can you go on a certain train but return in another?

This is an mn problem since the number of trains in which you go from station I to station II is $m = 6$, and the number of trains in which you return

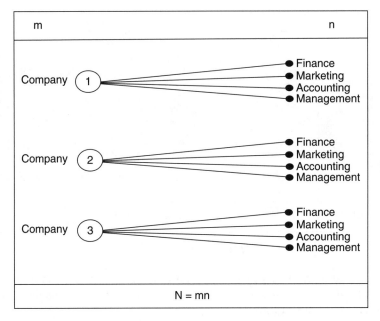

FIGURE E3.1.3

back from station II to station I is $n = 5$ because you exclude the train in which you went.

$$N = mn$$

$$= 6(5)$$

$$= 30$$

If you give the trains the colors red (R), green (G), blue (B), white (W), yellow (Y), and orange (O), we can visualize the total number of ways as shown in Figure E3.1.4.

How many ways you can return in the same train is still a matter of mn. The number of trains you go in is $m = 6$ and the number of trains you want to be back in is 1. Therefore,

$$N = mn$$

$$= 6(1)$$

$$= 6$$

3.2. PERMUTATION

Permutation is a way of arranging and ordering elements. It is the number of ways of ordering n objects taken r at a time. You can also look at it as a method

Ways to go back and forth in different trains	RG	GR	BR	WR	YR	OR
	RB	GB	BG	WG	YG	OG
	RW	GW	BW	WB	YB	OB
	RY	GY	BY	WY	YW	OW
	RO	GO	BO	WO	YO	OY
Excluding	RR	GG	BB	WW	YY	OO

FIGURE E3.1.4

of finding the number of ways to fill r positions with n objects. Permutation is denoted by nP_r, where n is the number of objects and r represents the way they are arranged at a given time.

Let's assume that there are four books with colored covers—red (R), green (G), blue (B), and yellow (Y)—and let's assume that there are four shelf compartments, 1 to 4. The number of ways that these books can be placed in the compartments one at a time is shown in Figure 3.1. For the first compartment, we could choose from all four books, and the choice was the green book (G); for the second compartment, we could choose from the remaining three books, and the choice was the blue book (B); for the third compartment, the choice was the yellow book (Y) from the only two books left; and for the last compartment we had only one book available, the red book (R). So the second row of compartments contains the number of choices available, which was descending, as we made our choices one at a time. We could have placed any of the books in the first compartment and any except green (which was chosen for the first compartment in the second compartment, and so on. Mathematically, there would be $n!$ ways of arranging these books:

$$n! = n(n-1)(n-2)(n-3)$$
$$4! = 4(4-1)(4-2)(4-3)$$
$$= 4 \cdot 3 \cdot 2 \cdot 1$$
$$= 24$$

But this is the permutation nP_r when $n = r$:

$$nP_r = n!$$
$$4P_4 = 4! = 24$$

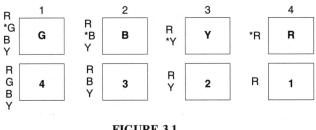

FIGURE 3.1

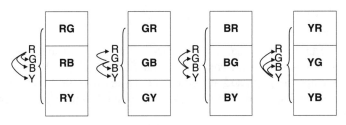

FIGURE 3.2

Now let's assume that we want to place two books at a time, one on top of the other, in each compartment. The order of colors between the top and the bottom makes a difference; that is, red and blue is a choice different from blue and red. In this case we would see the arrangement shown in Figure 3.2. This is a permutation nP_r where $r < n$. Here n represents the four books and r the way that we chose to arrange them, which is two books at a time. Mathematically, nP_r, where $n = 4$ and $r = 2$, can be obtained by

$$\boxed{nP_r = n(n-1)(n-2)\cdots(n-r+1)}$$

$$4P_2 = 4(4-1)$$

and we stop at $4 - 1$ because it is equal to the last term $(n - r + 1)$ according to the formula above.

$$4 - 1 = 3 = 4 - 2 + 1$$

Therefore, $4P_2$ is

$$4P_2 = 4(4-1)$$
$$= 4 \times 3$$
$$= 12$$

The most common formula for permutation is

$$\boxed{np_r = \frac{n}{(n-r)!}}$$

Example 3.2.1 If five people among 12 contestants are chosen to win a trip together by random selection, determine in how many ways the five winners would be arranged. $n = 12; r = 5$.

$$n P_r = \frac{n!}{(n-r)!}$$

$$12 P_5 = \frac{12!}{(12-5)!}$$

$$= 95,040$$

It is important to note that in the earlier examples, n always represented all different elements, such as different books and different people. Were n to include similar elements, the permutation formula would be different. Suppose that there are nine colored blocks with three colors distributed among the blocks in the following way:

3 red (R) blocks

4 green (G) blocks

2 blue (B) blocks

In this case, n is 9 but r comes in three groups: $r_1 = 3$, $r_2 = 4$, and $r_3 = 2$. The permutation can therefore be obtained by

$$n P_r = \frac{n!}{r_1! r_2! \cdots r_m!}$$

where m is the number of r groups. To solve this example we will have $m = 3$ because we have three groups of r that share the same characteristic (i.e., the color): $r_1 = 3$, $r_2 = 4$, and $r_3 = 2$.

$$P = \frac{9!}{3! \cdot 4! \cdot 2!}$$

$$= 1,260$$

3.3. COMBINATION

Just like permutation, **combination** is a way of arranging and ordering elements involving n objects and r ways of arrangement. The only difference is that with combination there is no consideration of either the order or the reverse order of the selection of r. For example, RGB is the same as GRB and BGR as long as all three elements are there. This means that the total number of ways of combinating is less than that of permutation. Combination is denoted by nC_r and is obtained by

$$nC_r = \frac{n!}{(n-r)!r!}$$

Example 3.3.1 If we want to form a five-member committee chosen from 10 professors, how many possible committees can be formed?

Here $n = 10, r = 5$, and the combination will be

$$nC_r = \frac{n!}{(n-r)!r!}$$
$$10C_5 = \frac{10!}{(10-5)!5!}$$
$$= 252$$

3.4. PROBABILITY

Probability is a measure of uncertainty: an estimation of the likelihood or chance that an uncertain event will occur. For example, we can never be certain of what we would get if we flip a coin or toss a die, since any outcome is subject to a probability. However, if the coin or die is fair, in a sense that it is physically perfect, all possible outcomes would be equally likely events, meaning that we would have a 50% chance of getting a head and a 50% chance of getting a tail when we flip a coin. Also, each of the six faces of a die would have $\frac{1}{6}$ of a chance on any toss.

Three major axioms describe the mathematical properties of probability. Let's assume that we call a certain event E and that there is an n-set of those events that are mutually exclusive and collectively exhaustive, such that the set is

$$E_1, \ E_2, \ E_3, \ldots, \ E_n$$

where E_i refers to any of those events. The three axioms follow.

> *Axiom 1.* Any event of that set (E_i) will either occur or not occur. If we assign a value of 1 to occurrence and a value of zero to nonoccurrence, we can express that by

$$0 \le P(E_i) \le 1$$

where $P(E_i)$ refers to the probability that event E_i occurs.

Axiom 2. Each event (E_i) in the set has a probability estimated by a fraction of 1, and therefore the probabilities of all events in the set together constitute a value of 1:

$$\sum P(E_i) = 1$$

Axiom 3. The probability of more than one individual event occurring separately is equal to the sum of the probabilities of these individual events:

$$P(E_1, E_2, E_3) = P(E_1) + P(E_2) + P(E_3)$$

Probability is often presented as the **relative frequency** of events that occur in a large number of trials. In this case it can be expressed mathematically as the limit of a relative frequency. According to the English logician John Venn (1834–1923), we can describe the major formula of probability as follows.

Suppose that an event E occurs f times among n trials. Then, the probability of such an event, $P(E)$, would approach the ratio f/n as n reaches infinity:

$$P(E) = \lim_{n \to \infty} \frac{f}{n}$$

Practically, n does not have to reach infinity for the formula to be valid. The point is that validity can be achieved with a large n. Therefore, the formula can be simplified for practical use as

$$P(E) = \frac{f}{n}$$

Example 3.4.1   What is the probability of drawing an ace from a full deck of cards?

The number of cards in a full deck is n, which is 52 cards. Since the deck includes only four aces, f is 4 and the probability is

$$P(E) = \frac{f}{n}$$

$$P(ace) = \frac{4}{52}$$

$$= \frac{1}{13}$$

TABLE E3.4.2a

Degree	Management	Nonmanagement	Total
Engineering	2	4	6
Nonengineering	8	6	14
Total	10	10	20

Example 3.4.2 A firm has a pool of potential managers to be considered for managerial positions. Eight persons have degrees in management, four have degrees in engineering, six have degrees in different fields, and two have both management and engineering degrees. They are organized as in Table E3.4.2a and the probabilities of the managers chosen, in terms of their degrees, are calculated.

- The probability that the chosen manager is among those who have only a management degree is

$$P(M) = \frac{8}{20} = 40\%$$

- The probability that the chosen manager is among those who have only an engineering degree is

$$P(E) = \frac{4}{20} = 20\%$$

- The probability that the chosen manager is among those who have a degree other than management or engineering is

$$P(N) = \frac{6}{20} = 30\%$$

- The probability that the chosen manager is among those who have a double degree in management and engineering is

$$P(E - M) = \frac{2}{20} = 10\%$$

- The probability that the chosen manager will be an engineer whether or not having a second degree is

$$P(AE) = \frac{6}{20} = 30\%$$

- The probability that the chosen manager will have a management degree whether or not having a second degree is

$$P(AM) = \frac{10}{20} = 50\%$$

TABLE E3.4.2b

Degree	Management	Nonmanagement	Total
Engineering	10%	20%	30%
Nonengineering	40%	30%	70%
Total	50%	50%	100%

- The probability that the chosen manager will be holding a nonmanagement degree is

$$P(NM) = \frac{10}{20} = 50\%$$

- The probability that the chosen manager will be holding a nonengineering degree is

$$P(NE) = \frac{14}{20} = 70\%$$

We can now record those probabilities in Table E3.4.2b to see how they add up to 100 in both directions.

3.5. MATHEMATICAL EXPECTATION AND EXPECTED VALUE

We stated that the probability is a relative frequency, which means that a probability distribution is a distribution of long-term frequencies. This also means that the mean of the probability distribution of a random variable reflects the centrality of the distribution. Therefore, both the value of the random variable and the probability of its occurrence are important to form a weighted average that would represent the mean of the distribution, which is also called the **expected value** of the random variable in the sense that it is the value that we expect to occur.

If we consider X to be the value of a discrete random variable and $P(x)$ its probability of occurrence, the expected value of such a variable is the sum of all values of the variable weighted by their own probabilities, which is also equal to the mean of the distribution, μ.

$$E(x) = \sum_{i=1}^{n} x_i P(x_i) = \mu$$

Example 3.5.1   Table E3.5.1a lists the possible rates of return on a certain stock and how probable each rate is. We can calculate the expected value of the rate,

TABLE E3.5.1a

Rate of Return X	Probability $P(x)$	$x(P_x)$
.10	.13	.013
.12	.19	.0228
−.08	.05	−.004
.095	.12	.0114
.14	.08	.0112
.11	.21	.0231
.125	.15	.01875
−.089	.07	−.00632
		$\sum xP(x) = .10035$

TABLE E3.5.1b

(1) X	(2) $f(x)$	(3) $P(x)$	(4) $xP(x)$	(5) $f(x)P(x)$
1	10.75	.20	.20	2.15
3	12.25	.30	.90	3.675
4.5	13.375	.10	.45	1.3375
6	14.5	.20	1.20	2.90
8.5	16.375	.10	.85	1.6375
10	17.5	.10	1.00	1.75
			4.60	13.45

which would also be the weighted average or mean of this distribution. So the expected value of the rate of return or the mean rate is a little more than 10%. In the same manner of obtaining the expected value of a single random variable, we can obtain the value expected for a function of a discrete random variable such as $f(x)$ if we know the probability of X:

$$E[fx] = \sum f(x)P(x)$$

Let's take, for example, a linear function of X such as

$$Y = f(x) = a + bx$$

If $a = 10$ and $b = .75$, the function is

$$f(x) = 10 + .75x$$

Suppose that the probabilities of X are as listed in the third column of Table E3.5.1b. The value expected for such a function can be calculated as the total of column (5) of the table. So the expected value of the function is

$$E[f(x)] = \sum f(x)P(x) = 13.45$$

Note that we can also obtain the expected value of the function by

$$E[f(x)] = E(a + bx)$$

$$\boxed{E[f(x)] = bE(x) + a}$$

Since $E(x)$ is equal to $xP(x) = 4.60$, and since $b = .75$ and $a = 10$,

$$E[f(x)] = .75(4.60) + 10$$

$$= 13.45$$

3.6. VARIANCE

Variance (σ^2) and standard deviation (σ) are probably the most important measures of variation. In the context of the probability distribution of a random variable, the **variance** is the expected squared deviation of that random variable (X) from its mean (μ). The probabilities of occurrence for the value of X still serve as weights in the calculation of the variance (σ^2):

$$\sigma^2 = E[(x - \mu)^2]$$

$$\boxed{\sigma^2 = \sum[(x - \mu)^2]P(x)}$$

Let's use the data from Table E3.5.1b to calculate the variance, as we do in Table 3.1. So the variance is 8.19. We can also calculate the variance by

$$\boxed{\sigma^2 = E(x^2) - [E(x)]^2}$$

where $E(x^2)$ is the expected value of the squared random variable (x^2), and $[E(x)]^2$ is the squared expected value of X. We can apply this formula by creating more columns Table 3.2. Given that

$$E(x^2) = \sum x^2 P(x)$$

We arrive at Table 3.2. Finding that $E(x) = 4.60$ and $E(x^2) = 29.35$, we can apply the formula

$$\sigma^2 = E(x^2) - [E(x)]^2$$
$$= 29.35 - (4.60)^2$$
$$= 8.19$$

The idea of this variance of a random variable is actually built on the original idea of the variance of a data set in its sample and population context. That is, the variance of a sample of n observations, which is called the **sample variance** (s^2), is calculated as

$$s^2 = \frac{\sum_{i=1}^{n}(x_i - \bar{x})^2}{n-1}$$

where x_i is the variable value, $\bar{x}$ is the sample mean, and n is the sample size. However, if we need to find the variance for the entire population, which is called the **population variance** (σ^2), we use the formula

$$\sigma^2 = \frac{\sum_{i=1}^{n}(x_i - \mu)^2}{N}$$

where μ is the population mean and N is the population size.

Example 3.6.1   Calculate the variance in the heights of a small sample of college students as listed in the first column of Table E3.6.1.

$$s^2 = \frac{\sum_{i=1}^{10}(x_i - \bar{x})^2}{n-1}$$
$$= \frac{2.1}{10-1}$$
$$= .233$$

TABLE 3.1

X	$P(x)$	$xP(x)$	$x - \mu$	$(x - \mu)^2$	$(x - \mu)^2 P(x)$
1	.20	.20	−3.6	12.96	2.592
3	.30	.90	−1.6	2.56	.768
4.5	.10	.45	−.10	.01	.001
6	.20	1.20	1.4	1.96	.392
8.5	.10	.85	3.9	15.21	1.521
10	.10	1.00	5.4	29.16	2.915
	$\sum xp(x) = \mu = 4.60$			$\sum(x-\mu)^2 P(x) = \sigma^2 = 8.19$	

TABLE 3.2

X	$P(x)$	$xP(x)$	x^2	$x^2 p(x)$
1	.20	.20	1	.20
3	.30	.90	9	2.7
4.5	.10	.45	20.25	2.025
6	.20	1.20	36	7.2
8.5	.10	.85	72.25	7.225
10	.10	1.00	100	10
		$E(x^2) = 4.60$		$E(x^2) = 29.35$

TABLE E3.6.1

x_i	$\overline{x}$	$x_i - \overline{x}$	$(x_i - \overline{x})^2$	x_i^2
5.9	5.8	.1	.01	34.81
5.2	5.8	−.6	.36	27.04
6.1	5.8	.3	.09	37.21
6.4	5.8	.6	.36	40.96
5.3	5.8	−.15	.25	28.09
6.0	5.8	.2	.04	36.00
6.5	5.8	.7	.49	42.25
5.1	5.8	−.7	.49	26.01
5.8	5.8	0	0	33.64
5.7	5.8	−.1	.01	32.49
58			$\sum_{i=1}^{10}(x_i - \overline{x})^2 = 2.1$	338.5

Another formula for the sample variance is

$$s^2 = \frac{\sum_{i=1}^{n} x_i^2 - (\sum_{i=1}^{n} x_i)^2/n}{n-1}$$

$$s^2 = \frac{338.5 - (58)^2/10}{10-1}$$

$$= .233$$

3.7. STANDARD DEVIATION

The **standard deviation** (σ) is the square root of the variance. It is still a measure of the variability and dispersion of the possible values of the random variable from their mean. The significance of the standard deviation stems from the fact that the higher the standard deviation, the greater the dispersion in the data and the higher the variability from the mean:

$$\sigma = \sqrt{\sigma^2}$$

$$\sigma = \sqrt{\sum(x - \mu)^2 p(x)}$$

Also,

$$\sigma = \sqrt{\frac{\sum_{i=1}^{N}(x_i - \mu)^2}{N}}$$

and for the standard deviation of the sample (s):

$$s = \sqrt{s^2}$$

$$s = \sqrt{\frac{\sum_{i=1}^{n}(x_i - \bar{x})^2}{n - 1}}$$

In the examples in Section 3.6, when the variance was 8.19, the standard deviation would be

$$\sigma = \sqrt{\sigma^2}$$
$$= \sqrt{8.19}$$
$$= 2.86$$

and when the sample variance s^2 was .233, the standard deviation (s) would be

$$s = \sqrt{s^2}$$
$$= \sqrt{.233}$$
$$= .483$$

3.8. COVARIANCE

The **covariance** between two variables, such as X and Y, is the expected value of the product of their deviations from their means:

$$\boxed{\text{Cov}(X, Y) = E[(X - \mu_x)(Y - \mu_y)]}$$

Only the sign of the covariance is meaningful:

- If $\text{Cov}(X, Y) > 0$: The two variables move together in the same direction.
- If $\text{Cov}(X, Y) < 0$: The two variables move opposite to each other.
- If $\text{Cov}(X, Y) = 0$: The two variables are not linearly related.

The magnitude of the covariance cannot refer to the strength of the linear association between variables unless it is related to the standard deviation of the two variables. Therefore, if we divide the covariance of X and Y by the product of the variables' standard deviations, we get the correlation coefficient (see Section 3.9).

Example 3.8.1   Table E3.8.1 shows the probability distribution of sales of two products (X and Y) for six months. Calculate the covariance between the two products.

The covariance is

$$\text{Cov}(X, Y) = \Sigma[(X - E(x))(Y - E(y))P]$$
$$= 12.3\%$$

The only important conclusion here is that the covariance is positive, which means that the two variables are associated linearly and positively with each other; that is, they move up and down together.

3.9. CORRELATION

The **correlation** between two variables, such as X and Y, is a measure of the magnitude of the linear relationship between them. It indicates how well and in

TABLE E3.8.1

(1)	(2)	(3)	(4)	(5)	(6)	(7)	(8) [X − E(x)]
X	Y	P	$XP(x)$ (1×3)	$X - E(x)$	$YP(y)$ (2×3)	$Y - E(y)$	$[Y - E(y)]P$ (5×7)
5.0	.49	.15	.75	−1.97	.0735	−.17	.05
7.5	.68	.20	1.5	.53	.136	.02	.002
6.2	.59	.30	1.86	−.77	.177	−.07	.016
8.2	.79	.15	1.23	1.23	.1185	.13	.024
7.8	.72	.10	.78	.83	.072	.06	.005
8.5	.83	.10	.85	1.53	.083	.17	.026
			6.97		.66		.123
			$E(X)$		$E(Y)$		$\text{Cov}(X, Y)$

what direction the variables move in a straight-line fashion and in association with each other. The correlation is calculated by dividing the covariance between the two variables by the product of their standard deviations.

$$\text{Corr}_{x,y} = \frac{\text{Cov}(x, y)}{\sigma_x \sigma_y}$$

The **correlation coefficient** at the sample level would be

$$r_{x,y} = \frac{SS_{x,y}}{\sqrt{SS_x}\sqrt{SS_y}}$$

The correlation coefficient would be interpreted as follows:

- If $\text{Corr}_{x,y} = 1$: X and Y are in a perfectly positive linear relationship. They move together in the same direction. If X increases, Y also increases, and vice versa.
- If $\text{Corr}_{x,y} = -1$: X and Y are in a perfectly negative linear relationship. They move opposite to each other. If X increases, Y decreases and vice versa.
- If $\text{Corr}_{x,y} = 0$: X and Y have no linear relationship. Therefore, any move in X does not affect Y, and any move in Y does not affect X.
- If $-1 < \text{Corr}_{x,y} < 0$: X and Y are in a negative relationship, reflected by the magnitude of the coefficient. So if $\text{Corr} = -.9$, it is strongly negative, but if $\text{Corr} = -.2$, it is weakly negative; and so on.
- If $0 < \text{Corr}_{x,y} < 1$: X and Y are in a positive relationship, reflected by the magnitude of the coefficient. So if $\text{Corr} = .9$, it is strongly positive; if it is equal to .2, it is weakly positive; and so on.

Example 3.9.1   Calculate the correlation between the two variables of Example 3.8.1. We had the covariance of X and Y. We need to calculate the standard deviation of both variables individually. For that we borrow columns (3), (5), and (7) from Table E3.8.1 and complete them in Table E3.9.1 with two more columns to calculate the variance and then the standard deviation.

$$\sigma_x = \sqrt{\sigma_x^2}$$
$$= \sqrt{1.346} = 1.16$$
$$\sigma_y = \sqrt{\sigma_y^2}$$
$$= \sqrt{.012} = .109$$
$$\text{Corr}_{x,y} = \frac{\text{Cov}(X, Y)}{\sigma_x \sigma_y}$$

TABLE E3.9.1

(1)	(2)	(3)	(4)	(5)	(6)	(7)
P	$X - E(x)$	$[X - E(x)]^2$	$[X - E(x)]^2 P$	$Y - E(y)$	$[Y - E(y)]^2$	$[Y - E(y)]^2 P$
15	−1.97	3.881	.582	−.17	.029	.00435
.20	.53	.281	.056	.02	.0004	.00008
.30	−.77	.593	.178	−.07	.005	.0015
.15	1.23	1.513	.227	.13	.017	.00255
.10	.83	.689	.069	.06	.0036	.00036
.10	1.53	2.34	.234	.17	.029	.0029

$$\sigma_x^2 = 1.346 \qquad\qquad\qquad \sigma_y^2 = .012$$

$$= \frac{.123}{1.16(.109)}$$

$$= .97$$

which means that the variables X and Y are almost perfectly positively related.

3.10. NORMAL DISTRIBUTION

The **normal distribution** of statistical data is famous for its *bell-shaped curve*, also known as the *Gaussian curve*. For more than a century, the discovery of the normal distribution was credited to Carl Friedrich Gauss (1777–1855) and the Marquis de Laplace (1749–1827). It was not until 1924 that Karl Pearson found the original pioneering paper by Abraham De Moivre (1667–1754), published in 1733. It contained the first analysis of the normal distribution and its equation.

The normal distribution is a continuous distribution that has a symmetrical dispersion around the mean or the expected value, which gives it a bell shape. Many variables in nature and in human life have numerical observations that tend naturally to cluster around their mean; thus, this curve provides a good model of data analysis. It is very useful in evaluating the accuracy of sampling outcomes. This is why the normal distribution can approximate the frequency distributions observed for many natural, physical, and human measurements, such as IQs, heights, weights, sales, returns, and many human and machine outputs. It can also estimate binomial probabilities, especially as the sample size increases.

The probability density function for the normal distribution depends greatly on the value of the distribution mean and its standard deviation:

$$f(x) = \frac{1}{\sigma\sqrt{2\pi}} \cdot e^{-(1/2)[(x-\mu)/\sigma]^2} \qquad \text{for} \quad -\infty \leq x \leq +\infty$$

where μ is the mean of the normal random variable, σ is the standard deviation of the distribution, e is 2.71828, and π is 3.14159.

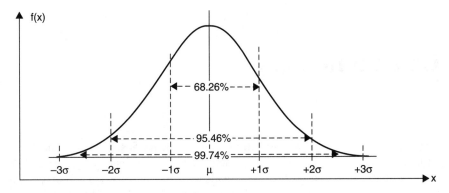

FIGURE 3.3

While the standard normal distribution has a mean of 0 and a standard deviation of 1, any other combinations of mean and standard deviation can identify a unique normal distribution. The total area under the curve is said to be equal to 100%, where it is divided between 50% above the mean and 50% below the mean. If the value of a random variable lies between two certain points, the probability for that to occur would be equal to the area under the curve between those points. The graph of the curve in Figure 3.3 shows that the actual value of a variable would be between ±1 standard deviation from the mean 68.26% of the time, between ±2 standard deviations from the mean 95.46% of the time, and between ±3 standard deviations from the mean 99.74% of the time.

A general conclusion can be made that the greater the standard deviation of a distribution, the greater the variability and dispersion of the data, and the farther the expected value of a variable from the mean of the distribution. In this case, the probability of an outcome that is very different from the mean or the expected value would be higher. The opposite would also hold: The smaller the standard deviation, the smaller the variability and dispersion and the closer the outcome to the value expected or the smaller the probability that the outcome is going to be different from the outcome expected.

Unit I Summary

The purpose of this unit was to refresh reactors memories of many fundamental mathematical concepts and terms often used to solve financial problems. We began with numbers and fractions and made a distinction between real and imaginary numbers and between rational and irrational numbers. We talked about proper and improper fractions and mixed and complex fractions. Decimals, repetends, and percentages were explained, as was the fact that the percentage concept is both the most frequently and the most incorrectly used.

Ratios, proportions, and aliquots were also, explained, as were exponents and logarithms, which it is essential to understand. These rules, functions, and uses are very closely related to the essence of many mathematical financial problems. Arithmetic, geometric, recursive, and infinite geometric progressions are also related to many financial issues and were discussed briefly.

Growth and decay functions and curves were part of the discussion because of their relevance as exponential functions. As they are more common in applications the growth and decay functions of the base of natural logarithms were also discussed. We then moved to more relevant concepts, such as the combinatorial rules and terms: specifically, permutation and combination.

Probability is a central topic in many financial and economic applications, especially because of its direct relevance to mathematical expectation and calculation of the most relevant concepts of variation, such as variance and standard deviation. As a continuation of the discussion of the tools used to measure the behavior of random variables and how they relate to each other, we described covariance and correlation. Finally, we explained the idea of the normal distribution of data, which represents one of the most important concepts for understanding and explaining the frequency distribution observed for many natural, physical, human, and certainly, financial measurements.

Mathematical Finance, First Edition. M. J. Alhabeeb.
© 2012 John Wiley & Sons, Inc. Published 2012 by John Wiley & Sons, Inc.

List of Formulas

Percentage amount:

$$P = B \cdot R$$

Base amount:

$$B = \frac{P}{R}$$

Percentage rate:

$$R = \frac{P}{B}$$

Extremes and means:

$$\frac{A}{B} = \frac{C}{D} \rightarrow \frac{A}{C} = \frac{B}{D} \rightarrow AD = BC$$

Laws of exponents:

$$\log X + \log Y = \log XY$$
$$X^a \cdot X^b = X^{a+b}$$
$$\frac{X^a}{X^b} = X^{a-b}$$
$$(X^a)^b = X^{ab}$$
$$(XY)^a = X^a X^b$$
$$\left(\frac{X}{Y}\right)^a = \frac{X^a}{Y^a}$$

Mathematical Finance, First Edition. M. J. Alhabeeb.
© 2012 John Wiley & Sons, Inc. Published 2012 by John Wiley & Sons, Inc.

$$\frac{1}{X^a} = X^{-a}$$

$$\sqrt[a]{X} = X^{1/a}$$

$$\sqrt[a]{X^b} = X^{b/a}$$

$$X^0 = 1 \qquad X \neq 0$$

Laws of the natural exponents:

$$e = \lim_{X \to \infty} \left(1 + \frac{1}{X}\right)^x$$

$$e^0 = 1$$

$$e^1 = 2.718281828$$

$$e^a \cdot e^b = e^{a+b}$$

$$(e^a)^b = e^{ab}$$

$$\frac{e^a}{e^b} = e^{a-b}$$

Laws of logarithms:

$$\log X + \log Y = \log XY$$

$$\log X - \log Y = \log \frac{X}{Y}$$

$$\log_X^a = a \log X$$

$$\log \sqrt[a]{X} = \log X^{1/a} = \frac{1}{a} \log X = \frac{\log X}{a}$$

$$\log 1 = 0$$

$$\log 10 = 1$$

Arithmetic progression:

$$a_n = a_i + (n - 1)d$$

Summation of arithmetic progression:

$$S_n = \frac{n}{2}(a_1 + a_n)$$

Geometric progression:

$$a_n = a_1 r^{n-1}$$

Summation of geometric progression:

$$S_n = \frac{a_1(1 - r^n)}{1 - r}$$

Infinite geometric progression:

$$S_n = a_1 \frac{1 - r^n}{1 - r}$$

$$S_n = \frac{a_1}{1 - r} - \frac{a_1 r^n}{1 - r}$$

Growth–decay function:

$$Y = ab^x$$
$$Y = ae^x$$
$$Y = ae^{-x}$$

n-Factorial:

$$n! = n(n - 1)(n - 2)(n - 3)$$

mn Rule:

$$N = mn$$

Permutation:

$$nP_r = \frac{n!}{(n - r)!}$$

$$nP_r = \frac{n!}{r_1! r_2! \cdots r_m!}$$

Combination:

$$nC_r = \frac{n!}{(n - r)! r!}$$

Probability:

$$P(E) = \lim_{n \to \infty} \frac{f}{n}$$

Mathematical expectation:

$$E(x) = \sum_{i=1}^{n} x_i P(x_i) = \mu$$

$$E[f(x)] = \sum f(x)P(x)$$
$$E[f(x)] = bE(x) + a$$

Variance:

$$\sigma^2 = E[(x - \mu)^2]$$
$$\sigma^2 = \Sigma[x - \mu]^2 P(x)$$
$$\sigma^2 = E[(x^2)] - [E(x)]^2$$
$$s^2 = \frac{\sum_{i=1}^{n}(x_i - \overline{x})^2}{n - 1}$$
$$\sigma^2 = \frac{\sum_{i=1}^{N}(x_i - \mu)^2}{N}$$
$$s^2 = \frac{\sum_{i=1}^{n} x_i^2 - \frac{(\sum_{i=1}^{n} x_i)^2}{n}}{n - 1}$$

Standard deviation:

$$\sigma = \sqrt{\sigma^2}$$

$$\sigma = \sqrt{\sum_{i=1}^{n}(x - \mu)^2 P(x)}$$

$$\sigma = \sqrt{\frac{\sum_{i=1}^{n}(x_i - \mu)^2}{N}}$$

$$s = \sqrt{s^2}$$

$$s = \sqrt{\frac{\sum_{i=1}^{n}(x_i - \overline{x})^2}{n - 1}}$$

Covariance:

$$Cov(X, Y) = E[(X - \mu_x)(Y - \mu_y)]$$

Correlation:

$$Corr_{x,y} = \frac{Cov(x, y)}{\sigma_X \sigma_Y}$$

$$r_{x,y} = \frac{SS_{x,y}}{\sqrt{SS_x}\sqrt{SS_y}}$$

Normal distribution function:

$$f(x) = \frac{1}{\sigma\sqrt{2\pi}} \cdot e^{-(1/2)[(x-\mu)/\sigma]^2}$$

Exercises for Unit I

1. Rewrite $\frac{12}{16}$ in a lower-term fraction, and $\frac{3}{5}$ in a first-step-higher fraction.

2. Rewrite $\frac{5}{4}$ in mixed number format, and turn $1\frac{2}{3}$ into regular fraction format.

3. Turn $\frac{5}{27}$ into a decimal and round it to four decimal places.

4. Rewrite 9.4 as a regular fraction.

5. If your real estate taxes are \$4,500 annually and you know that the rate is 3%, what is the value of your house?

6. Jack has two jobs, his annual income from the first job being \$37,600 and that of the second job being \$9,400. write the relation of the second income to the first in ratio format.

7. Prove that $x^K/x^L = 1/x^{L-K}$.

8. Simplify $(2xx^2/y^2y)^3$.

9. Simplify $(\sqrt{x^4}\sqrt[3]{y^2}/xy^3)^{-3}$.

10. Solve for $M : M = \log_3 \sqrt[5]{3}$.

11. Solve for $G : \log_9 G = \frac{3}{2}$.

12. Solve for $x : 20(2.25)^x = 120$.

13. Given the arithmetic progression $-\frac{1}{6}, \frac{1}{12}, \frac{1}{3}, \ldots$, find the 10th term and the sum of the 12 terms.

14. Given the geometric progression $625, 125, 25, \ldots$ find the sixth term and the sum of the 15 terms.

15. Draw the function $y = 230(1.34)^x$.

16. Find the number of permutations of the letters v, w, x, y, z taken five at a time.

17. Write the letters M, N, O in six different arrangements of three each.

18. How many teams of five students each can be formed in a class of 15 students?

19. Calculate $4C_2$ and $5C_3$.

Mathematical Finance, First Edition. M. J. Alhabeeb.
© 2012 John Wiley & Sons, Inc. Published 2012 by John Wiley & Sons, Inc.

20. If you are offered $20 to draw a red ball from a bag that contains six green balls, two blue balls, and two red balls, how much would you be willing to pay for a ticket to draw?

21. Calculate the variance and standard deviation of x, which has the following values at the following y probabilities:

x_i	3.5	4.8	−2.8	5.5	6.1	3.7
P	.15	.20	.25	.10	.20	.10

22. Calculate the covariance of x of Exercise 21 with the following y:

$$y_i : 3.9 \quad 4.2 \quad -2.5 \quad 5.0 \quad 6.7 \quad 3.5$$

23. What would the correlation coefficient be between x and y above?

UNIT II

Mathematics of the
Time Value of Money

Introduction

The **time value of money**, a key theoretical concept and fundamental tool in finance, refers to the bidirectional nature of the value of money as it fluctuates up and down over time. Generally, in the absence of an interest rate, money tends to have a higher value in the present and a lower value in the future. Three factors may explain this fact: inflation, consumer impatience, and life uncertainty. **Inflation** is a steady rise in the general level of the price of goods and services. When these prices increase, the purchasing power of money decreases, simply because more money will be needed after inflation begins to make the same purchases as were made before. However, even if there is no inflation, certain noninflationary factors, such as consumer impatience and the uncertain nature of life, would still contribute to decrease the value of money in the future compared to its value at present. **Consumer impatience** refers to people's general preference for today's satisfaction over tomorrow's. Almost anyone would prefer to purchase a favorite car or stereo set today as opposed to next year or next month. So the immediate utility derived from the goods and services purchased immediately gives the money its higher current value compared to a lower value at a later time, which would yield a delayed utility. **Life uncertainty** poses a great risk to the extension of time to utilize money. For example, if a prizewinner would probably be very disappointed were the delivery of his or her prize to be postponed for a year. The rules and regulations governing the prize might change unfavorably; the obligations and liabilities might become greater; taxes on the prize and the fees associated with collection might all rise; and many other things might occur and jeopardize the collection or at least reduce the benefits of the prize. So, collecting the prize sooner gives the money higher current value, and delaying the collection lowers the value of the prize.

For all of these reasons—inflation, impatience, and uncertainty—people would be better off to utilize their money as soon as possible. However, if they have to wait and forgo the immediate satisfaction brought about by spending the money, a reward to compensate for the sacrifice should be due. We call this reward **interest**, defined as the price of money services, the focal point of the time value of money concept and the theoretical and practical core of finance. The **rate of interest** is the reward that would have to be paid by a borrower to a lender for the use of money borrowed. This rate is usually expressed as a percentage of the original amount of money borrowed. Since money is generally

Mathematical Finance, First Edition. M. J. Alhabeeb.
© 2012 John Wiley & Sons, Inc. Published 2012 by John Wiley & Sons, Inc.

characterized as being a store of value or worth, the rate of interest may also reflect the opportunity cost of holding that value or wealth that could be earned on other financial alternatives.

Interest rates have become a major indicator of the economic performance of a country. They are often characterized as the most important regulator of the pace of business and the prosperity of nations. Interest rates are crucial factors for consumers, whether they are borrowing money to purchase a home, saving or investing part of their income, or building funds for their retirement or for their children's education. They are also crucial determinants for businesses in their quest to expand their operations, venture into new projects, or develop product innovations. As to the interest rate involved, calculations of the time value of money can be bidirectional. First, the money value would increase as we move forward from the present to the future. Such an increase in value is due to the compensatory effect of an interest rate in both its **simple** and **compounded** accumulative processes. Second, the money value decreases as we move backward from the future to the present. This decline in value is due to the depreciative effect of the discount rate in a reducing process called **discounting**.

In all calculations of the time value of money, five key terms are involved:

1. The worth of money in the current period, as it is represented by the initial amount of money saved or invested, often called the **principal**, **current value**, or **present value**.
2. The worth of money in the future, often called the **future value**, which is the current value after it grows due to the accumulation of interest.
3. The prevailing interest rate, which is the rate that is applied to the current value to turn it into a future value. It can also work backward to return the future value to its original value. In this case it would be called a **discount rate**.
4. The time of maturity, the time span between the current and future values.
5. The periodic payment.

Commonly, most calculations involve solving for one variable in terms of the other variables. Calculation by mathematical formula is the focus of this book, although the table method is introduced in a few examples. Financial calculator and computer methods are not discussed. Calculation of the financial variables can be made based on either the simple or compound processes of interest accumulation. We begin with simple interest and discount, then move to the compound method for both a single amount of money and a stream of periodic payments.

1 Simple Interest

1.1. TOTAL INTEREST

In the simple interest method, interest is assessed on the principal only. Since
we emphasize fluctuation of the money value across time, the principal is called
the **current value** (CV) to reflect the money value of the initial fund as it stands
at any moment in the present time, while the **future value** (FV) represents the
money value of what that principal would grow to over a certain period of time
(n) under the effect of a certain interest rate (r). Therefore, the total simple
interest (I) that is accumulated for a certain principal is calculated as

$$I = CV \cdot r \cdot n$$

Example 1.1.1 What would the total interest be on a deposit of $2,000 for
5 years in a savings account yielding $3\frac{1}{2}\%$ annual simple interest?

$$I = CV \cdot r \cdot n$$
$$= 2,000(.035)(5)$$
$$= 350$$

1.2. RATE OF INTEREST

We can obtain the rate of interest (r) if the other three variables [total interest
earned (I), time of maturity (n), and initial amount of money invested (CV)] are
all given:

$$r = \frac{I}{CV \cdot n}$$

Mathematical Finance, First Edition. M. J. Alhabeeb.
© 2012 John Wiley & Sons, Inc. Published 2012 by John Wiley & Sons, Inc.

Example 1.2.1   Jill saved $1,500 in her local bank for 3 years and earned $236.25. What would the rate of interest have been?

$$r = \frac{236.25}{1,500(3)}$$

$$= .0525 = 5.25\%$$

1.3. TERM OF MATURITY

In the same way, we can obtain the term of maturity (n) given CV, I, and r.

$$\boxed{n = \frac{I}{CV \cdot r}}$$

Example 1.3.1   Suppose that another local bank offers Jill $5\frac{1}{2}\%$ interest if she doubles her deposit with them, and that she can earn a total of $825 in interest after a certain time. How long would Jill have to keep her money in that bank to earn the $825 interest?

$$1,500 \times 2 = 3,000$$

$$n = \frac{825}{3,000(.055)}$$

$$= 5$$

1.4. CURRENT VALUE

We can obtain CV given I, r, and n.

$$\boxed{CV = \frac{I}{r \cdot n}}$$

Example 1.4.1   How much should Jill deposit at 6% interest to collect $1,000 in total interest at the end of 4 years?

$$CV = \frac{1,000}{.06(4)}$$

$$= 4,167$$

Note that **current value** is used in this book as a synonym of **present value**, as is used in many other books. Generally in the financial literature, the two terms are, used interchangeably, but for greater accuracy, the present value should reflect more than one current value, as we will see later.

1.5. FUTURE VALUE

If we add the total interest earned (I) to the initial amount saved or invested (CV), we obtain the maturity amount, or future value (FV), to which the current value will grow.

$$FV = CV + I$$
$$FV = CV + CV \cdot r \cdot n$$
$$\boxed{FV = CV(1 + rn)}$$

Example 1.5.1   Amy borrowed \$900 for 16 months at an annual simple interest of 7%. How much would the payoff be?

As 16 months $= 1\frac{1}{3}$ years $= 1.33$,

$$FV = CV(1 + rn)$$
$$= 900[1 + .07(1.33)]$$
$$= 983.79$$

1.6. FINDING n AND r WHEN THE CURRENT AND FUTURE VALUES ARE BOTH KNOWN

In the equation $FV = CV(1 + rn)$, we can get either r or n in terms of other variables in the equation.

$$\boxed{r = \frac{(FV/CV) - 1}{n}}$$

and

$$\boxed{n = \frac{(FV/CV) - 1}{r}}$$

Example 1.6.1   Tom will need to have \$7,500 in $2\frac{1}{2}$ years. At present he has only \$6,218 in his savings account. What simple interest rate would allow him to collect \$7,500 after $2\frac{1}{2}$ years?

$$r = \frac{(FV/CV) - 1}{n}$$
$$= \frac{(7,500/6,218) - 1}{2.5}$$
$$= .0825$$

The interest rate must be $8\frac{1}{4}\%$.

Example 1.6.2 If Tom needs \$9,000 and the interest rate drops to 8%, how long would he need to keep the \$6,218 in the savings account?

$$n = \frac{(FV/CV) - 1}{r}$$

$$= \frac{(9,000/6,218) - 1}{.08}$$

$$= 5.5$$

He needs to keep the money for at least $5\frac{1}{2}$ years.

1.7. SIMPLE DISCOUNT

If we reverse the process of obtaining the future value (FV) as shown in Section 1.5, we would obtain CV from the future value, interest rate, and time. This process of getting the current value is called the **simple discount**, for it brings the future value back from its maturity date to the current time. Therefore, the current value would be calculated as

$$\boxed{CV = \frac{FV}{1 + rn}}$$

We can use this simple discount to get the current value of a future value in the two types of debt: non-interest- and the interest-bearing debt.

Simple Discount of a Non-Interest-Bearing Future Amount

Example 1.7.1 Determine the simple discount of a debt of \$5,500 that is due in 8 months when the interest rate is 6.5%. FV = \$5,500; n = 8 months; r = 6.5% (see Figure E1.7.1).

$$CV = \frac{\$5,500}{1 + .065(8/12)}$$

$$= \$5,271.56$$

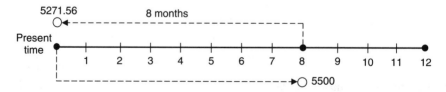

FIGURE E1.7.1

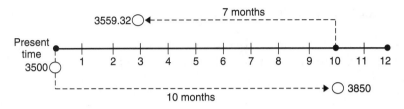

FIGURE E1.7.2

The difference between the original and discounted amount is the simple discount. Note that the interest rate, 6.5%, is the same as the discount rate because the process is reversed. If it happens that we know an amount of \$5,271.56 in the present time and we want to see how much it would grow to in 8 months at an interest rate of 6.5%, we would get \$5,500.

$$FV = CV(1 + rn)$$
$$= \$5,271.56\left[1 + .065\,(8/12)\right]$$
$$= \$5,500$$

Simple Discount of an Interest-Bearing Future Amount

Example 1.7.2 Dina borrowed \$3,500 at 12% for 10 months, but 3 months later she had to go abroad, and her lender agreed to settle by discounting the loan at 14%. How much did she have to pay?

Since this is an interest-bearing debt, we have to calculate the original payoff amount. $CV = \$3,500$; $n = 10$ months; $r = 12\%$ (see Figure E1.7.2).

$$FV = CV(1 + rn)$$
$$= \$3,500\left[1 + .12\left(\frac{10}{12}\right)\right]$$
$$= \$3,850$$

Next we discount this amount for 7 months at a discount rate of 14%. $CV = \$3,850$; $n = 7$ months; $r = 14\%$.

$$CV = \frac{FV}{1 + rn}$$
$$= \frac{\$3,850}{1 + .14(7/12)}$$
$$= \$3,559.32$$

1.8. CALCULATING THE TERM IN DAYS

When the term of maturity is expressed in days, there are two methods that can be used to calculate the number of days between the term starting day and the due day.

Exact Time

The term would include all days in the time segment except the first day. The exact number of days would be obtained by subtracting the serial number of the starting day from the serial number of the due day. **Serial numbers** refer to the accumulative number of calendar days at any day of the year, as shown in Table 4 in the Appendix.

Example 1.8.1   If the term of a loan starts on May 15 and its maturity day is November 23, how many days will the term include using the exact time method?

Look at the serial table for the numbers associated with the term dates:

$$May\ 15: 135$$

$$November\ 23: 327$$

The number of days in this term is $327-135 = 192$ days exactly.
If a serial table is not available, the exact time can be calculated this way:

$$May:\ 16\ days(31 - 15 = 16)$$
$$June:\ 30$$
$$July:\ 31$$
$$August:\ 31$$
$$September:\ 30$$
$$October:\ 31$$
$$November:\ \underline{23}$$
$$Total:\ 192$$

Approximate Time

In this method, the number of months is calculated first. It is obtained by observing how many whole months are between the starting day and the same day of the month where the due day would be. All months of the year are assumed to have a standard 30 days each. If there are days remaining in the term, they would be added.

Example 1.8.2   Calculate the days of the term in Example 1.8.1 using the approximate time method.

Number of months from May 15 to November 15 is 6

Number of days between May 15 and November 15 is $6 \times 30 = 180$

Number of days between November 15 and November 23 is $23 - 15 = 8$

Total approximate number of days $= 180 + 8 = 188$

1.9. ORDINARY INTEREST AND EXACT INTEREST

The assumption of assigning 30 days to each month of the year means that the year would have 360 days (12×30). When interest is calculated using the per day interest, and when the 360 days are used as a denominator for the proportion of the term, the interest obtained would be called **ordinary interest**. But when 365 days are used in the denominator, the interest obtained is called **exact interest**. However, for a leap year, 366 days is used.

Example 1.9.1   What would be the simple interest on $8,500 for 90 days at 7.25% annual simple interest? Use both the ordinary and exact interest methods.

Using ordinary interest:

$$I_0 = CV \cdot r \cdot n$$

$$= 8,500(.0725) \left(\frac{90}{360} \right) = \$154.06$$

Using exact interest:

$$I_e = 8,500(.0725) \left(\frac{90}{365} \right) = \$151.95$$

If a combination of exact time and ordinary interest is used, the method is called the **banker's rule**, which is followed by most commercial banks. But most credit card companies use a combination of exact time and exact interest.

1.10. OBTAINING ORDINARY INTEREST AND EXACT INTEREST IN TERMS OF EACH OTHER

$$I_0 = CV \cdot r \cdot \frac{n}{360} \quad \text{or} \quad I_0 = \frac{CV \cdot r \cdot n}{360}$$

$$I_e = CV \cdot r \cdot \frac{n}{365} \quad \text{or} \quad I_e = \frac{CV \cdot r \cdot n}{365}$$

Dividing I_0 by I_e gives us:

$$\frac{I_0}{I_e} = \frac{(CV \cdot r \cdot n)/360}{(CV \cdot r \cdot n)/365}$$

$$= \frac{\cancel{CV \cdot r \cdot n}}{360} \times \frac{365}{\cancel{CV \cdot r \cdot n}}$$

$$= \frac{365}{360}$$

Dividing by 5:

$$= \frac{73}{72} = \frac{72 + 1}{72}$$

$$= \frac{72}{72} + \frac{1}{72}$$

$$\frac{I_0}{I_e} = 1 + \frac{1}{72}$$

$$\boxed{I_0 = I_e \left(1 + \frac{1}{72}\right)} \quad \text{or} \quad \boxed{I_0 = 1.014 I_e}$$

Similarly, we obtain

$$\boxed{I_e = I_0 \left(1 + \frac{1}{73}\right)} \quad \text{or} \quad \boxed{I_e = \frac{I_0}{1.014}}$$

Example 1.10.1   What would the ordinary interest be if the exact interest is $27.70?

$$I_0 = 27.70 \left(1 + \frac{1}{72}\right)$$

$$= 27.70 + \frac{27.70}{72}$$

$$= 28.08$$

Example 1.10.2   Obtain the exact interest corresponding to an ordinary interest of $502.66.

$$I_e = I_0 \left(1 - \frac{1}{73}\right)$$

$$= 502.66 \left(1 - \frac{1}{73}\right)$$

$$= 502.66 - \frac{502.66}{73}$$

$$= 495.77$$

1.11. FOCAL DATE AND EQUATION OF VALUE

Given that the money value fluctuates across time, it is imperative in finance that funds be stated in the same time terms so that monies can be compared fairly, added up, or reconciled. For this purpose, a **time line diagram** can prove to be very helpful in illustrating the values of funds at various points in time. The **focal date** is the date at which various funds are chosen to be evaluated. Most often, a focal date is at the present time, with values of funds maturing at different times all pulled back to the current time. In other words, this process is to evaluate future funds as if they are to be cashed in today. However, a focal date can also be at a future date.

Example 1.11.1   Jimmy received a loan that he was to pay off in three installments: $500 in a year, $1,200 in 20 months, and $1,500 in 2 years (see Figure E1.11.1). What would be the amount of loan received if the annual simple interest rate is $9\frac{1}{2}\%$?

We have three future values, which need individually to be brought back to the present—to a focal date that is today.

$$CV = \frac{FV_1}{1+rn} + \frac{FV_2}{1+rn} + \frac{FV_3}{1+rn}$$

$$= \frac{500}{1+.095(1)} + \frac{1,200}{1+.095(1\frac{2}{3})} + \frac{1,500}{1+.095(2)}$$

$$= 456.62 + 1,036 + 1,260.50$$

$$= 2,753.12$$

Example 1.11.2   Jill signed a loan contract that charges an annual simple interest of $8\frac{1}{2}\%$. The payoff term required two payments: $5,500 in 9 months and $6,750 in 30 months (see Figure E1.11.2). But later Jill decided to pay it all off in only $1\frac{1}{2}$ years. How much should Jill expect to pay?

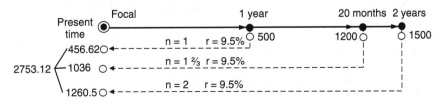

FIGURE E1.11.1

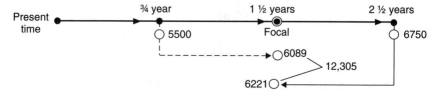

FIGURE E1.11.2

The focal date here is $1\frac{1}{2}$ years from now, and the two payments have to be evaluated at that date. Note that the value of the first payment has to move from its time forward to the focal day, and the value of the second payment has to be brought back from a future day to the focal day.

$$FV = CV(1 + rn)$$
$$= 5,500\left[1 + .085\left(1\tfrac{1}{4}\right)\right]$$
$$= 6,084.37$$

$$CV = \frac{FV}{1 + rn} = \frac{6,750}{1 + .085(1)} = 6,221.19$$

The payment at $1\frac{1}{2}$ years $= 5,850.62 + 6,221.19 = 12,071.81$.

This method of obtaining one payment at the focal day for two obligations that are due on different dates is, called **equating the values**. We essentially equate the unknown single payment (X) to the sum of the growing first payment (FV) and the discounted second payment (CV).

$$X = FV + CV$$

An **equation of value** is therefore a mathematical equation that expresses the equivalence in value of a number of original obligations due on specific dates with a new payment after the value of all have been brought to the focal date, using a given interest rate.

Example 1.11.3   A man has two loans:

1. $1,500 that is due 2 months from now with 7% annual simple interest
2. $750 that is due 5 months from now (see Figure E1.11.3).

If he wants to mix them in a single payment 10 months from now, how much would he pay given that the interest rate is 5%?

First we calculate the first debt, as it is due in 2 months with a 7% annual interest.

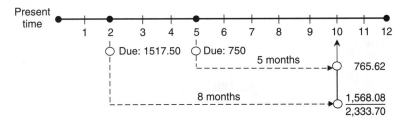

FIGURE E1.11.3

$$FV = CV(1 + rn)$$

$$= \$1,500\left[1 + .07\left(\frac{2}{12}\right)\right]$$

$$= \$1,517.50$$

and if this debt is pushed to be paid in 8 months at 5% interest, then

$$FV = CV(1 + rn)$$

$$= \$1,517.50\left[1 + .05\left(\frac{8}{12}\right)\right]$$

$$= \$1,568.08$$

The second debt ($750) is going to be pushed to be paid in 5 months at 5% interest:

$$FV = CV(1 + rn)$$

$$= \$750\left[1 + .05\left(\frac{5}{12}\right)\right]$$

$$= \$765.62$$

The single payment 10 months from now would be

$$\$1,568.08 + \$765.62 = \$2,333.70$$

which is the equivalent value of the two debts.

Example 1.11.4   A man has a debt of $1,300 that is due 7 months from now (see Figure E1.11.4). If he has the option to pay it either earlier, such as 3 months from now, or later, at the end of the year (12 months from now), how much would he pay in both cases given that the interest rate is 12%?

In the first option, the debt would be paid 4 months earlier. In this case, the original debt of $1,300 would be considered a future value that we need to bring back to its current value on the third month:

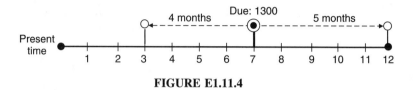

FIGURE E1.11.4

$$CV = \frac{FV}{1 + rn}$$

$$= \frac{\$1,300}{1 + .12(4/12)]}$$

$$= \$1,250$$

In the second option, the debt of $1,300 is treated as a current value at its original due date in the seventh month. Here we want to obtain its future value at the end of the year, 5 months later:

$$CV = CV(1 + rn)$$

$$= \$1,300 \left[1 + .12 \left(\frac{5}{12} \right) \right]$$

$$= \$1,365$$

1.12. EQUIVALENT TIME: FINDING AN AVERAGE DUE DATE

If several obligations are due on different maturity dates, and if there is a desire to pay them all off with interest in a single payment, a new date when that single payment will be due has to be found. The date on which the single payment would discharge all debts is called the **average due date** or **equated date**. It is the corresponding date of the last day of the average term, which can be obtained as the weighted average of all maturity terms of the various obligations. It is called the **equivalent time** ($\bar{n}$).

$$\bar{n} = \frac{\sum P_i n_i}{\sum P_i} = \frac{P_1 n_1 + P_2 n_2 + \cdots + P_k n_k}{P_1 + P_2 + \cdots + P_k}$$

where $i = 1, 2, 3, \ldots, k$; $P_1, P_2, P_3, \ldots, P_k$ are payments; and $n_1, n_2, n_3, \ldots, n_k$ are the due dates of those payments, respectively.

Example 1.12.1   A small store owner has to pay his supplier three payments: $200 in 30 days, $400 in 60 days, and $600 in 90 days. If the interest is 8%, what single payment would discharge all three payments? What would be the equated date?

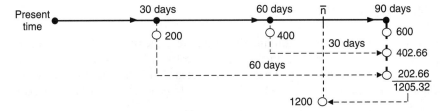

FIGURE E1.12.1

First, let's go through the traditional solution with a time line diagram (Figure E1.12.1), making 90 days the focal date. We obtain future values of $200 for 60 days and $400 for 30 days.

$$FV = CV(1 + rn)$$

$$= 200\left[1 + .08\left(\frac{60}{360}\right)\right] = 202.66$$

$$= 400\left[1 + .08\left(\frac{30}{360}\right)\right] = 402.66$$

We can obtain the total future value at the end of 90 days by adding these values to $600:

$$\text{total FV} = 600 + 402.66 + 202.66 = 1,205.32$$

Since the total of the original payments is $1,200, we can conclude that $1,200, which is less than $1,205.32, would be due at a time earlier than the 90th day. If we assume that this day is $\bar{n}$ somewhere before the 90th day, we can discount the $1,205.32 to $1,200 at $\bar{n}$, obtaining the term $90 - \bar{n}$.

$$CV = \frac{FV}{1 + rn}$$

$$1,200 = \frac{1,205.32}{1 + .08[(90 - \bar{n})/360]}$$

Solving for $\bar{n}$, we get

$$\bar{n} = 70 \text{ days}$$

Now we can apply the $\bar{n}$ formula to get the same result.

$$\bar{n} = \frac{P_1n_1 + P_2n_2 + P_3n_3}{P_1 + P_2 + P_3}$$

$$= \frac{200(30) + 400(60) + 600(90)}{200 + 400 + 600} = 70$$

1.13. PARTIAL PAYMENTS

If part of a debt is paid before the due date, the principal should be discounted properly so that a reduction in the total interest can be assured. But all of the details governing interest and how it is calculated should be put down in the loan contract. Generally, two methods are available to calculate the balance due after a partial payment is made.

1. *Merchant's rule.* In this method, the focal date is the final due date, and therefore each partial payment earns interest from the time it is made to the focal date. The balance due is therefore the difference between the amount of the debt and the sum of the partial payments made.
2. *U.S. rule.* In this rule, the outstanding principal would be adjusted each time a partial payment is made. Any partial payment exceeding the interest would be discounted from the outstanding principal, and any partial payment that is less than the interest would be held without interest until another partial payment is made and until the combined payment exceeds the interest and results in reduction of the principal.

Example 1.13.1   A loan of $1,300 with 7% interest is due in a year. The borrower made a $300 payment after 3 months and $500 after 8 months. How much would the final balance be at the end of the year? Use both the merchant's rule and the U.S. rule.

By the merchant's rule (see Figure E1.13.1a):

$$FV_1 = CV(1 + rn)$$

$$= 1,300\left[1 + .07\left(\frac{12}{12}\right)\right] = 1,391$$

$$FV_2 = 300\left[1 + .07\left(\frac{9}{12}\right)\right] = 315.75$$

$$FV_3 = 500\left[1 + .07\left(\frac{4}{12}\right)\right] = 511.66$$

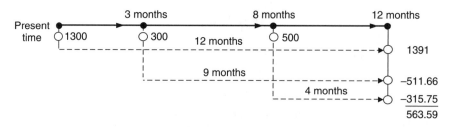

FIGURE E1.13.1a

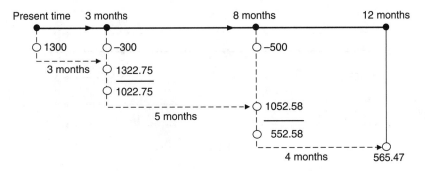

FIGURE E1.13.1b

The balance due $= 1{,}391 - (315.75 + 511.66) = 563.39$.
   By the U.S. rule (see Figure E1.13.1b):

$$FV_1 = CV(1 + rn)$$

$$= 1{,}300\left[1 + .07\left(\frac{3}{12}\right)\right] = 1{,}022.75$$

$$FV_2 = 1{,}022.75\left[1 + .07\left(\frac{5}{12}\right)\right] = 1{,}052.58$$

$$FV_3 = 552.58\left[1 + .07\left(\frac{4}{12}\right)\right] = 565.47 \qquad \text{final balance}$$

1.14. FINDING THE SIMPLE INTEREST RATE BY THE DOLLAR-WEIGHTED METHOD

The *dollar-weighted method* is used to obtain the simple interest rate when the beginning and end of a balance are known as well as the transactions of deposits and withdrawals into and out of the fund. Let's assume that a fund has B as the starting balance and E as the ending balance for a certain period of time t (see Figure 1.1). Let's also assume that throughout period t there are, for simplicity, two transactions: a deposit D at time t_1 and a withdrawal W at time t_2. With this information we can calculate the simple interest r by averaging the amounts of the transactions after weighing them by their own standing periods.

$$r = \frac{E - [(B + D) - W]}{Bt + D(t - t_1) - W(t - t_2)}$$

This method is practical only for short-term fund activities that have a maturity date not exceeding a year.

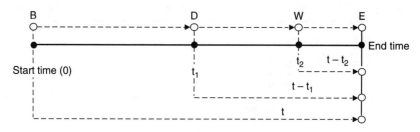

FIGURE 1.1

Example 1.14.1 Sevina entered a short-term investment fund with $3,250 on February 1. She deposited $750 on April 1 and withdrew $1,200 on August 1 (see Figure E1.14.1). What would the simple interest rate of this fund be if on November 1 she had a balance of $3,100?

$B = \$3,250$ at the start time $=$ February1.

$E = \$3,100$ at time $t =$ from February 1 to November 1 $= 9$ months

$D = \$750$ at time $t_1 =$ from April 1 to November 1 $= 7$ months

$W = \$1,200$ at time $t_2 =$ from August 1 to November 1 $= 3$ months

$$r = \frac{E - [(B + D) - W]}{Bt + D(t - t_1) - W(t - t_2)}$$
$$= \frac{\$3,100 - [(\$3,250 + \$750) - \$1,200]}{\$3,250 + \$750(7) - \$1,200(3)}$$
$$= \frac{300}{30,900} = .0097$$

This is the monthly rate, which would be an annual rate of

$$.0097 \times 12 = 11.6\%$$

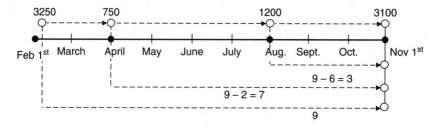

FIGURE E1.14.1

2 Bank Discount

When a lender collects the interest due from the borrower up front, and at the time the loan amount is finalized, that interest paid is called the **discount**. The lender is actually deducting the amount of interest directly from the loan amount and giving the borrower the rest of the loan money, which is called the **proceeds**. Similar to the interest rate, the rate at which the discount is collected is called the **discount rate** and the term of the loan is called the **discount term**. The entire process and its elements are the same as those of the simple interest rate except for one crucial difference: the value on which the interest rate and discount rate are based. The interest rate is based on the current value, and the discount rate is based on the future value.

Example: If the discount rate is 10% on a loan of $1,000, the bank would deduct its discount of $100 ($1,000 × .10) and give the borrower $900 as the proceeds. Since the borrower would have to pay back the $1,000, this $1,000 would be considered a future value (FV) to a current value of $900. The discount rate is 10% (100/1,000) and is not to be confused with the interest rate, which would be based on what is actually received, the proceeds, or the current value ($900), and therefore it would be equal to 11% (100/900).

Similar to the method of calculating the total interest as:

$$I = \text{CV} \cdot r \cdot n$$

the total discount (D) can be obtained by:

$$D = \text{FV} \cdot d \cdot n$$

where D is the total discount (bank discount), FV is the maturity value of a loan, d is the discount rate, and n is the term of the discount. Since the lender would deduct the discount amount (D) from the amount of the loan that is going to be paid back (FV), the borrower would get the proceeds (C), which would be obtained by

$$C = \text{FV} - D$$

Mathematical Finance, First Edition. M. J. Alhabeeb.
© 2012 John Wiley & Sons, Inc. Published 2012 by John Wiley & Sons, Inc.

But as D is equal to $FV \cdot d \cdot n$,

$$C = FV - FV \cdot d \cdot n$$

$$\boxed{C = FV(1 - dn)}$$

2.1. FINDING FV USING THE DISCOUNT FORMULA

Just as we did earlier, we can rearrange the discount formula to find the future value when the proceeds, discount rate, and discount term are known.

$$\boxed{FV = \frac{C}{1 - dn}}$$

Example 2.1.1   Megan would like to borrow \$1,500 from her local bank for $1\frac{1}{2}$ years at a discount rate of $9\frac{3}{4}\%$. How much would Megan receive as proceeds? How much would the bank get as a discount?

$$D = FV \cdot d \cdot n$$

$$= 1,500(.0975)(1.5) = 219.37 \quad \text{the discount}$$

$$C = FV - D$$

$$= 1,500 - 219.37 = 1,289.63 \quad \text{the proceeds}$$

Example 2.1.2   If Megan needs to get a net of \$1,500, what size loan should she apply for?

In this case the \$1,500 would be the proceeds, and we need to find the application amount (maturity value or future value).

$$FV = \frac{C}{1 - dn} = \frac{1,500}{1 - .0975(1.5)} = 1,756.95$$

Megan has to apply for \$1,756.95 in order to receive a net of \$1,500 because the bank discount would be \$256.95.

2.2. FINDING THE DISCOUNT TERM AND THE DISCOUNT RATE

Considering the proceeds formula $C = FV(1 - dn)$, we can obtain both n and d in terms of other variables.

$$d = \frac{1 - (C/FV)}{n}$$ discount rate formula

$$n = \frac{1 - (C/FV)}{d}$$ discount term formula

Example 2.2.1   What would be the discount rate for a $700 loan for 60 days if the borrower gets $679?

$$d = \frac{1 - (C/FV)}{n}$$

$$= \frac{1 - (679/700)}{(60/360)}$$

$$= .18 \text{ or } 18\%$$

Example 2.2.2   What would be the term of discount for Paul if he receives proceeds of $985 for a 6% loan of $1000?

$$n = \frac{1 - (C/FV)}{d}$$

$$= \frac{1 - (985/1,000)}{.06}$$

$$= .25 \text{ or } \tfrac{1}{4} \text{ of a year, which is 90 days}$$

2.3. DIFFERENCE BETWEEN A SIMPLE DISCOUNT AND A BANK DISCOUNT

In addition to the procedural difference of collecting the discount amount in advance, a bank discount has a slight computational difference from a simple discount. Let's take an example to observe the difference.

Example 2.3.1   Let us discount $5,000 for 6 months at 9% using the simple discount method and the bank discount method.

Using a simple discount, we obtain

$$CV = \frac{FV}{1 + rn}$$

$$= \frac{\$5,000}{1 + .09(6/12)}$$

$$= \$4,784.69$$

Simple discount $= \$5,000 - \$4,784.69 = \$215.31$

Using a bank discount, we obtain

$$D = FV \cdot d \cdot n$$

$$= \$5,000(.09)\left(\frac{6}{12}\right)$$

$$= \$225$$

$$C = FV - D$$

$$= \$5,000 - \$225$$

$$= \$4,775$$

$$\text{Bank discount} = \$5,000 - \$4,775 = \$225$$

So the bank discount produced a larger discount ($225) than the simple discount method, which produced $215.31. This result came when we used the same interest rate r and discount rate d (.09).

Now, let's reverse the logic in this example and try to obtain the discount rate (d) by the bank discount process and the interest rate (r) by the simple discount process.

$$d = \frac{D}{FV \cdot n}$$

$$= \frac{\$225}{\$5,000(6/12)}$$

$$= .09 = 9\%$$

and

$$r = \frac{\$225}{\$4,784.69(6/12)}$$

$$= .094 = 9.4\%$$

So the interest rate of the simple discount process is larger than the discount rate of the bank discount process if we equate the simple discounts of both processes. We can therefore conclude that:

1. If we use the same rate of interest (r) and discount rate (d), the simple discount amount of the bank discount method (D_B) would be larger than the simple discount amount of the simple method (D_S):

$$\boxed{D_B > D_s \qquad \text{if} \quad r = d}$$

2. If we use the same simple discount in both the bank and simple methods, the interest rate of the simple method (r) would be larger than the discount rate of the bank method (d):

$$\boxed{r > d \qquad \text{if} \quad D_B = D_S}$$

In the following section we compare the discount rate (d) to the interest rate (r).

2.4. COMPARING THE DISCOUNT RATE TO THE INTEREST RATE

If a borrower gets to choose between an interest rate and a discount rate, going by the bank discount would cost the borrower slightly more than going by the interest rate. This is one important reason for getting to know how to compare the two rates.

Let's take the two equations of the discount rate and the interest rate:

$$C = FV(1 - dn) \tag{1}$$

and

$$CV = \frac{FV}{1 + rn} \tag{2}$$

Since the proceeds in (1) stand for the current value of the maturity amount in (2), we equate the two terms:

$$FV(1 - dn) = \frac{FV}{1 + rn} \tag{3}$$

Dividing by FV yields:

$$1 - dn = \frac{1}{1 + rn} \tag{4}$$

This can also be written as

$$\frac{1}{1 - dn} = 1 + rn \tag{5}$$

$$\frac{1}{1 - dn} - 1 = rn$$

$$\frac{1 - (1 - dn)}{1 - dn} = rn$$

$$\frac{1 - 1 + dn}{1 - dn} = rn$$

$$\frac{dn}{1 - dn} = rn$$

$$\boxed{\frac{d}{1 - dn} = r} \qquad \text{interest rate in terms of discount rate}$$

To get the d value, we rearrange the equation:

$$d \cdot \frac{1}{1 - dn} = r$$

Substitute for $1/(1 - dn)$ from equation (5):

$$d(1 + rn) = r$$

$$\boxed{d = \frac{r}{1 + rn}}$$ discount rate in terms of interest rate

Example 2.4.1   If a bank discounts a note at a rate of 3% for 120 days, what would be the equivalent interest rate?

$$r = \frac{d}{1 - dn}$$

$$= \frac{.03}{1 - .03(120/360)}$$

$$= .02 \text{ or } 2\% \text{ interest rate}$$

Example 2.4.2   Kathy was offered an interest rate of 9.6% on a loan that she planned to pay back in 8 months, but she was curious to know what the discount rate would be.

$$d = \frac{r}{1 + rn}$$

$$= \frac{.096}{1 + .096(8/12)}$$

$$= 9\%$$

2.5. DISCOUNTING A PROMISSORY NOTE

One of the most common applications for which the interest rate and the discount rate work together is that of cashing a promissory note. When a lender holding a promissory note needs some cash before the maturity date on the note, he may go to a bank to cash the note with a discount. In this case the discount has to be assessed based on the maturity value of the note, which is to be determined based on the original amount loaned (the face value).

Example 2.5.1   Michael borrowed $14,750 from Brian at 7% interest to be matured in 90 days, and he signed a promissory note to that effect (see Figure E2.5.1). But Brian needed cash for an emergency only 30 days later, and he could not wait until Michael could pay him back. He took the promissory note to his local bank and discounted it at $8\frac{1}{4}\%$ discount rate. How much

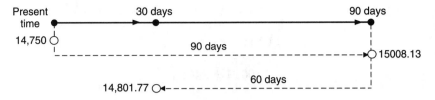

FIGURE E2.5.1

cash would the bank give him, and how much would the bank make from this transaction?

$$FV = CV(1 + rn)$$

$$= 14,750 \left[1 + .07 \left(\frac{90}{360} \right) \right]$$

$$= 15,008.13 \qquad \text{maturity value}$$

$$C = FV(1 - dn)$$

$$= 15,008.13 \left[1 - .0825 \left(\frac{60}{360} \right) \right]$$

$$= 14,801.77 \qquad \text{what Brian receives in cash}$$

$$15,008.13 - 14,801.77 = 206.36 \qquad \text{what the bank receives}$$

Note that what the bank earned is the same amount that Brian lost.

Example 2.5.2 Sylvia signed a promissory note of \$7,800 on July 23 at $6\frac{1}{4}\%$ interest and a maturity date of May 20 of the following year (see Figure E2.5.2). The bearer of the promissory note sold it to his bank on November 15 at a discount rate of $7\frac{1}{2}\%$. (1) How much cash would the bearer of the note receive? (2) Would he ever make money on what he lent to Sylvia? If so, how much would he make and at what interest rate? (3) How much would the bank make out of the discount transaction?

Looking at the serial table, we obtain the date numbers:

$$\text{July 23: 204} \qquad \text{May 20: 140} \qquad \text{Nov. 15: 319}$$

$$\text{Maturity term: } 365 - 204 = 161 + 140 = 301$$

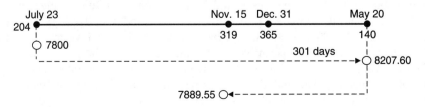

FIGURE E2.5.2

Using the banker's rule, the maturity value would be

$$FV = CV(1 + rn)$$

$$= 7{,}800 \left[1 + .0625 \left(\frac{301}{360} \right) \right]$$

$$= 8{,}207.60$$

Discount term: $140 + (365 - 319) = 186$

$$C = FV(1 - dn)$$

$$= 8{,}207.60 \left[1 - .075 \left(\frac{186}{360} \right) \right]$$

$$= 7{,}889.55 \quad \text{the proceeds; what the bearer of the note receives}$$

$$8{,}207.60 - 7{,}889.55 = 318.05 \quad \text{what the bank would make}$$

The lender lent Sylvia \$7,800 on July 23 and ended up receiving \$7,889.55 on November 15. So he made \$89.55 in 115 days. That is a rate of interest of

$$r = \frac{I}{CV \cdot n}$$

$$= \frac{.89.55}{7{,}800(115/360)}$$

$$= .036 \text{ or } 3.6\%$$

2.6. DISCOUNTING A TREASURY BILL

Investing in U.S. Treasury bills means that an investor pays less in the present for more in a short-term future. In a financial sense, the federal government accepts a discounted amount for a future value that matures within a year. In other words, the government sells notes to individual and institutional investors. The denominations and maturity terms are designated by the government. The notes come with face values of \$1,000, \$10,000, \$15,000, \$50,000, \$100,000, and \$1,000,000. Those amounts would be the maturity or future value of the notes. The maturity terms are in days: 28, 91, 182, and 364. Treasury bills are usually sold at public auctions, and because the return on investment is the difference between what the investor pays and what he or she gets in the near future, a Treasury bill is technically considered a short-term, non-interest-bearing, negotiable security. Although the discount rate can be obtained by the traditional method that we have come to know so far, a formula can be developed to get the discount rate in terms of the discount term and the bid percentage.

Let's consider the bid price as the proceeds, and since it is a percentage, the future value of the proceeds would be $100. Therefore:

$$C = FV(1 - dn)$$

$$B = 100\left[1 - d\left(\frac{n}{360}\right)\right] \qquad \text{where} \quad B = \text{bid\%}$$

Divide by 100:

$$\frac{B}{100} = 1 - d\left(\frac{n}{360}\right)$$

Rearrange:

$$d\left(\frac{n}{360}\right) = 1 - \frac{B}{100}$$

$$d\left(\frac{n}{360}\right) = \frac{100 - B}{100}$$

$$d = \frac{(100 - B)/100}{n/360}$$

$$d = \frac{100 - B}{100.} \cdot \frac{360}{n}$$

Divide by 100:

$$d = \frac{100 - B}{1} \cdot \frac{3.60}{n}$$

$$= \frac{3.6(100 - B)}{n} = \frac{360 - 3.6B}{n}$$

$$\boxed{d = \frac{360 - 3.6B}{n}}$$

Example 2.6.1 A 182-day Treasury bill of $15,000 was purchased for a 95.350% bid. Find the discount rate the traditional way and by using the d formula.

By the traditional method the discount rate (d) is the discount amount (D) divided by the product of the maturity value (FV) and the maturing term (n). The discount amount is the difference between the maturity value (FV) and the purchase price (CV), which can be obtained by multiplying the bid % by the future value.

$$CV = B \cdot FV$$

$$= .95350(15,000)$$

$$= 14,302.50$$

$$D = \text{FV} - \text{CV}$$

$$= 15{,}000 - 14{,}302.50 = 697.50$$

$$d = \frac{D}{\text{FV} \cdot n} = \frac{697.50}{15{,}000(182/360)} = .092 \text{ or } 9.2\%$$

We can also use the d formula:

$$d = \frac{360 - 3.6B}{n}$$

$$= \frac{360 - 3.6(95.350)}{182}$$

$$= .092 \text{ or } 9.2\%$$

Example 2.6.2   On March 3, Charles Tires Co. had a bid of $96.438 on a 271-day Treasury bill of $500,000. What are the purchase price, the total discount, and the rate of return in both the discount rate and interest rate terms?

$$\text{purchase price} = \text{CV} = B \cdot \text{FV}$$

$$= .96438(500{,}000)$$

$$= 482{,}190$$

$$\text{total discount } D = \text{FV} - \text{CV}$$

$$= 500{,}000 - 482{,}190$$

$$= 17{,}810$$

$$d = \frac{D}{\text{FV} \cdot n}$$

$$= \frac{17{,}810}{500{,}000(271/360)}$$

$$= 4.73\% \text{ discount rate}$$

Also,

$$d = \frac{360 - 3.6B}{n}$$

$$= \frac{360 - 3.6(96.438)}{271}$$

$$= 4.73\% \text{ discount rate}$$

$$r = \frac{d}{1 - dn}$$

$$= \frac{.0473}{1 - .0473(271/360)}$$

$$= .049 = 4.9\%$$

3 Compound Interest

Unlike the simple interest method, where interest is earned only on the principal, in the compound interest method, interest is earned on the principal as well as on any interest earned. That is why compound interest is called "interest on interest." Let's consider an example where $1,000 would earn 10% interest for 3 years by both the simple and compound methods.

In the simple method:

Time, n (years)	Principal, CV ($)	Interest, $r(10\%)\$$
1		100
2	$1,000	100
3		100
	1,000	300

$$FV = 1,000 + 300$$
$$= 1,300$$

In the compound method:

Time, n (years)	Principal, CV ($)	Interest, $r(10\%)\$$
1	1,000	100
2	1,100	110
3	1,210	121
	1,331	331

$$FV = 1,000 + 331$$
$$= 1,331$$

Mathematical Finance, First Edition. M. J. Alhabeeb.
© 2012 John Wiley & Sons, Inc. Published 2012 by John Wiley & Sons, Inc.

TABLE 3.1   Simple and Compound Interest Accumulation

				r		
		5%	10%	15%	20%
CV = $100	n			FV	
Simple	10	150	200	250	300
Compound		162.89	259.37	404.56	619.17
Simple	20	200	300	400	500
Compound		265.33	672.75	1,636.65	3,833.76
Simple	30	250	400	550	700
Compound		432.19	1,744.94	6,621.17	23,737.63
Simple	40	300	500	700	900
Compound		703.99	4,525.92	26,786.35	14,697.15
Simple	50	350	600	850	1,100
Compound		1,146.74	11,739.08	108,365.74	810,043.82

The multiple interest earned is the reason behind the dramatic accumulation that occurs by compounding, which led the compound interest method to be described as a "wonder."[†]

In Table 3.1 and Figures 3.1 and 3.2 we can follow how $100 grows by simple and compound interest at four different rates and five different maturity times. We can see that as both r and n increase, the linear function of the simple interest accumulation is represented by proportionately ascending straight lines, while the exponential function of the compound accumulation is represented by steeper and steeper upward curves.

3.1. THE COMPOUNDING FORMULA

Let's turn to our recent example of calculating 10% compound interest on $1000 for 3 years into an example of general terms. Let's consider the amount of $1000

[†]The formula for compounding interest is the most critical formula in finance. The impressive mathematical multiplication of this formula is behind what has been called The "wonder of compounding." The British economist John Maynard Keynes called compound interest "magic," and Baron Rothschild, a banker with international stature, called compounding the "eighth wonder of the world." Mathematicians and economists have been circulating several interesting historical scenarios to illustrate the wonder of compounding. For example, what has become the heart of New York City, the Island of Manhattan, was bought in 1624 from the native Indians for the sum of $24. If this amount was invested at, say, 5% interest, it would have grown to $1,922,293,931 today. When Benjamin Franklin died in 1790, he left to the city of Boston a bequest that was equivalent to $4,570. In his will, he wanted the money to earn interest for a century. He stipulated that part of the fund was to be spent on public projects, and the remaining part was to be invested for another century. By 1890 the bequest had grown to $322,000. About $438,742 was spent in 1907 on the Franklin Institute; the remainder of the fund was left to accumulate to $3,458,797 by 1980. In 1810, Francis Bailey, an English mathematician and astronomer, calculated that if, at the birth of Christ, a British penny was invested at 5% interest compounded annually, it would have grown to what could buy enough gold to fill 357,000,000 Earths.

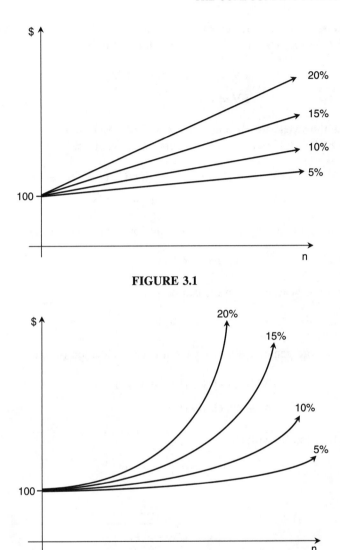

FIGURE 3.1

FIGURE 3.2

as the current value (CV), the 10% compound interest as r, and the maturity term of 3 years as n. The future value (FV) would be calculated according to the following facts:

- In the first year, CV would earn $r \cdot \text{CV}$, and the future value becomes

$$\text{FV} = \text{CV} + r \cdot \text{CV}$$
$$\text{FV} = \text{CV}(1 + r) \tag{1}$$

- In the second year, the current value would be $CV(1 + r)$ and it would earn $r[CV(1 + r)]$. Therefore, the future value would be

$$FV = CV(1 + r) + r[CV(1 + r)]$$
$$FV = CV(1 + r)(1 + r) \qquad (2)$$

- In the third year, the current value would be $CV(1 + r)(1 + r)$ and it would earn $r \cdot CV(1 + r)(1 + r)$. The future value would be

$$FV = CV(1 + r)(1 + r) + r[CV(1 + r)(1 + r)]$$
$$FV = CV(1 + r)(1 + r)(1 + r) \qquad (3)$$

If we notice that in the second year, equation (2) can be written

$$FV = CV(1 + r)^2$$

and in the third year, equation (3) can be written

$$FV = CV(1 + r)^3$$

We can therefore generalize that for the nth year, equation (n) can be written

$$FV = CV(1 + r)^{n-1} + rCV(1 + r)^{n-1}$$
$$FV = CV(1 + r)^{n-1}(1 + r)$$

This equation can be written

$$FV = CV(1 + r)^{n-1+1}$$

and finally,

$$\boxed{FV = CV(1 + r)^n}$$

which is the general formula for compounding interest on a single amount (CV).

Example 3.1.1   Heather invested \$12,000 for 5 years in an account earning $7\frac{1}{3}\%$ annual compound interest. How much would she expect to collect at the end of 5 years?

$$FV = CV(1 + r)^n$$
$$= 12,000(1 + .0733)^5$$
$$= 17,091.76$$

The term $(1 + r)^n$ is calculated as a table value when CV = \$1.00. This table value is called s and can be used as a multiple to calculate the future values, and the formula above can therefore be adjusted to

$$\boxed{FV = CV \cdot s}$$

Example 3.1.2 Using the table method, calculate the future value of an investment amount of \$930 for 6 years if the annual compound interest rate is $7\frac{1}{2}\%$.

We look at Table 5 in the Appendix to get the value of s across $7\frac{1}{2}\%$ interest and 6 years:

$$FV = CV \cdot s$$
$$= \$930(1.543302)$$
$$= \$1,435.27$$

The result is the same if we use the formula

$$FV = CV(1 + r)^n$$
$$= \$930(1 + .075)^6$$
$$= \$1,435.27$$

3.2. FINDING THE CURRENT VALUE

Just as with simple interest, the future value in the compounding formula can be discounted into a current value. That is, the money value can be brought back from the future to the present.

$$\boxed{CV = \frac{FV}{(1 + r)^n}}$$

This formula can be rewritten as

$$CV = \frac{1}{(1 + r)^n} \cdot FV \tag{4}$$

or

$$CV = FV(1 + r)^{-n} \tag{5}$$

The term $1/(1 + r)^n$ or $(1 + r)^{-n}$ is a table value called v^n (Table 6 in the Appendix). This value has been calculated based on the current or present value of $1.00. It is therefore used as a multiple to calculate any future value. The CV formula can be rewritten as

$$\boxed{CV = FV \cdot v^n}$$

Example 3.2.1 An amount of $5,000 is to be inherited in 4 years. How much would it be if it is cashed in now given that the interest rate is $6\frac{3}{4}\%$ compounded annually?

$$CV = \frac{FV}{(1+r)^n}$$

$$= \frac{5,000}{(1+.0675)^4}$$

$$= 3,850.33$$

Example 3.2.2 Use Table 6 in the Appendix to calculate the current value of $2,700 at 5.5% for 9 years.

$$CV = FV \cdot v^n$$

$$= \$2,700(.617629)$$

$$= \$1,667.60$$

Verifying this with a formula method, we get

$$CV = \frac{2700}{(1+.055)^9}$$

$$= \$1,667.60$$

3.3. DISCOUNT FACTOR

A future value can be logically discounted to a current value by being multiplied by a certain **discount factor**, which is determined by both the interest rate and the term of time during which the future value is pulled back from the future to the present time. We can use the format of the current value formula in equation (4) to point out the discount factor:

$$CV = FV \cdot \frac{1}{(1+r)^n}$$

This multiplier of the future value, $1/(1+r)^n$ is the discount factor:

$$DF = \frac{1}{(1+r)^n}$$

It is significant to see how the current value changes depending on the discount factor, which, in turn, depends on the fluctuations of both r and n. In Example 3.2.1, $DF = 1/(1 + .0675)^4 = .77$. Therefore, CV could be obtained by

$$CV = FV \cdot DF$$

$$= 5000(.77) = 3850$$

Example 3.3.1 Next we look at the inheritance value in Example 3.2.1 if it were to be cashed in under the following conditions (see Figure E3.3.1):

(a) The interest rate goes up to 8% but the time stays the same.
(b) The time extends to 6 years but the interest rate stays the same.
(c) The interest rate goes up to 8% and the time is extended to 6 years.

We adjust the discount factor according to the changes given:

(a)
$$DF = \frac{1}{(1 + .08)^4} = .735$$

$$CV = DF \cdot FV$$

$$= .735(5,000) = 3,675$$

Notice that the discount factor became smaller and the current value decreased as we increased the interest rate (r).

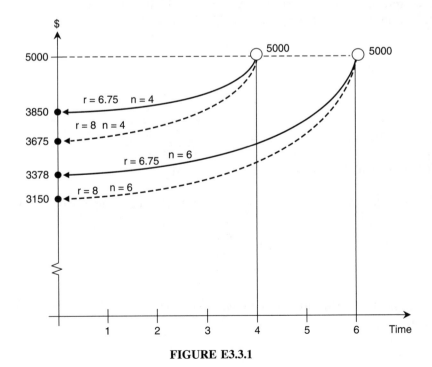

FIGURE E3.3.1

(b)
$$DF = \frac{1}{(1+.0675)^6} = .676$$
$$CV = .676(5,000) = 3.378$$

Here, too, the discount factor got smaller and the current value decreased as we extended the time.

(c)
$$DF = \frac{1}{(1+.08)^6} = .63$$
$$CV = .63(5,000) = 3,150$$

We can conclude that while the discounted value (CV) changes positively with the DF, it has a negative relationship with both the interest rate (r) and the term of discounting (n).

3.4. FINDING THE RATE OF COMPOUND INTEREST

The compound interest rate can be found in terms of all other remaining factors in the compounding formula: FV, CV, and n.

$$r = \sqrt[n]{\frac{FV}{CV}} - 1$$

Example 3.4.1 If you want to buy a car for \$15,485 in 5 years and you want to start investing what you have now, \$7,700, what should the interest rate be?

$$r = \sqrt[n]{\frac{FV}{CV}} - 1$$
$$= \sqrt[5]{\frac{15,485}{7,700}} - 1$$
$$= 15\%$$

3.5. FINDING THE COMPOUNDING TERM

The term (n) can also be obtained in relation to all other compounding factors:

$$n = \frac{\ln(FV/CV)}{\ln(1+r)}$$

Example 3.5.1 How long would it take to collect \$13,000 if you invest \$6,777 at an annual interest rate of $9\frac{3}{8}\%$?

$$n = \frac{\ln(FV/CV)}{\ln(1+r)}$$

$$= \frac{\ln(13{,}000/6{,}777)}{\ln(1+.09375)}$$

$$= 7.27 \text{ years}$$

3.6. THE RULE OF 72 AND OTHER RULES

The maturity formula above can almost accurately estimate the time required for a principal value to grow to any future value. However, if a general approximation is needed to find the time required for a principal sum to grow in certain multiple times, there are simpler formulas. The most common rule in this regard is the **rule of 72**. This rule estimates the number of years (n) for which a principal amount would double given a certain interest rate (r):

$$\boxed{n = \frac{72}{r}} \qquad \text{for 2CV}$$

For the principal to double, it means that we obtain 2CV. Similarly, mathematicians postulated the rule of 114, which estimates the number of years required to triple a principal (getting 3CV), and the rule of 167, to let the principal grow fivefold (getting 5CV), given certain interest rates.

$$\boxed{n = \frac{114}{r}} \qquad \text{for 3CV}$$

and

$$\boxed{n = \frac{167}{r}} \qquad \text{for 5CV}$$

Example 3.6.1   At a time when the interest rate is 10%, it would take 7.2 years to double a principal sum, 11.4 years to triple it, and 16.7 years to have it grow fivefold.

$$\frac{72}{10} = 7.2 \qquad \frac{114}{10} = 11.4 \qquad \frac{167}{10} = 16.7$$

Note: Please notice that the interest rate is used as a whole number, not in a percent format.

Mathematically, we can approximate the time required for an investment amount to grow by any factor if we can break that factor down into its multiples, which would correspond to any of the three rules above and add the rules up in the numerator of the formula.

Example 3.6.2 How long would it take to have an investment amount grow 30-fold if the interest rate is 12%?

First, we analyze 30 into factors corresponding to our rules:

$$30 = 2 \times 3 \times 5$$

Therefore,

$$n = \frac{\text{rule to double} + \text{rule to triple} + \text{rule to grow fivefold}}{r}$$

$$= \frac{72 + 114 + 167}{12} = \frac{353}{12}$$

$$= 29.4 \text{ years}$$

Example 3.6.3 How many years would it take for a principal sum to grow by a factor of 12 given an interest rate of 8.5%?

$$12 = 2 \times 2 \times 3$$

$$n = \frac{72 + 72 + 114}{8.5} = \frac{258}{8.5}$$

$$= 30.3 \text{ years}$$

3.7. EFFECTIVE INTEREST RATE

The interest rate stated on all financial transactions and documents is normally a nominal rate that may also be referred to as the **annual percentage rate** (APR). However, the frequency of the compounding process makes a difference in the amount of interest assessed. The more frequent the compounding, the more interest would accumulate. Therefore, the conversion period becomes important in assessing interest more accurately. The **conversion period** is the time between successive computations of interest. It is the basic unit of time in the compounding process. The best known conversion periods are:

1: for annual compounding

2: for semiannual compounding

4: for quarterly compounding

12: for monthly compounding

TABLE E3.7.1

Nominal Rate, r	Term of Compounding	Conversion Period, m	Effective Rate, R (%)
6%	Annually	1	6
	Semiannually	2	6.09
	Quarterly	4	6.13636
	Monthly	12	6.16778
	Weekly	52	6.17782
	Daily	365	6.18383

52: for weekly compounding

365: for daily compounding

The **effective interest rate** is the rate obtained according to the conversion period used in the compounding process. If we denote the conversion period by m, we can calculate the effective interest rate (R) for any quoted, nominal rate of interest (r), using the following formula:

$$R = \left(1 + \frac{r}{m}\right)^m - 1$$

Example 3.7.1 If the nominal rate of interest stated is 6% (see Table E3.7.1), what would the effective interest rate be if the compounding occurs annually, semiannually, quarterly, monthly, weekly, and daily?

Annually: $R = \left(1 + \frac{.06}{1}\right)^1 - 1 = .06$ or 6%

Semiannually: $R = \left(1 + \frac{.06}{2}\right)^2 - 1 = .0609$ or 6.09%

Quarterly: $R = \left(1 + \frac{.06}{4}\right)^4 - 1 = .0613636$ or 6.136%

Monthly: $R = \left(1 + \frac{.06}{12}\right)^{12} - 1 = .0616778$ or 6.168%

Weekly: $R = \left(1 + \frac{.06}{52}\right)^{52} - 1 = .0617782$ or 6.178%

Daily: $R = \left(1 + \frac{.06}{365}\right)^{365} - 1 = .0618383$ or 6.184%

3.8. TYPES OF COMPOUNDING

The following example illustrates how the compounding process can be adjusted according to the conversion period used. The interest yield would also be different, based on the type of compounding used.

Example 3.8.1   If you invest $1,000 at $7\frac{1}{4}\%$ interest for $3\frac{1}{2}$ years, how much would you accumulate if the compounding occurs annually; semiannually; quarterly; monthly; weekly; daily?

Annually: $FV = CV(1 + r)^n$

$\qquad = 1,000(1 + .0725)^{3\ 1/2}$

$\qquad = 1277.58$

Semiannually: $FV = 1,000(1 + .03625)^7$

$\qquad = 1,283.07 \quad \begin{cases} \text{semiannual rate} = \dfrac{.0725}{2} = .03625 \\ \text{semiannual terms} = 3.5 \times 2 = 7 \end{cases}$

Quarterly: $FV = 1,000(1 + .018125)^{14}$

$\qquad = 1,285.92 \quad \begin{cases} \text{quarterly rate} = \dfrac{.0725}{4} = .018125 \\ \text{quarterly terms} = 3.5 \times 4 = 14 \end{cases}$

Monthly: $FV = 1,000(1 + .006)^{42}$

$\qquad = 1,287.86 \quad \begin{cases} \text{monthly rate} = \dfrac{.0725}{12} = .006 \\ \text{monthly terms} = 3.5 \times 12 = 42 \end{cases}$

Weekly: $FV = 1,000(1 + .00139)^{182}$

$\qquad = 1,288.62 \quad \begin{cases} \text{weekly rate} = \dfrac{.0725}{52} = .00139 \\ \text{weekly terms} = 3.5 \times 52 = 182 \end{cases}$

Daily: $FV = 1,000(1 + .00198)^{1277}$

$\qquad = 1,288.69 \quad \begin{cases} \text{weekly rate} = \dfrac{.0725}{365} = .000198 \\ \text{weekly terms} = 3.5 \times 365 = 1277 \end{cases}$

Notice that the difference in the total interest earned got smaller and smaller as the conversion periods increased dramatically, and particularly, that there is no significant difference between weekly and daily compounding, which is one reason that continuous compounding is not very important.

3.9. CONTINUOUS COMPOUNDING

We have seen that compounding can be carried out for more conversion periods from one period in annual compounding to 365 in daily compounding. We have also seen that the interest rate earned would continue to increase, but not significantly. We can therefore imagine that compounding can be done beyond daily: that is, minute by minute or second by second. It is continuous compounding, which actually is more theoretical than practical because the increase in interest would be even less significant. To understand the concept of continuous compounding, let's go back to the effective interest rate formula:

$$R = \left(1 + \frac{r}{m}\right)^m - 1$$

We can rewrite the formula as

$$R = \left[\left(1 + \frac{r}{m}\right)^{m/r}\right]^r - 1$$

Let $m/r = K$, so that r/m would be $1/K$.

$$R = \left[\left(1 + \frac{1}{K}\right)^K\right]^r - 1$$

Now if we assume that K increases infinitely, the limit of $(1 + 1/K)^K$ would approach the value of e, since

$$e = \lim_{n \to \infty} \left(1 + \frac{1}{n}\right)^n$$

$$e = \lim_{n \to \infty} \left[\left(1 + \frac{1}{K}\right)^K\right]^r - 1$$

Therefore,

$$\boxed{R = e^r - 1}$$

which is the formula for the effective rate when the compounding is continuous. Now we can use the effective rate value R instead of r in the compounding formula.

$$FV = CV(1 + R)^n$$
$$= CV(1 + e^r - 1)^n$$
$$= CV(e^r)^n$$

Hence, the future value under the continuous compounding would be

$$FV = CV \cdot e^{rn}$$

We can also reverse it to obtain the current value:

$$CV = FV \cdot e^{-rn}$$

Example 3.9.1   Use continuous compounding to find the future value in Example 3.8.1.

$$FV = 1,000e^{.0725(3.5)}$$

$$= 1,000(2.71828)^{.25375}$$

$$= 1,288.85$$

Example 3.9.2   A fund of $7,250, compounded continuously at a rate of $5\frac{1}{2}$ %, will be received in 30 months. How much would it be if it were to be cashed today?

$$CV = FV \cdot e^{-rn}$$

$$= 7,250(2.71828)^{-(.055)(2.5)} \quad \left(\text{where } n = \frac{30}{12} = 2.5 \right)$$

$$= 6,318.62$$

3.10. EQUATIONS OF VALUE FOR A COMPOUND INTEREST

Just as we did with simple interest, many obligations and payments assessed at compound interest may need to be reconciled for a final settlement. The same method of equations of value would be used where an appropriate focal day is chosen to achieve an equitable settlement for both sides: obligations and payment.

Example 3.10.1   Bill borrowed $10,500 that will be due in 2 years at $7\frac{3}{4}$ % interest compounded monthly (see Figure E3.10.1). But he made a payment of $3,000 in 8 months and another payment of $2,500 in 15 months. How much will he still owe on the maturity day?

First we calculate the amount due originally:

$$\text{monthly rate} = \frac{.0775}{12} = .0064$$

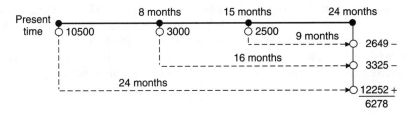

FIGURE E3.10.1

$$FV = CV(1 + r)^n$$
$$= 10,500(1 + .00645)^{24}$$
$$= 12,252$$

Second, we take the two payments to the due date as a focal date.

$$FV_1 = 3,000(1 + .00645)^{16} = 3,325$$
$$FV_1 = 2,500(1 + .00645)^9 = 2,649$$
$$12,252 - (3,325 + 2,649) = 6,278$$

what Bill still owes at the end of 2 years

Example 3.10.2   An inheritance of \$993,715 is to be distributed among heirs in the following way (see Figure E3.10.2):

- The wife will take $\frac{1}{2}$ in 5 years.
- The 11-year-old son will take $\frac{1}{4}$ when he is 19.
- The 7-year-old daughter will take $\frac{1}{4}$ when she is 19.

How much would each heir receive if the fund is invested at $6\frac{1}{2}\%$ interest compounded quarterly?

The money for each heir will mature within a different time period. The best way to find the answer is to bring the values of all shares from their maturity

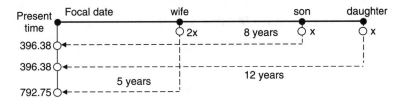

FIGURE E3.10.2

date to the present day, which would be the focal date: $.065/4 = .01625$ is the quarterly rate, $5 \times 4 = 20$ is the wife's maturity term, $8 \times 4 = 32$ is the son's maturity term, and $12 \times 4 = 48$ is the daughter's maturity term. Let's assume that one-fourth of the inheritance is X. Therefore, one-half would be $2X$. These are future values that would be discounted to their current values.

$$CV = \frac{FV}{(1+r)^n}$$

$$\frac{2X}{(1+.01625)^{20}} + \frac{X}{(1+.01625)^{32}}$$

$$+ \frac{X}{(1+.01625)^{48}} = 993{,}715$$

$$\frac{12.56X}{5.01} = 993{,}715$$

$$X = 396.38 \qquad \text{for each son and daughter}$$

$$2X = 396.38 \times 2 = 792.75 \qquad \text{for the wife}$$

3.11. EQUATED TIME FOR A COMPOUND INTEREST

Just as with simple interest, here we can pay the sum of the maturity values of different obligations at a certain date called the **equated date**. The different obligations usually have different due dates, but we can calculate what is called the **equated time**, which is the time extended between the equated date and the present time, where the present time usually serves as a focal date.

Example 3.11.1   A loan of $10,000 is to be paid off in three installments: $2,000 in 2 years, $5,000 in 4 years, and $3,000 in 5 years (see Figure E3.11.1). If the interest is 5%, when would the borrower discharge his debt in a single payment of $10,000?

First we have to discount all of these installments to the present time, which would serve as the focal date. Then we equate the total of all discounted installments with the discounted $1,000 in an equation of value that would be solved for n, which would be the equated time.

$$\frac{2{,}000}{(1+.05)^2} + \frac{5{,}000}{(1+.05)^4} + \frac{3{,}000}{(1+.05)^5} = \frac{10{,}000}{(1+.05)^n}$$

$$\$1{,}814 + \$4{,}113 + \$2{,}350 = \$10{,}000(1+.05)^{-n}$$

$$\$8{,}277 = \$10{,}000(1+.05)^{-n}$$

$$\frac{\$8{,}277}{\$10{,}000} = (1+.05)^{-n}$$

$$.8277 = (1+.05)^{-n}$$

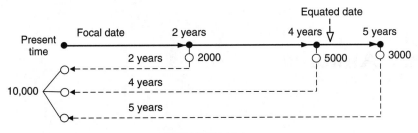

FIGURE E3.11.1

Now we can interpolate the n value using Table 6 in the Appendix.

$$\underline{n} \quad (1 + .05)^{-n}$$

$$3 - 4 \left\{ \begin{array}{ll} 3 & .8638 \\ n & .8277 \end{array} \right\} \quad .8638 - .8227 = .0411$$
$$n - 4 \left\{ \begin{array}{ll} n & .8277 \\ 4 & .8227 \end{array} \right\} \quad .8277 - .8227 = .005$$

$$\frac{n - 4}{3 - 4} = \frac{.005}{.0411}$$

$$n - 4 = \frac{-.005}{.0411} = .1216$$

$$n = .1216 + 4$$

$$= 4.1216 \text{ or } 4 \text{ years and } 44 \text{ days}$$

4 Annuities

Although the word *annuity* was meant originally to refer to an annual payment, the concept has been expanded to represent more than simply an annual payment. An **annuity** is any set of equal payments made at equal intervals of time. In this sense, all periodic saving or investment deposits, purchases on installment, mortgages, auto loans, life insurance premiums, and even Social Security deductions are types of annuities.

4.1. TYPES OF ANNUITIES

Annuities are classified according to three criteria: time of payments, term of annuity, and the timing of the compounding process. The following annuities are based on time of payments:

1. *Ordinary annuity:* an annuity for which payments are made at the end of the interval period. If the annuity is monthly, payments are due at the end of each month. This annuity is also called an **annuity immediate**.
2. *Annuity due:* an annuity for which payments are made at the beginning of each interval period. So if it is a monthly annuity, payments are due at the beginning of each month.
3. *Deferred annuity:* an ordinary annuity but with a deferred term. In this case, the entire payment process would not begin until after a certain designated time has passed. An example is a debt paid off by a certain set of equal payments made at the end of equal intervals, but with the first payment due 3 years from now. So it is an ordinary annuity with a 3-year deferment.

The **term of annuity** is the time between the beginning of the first interval and the end of the last interval. For example, a 3-year term would be between January 1, 2009 and December 31, 2011.

The following annuities are based on the term of the annuity:

1. *Annuity certain:* an annuity for which the beginning and end of the term are designated and recognized in advance. A mortgage and an auto loan are examples of an annuity certain.

Mathematical Finance, First Edition. M. J. Alhabeeb.
© 2012 John Wiley & Sons, Inc. Published 2012 by John Wiley & Sons, Inc.

2. *Contingent annuity:* an annuity for which the beginning of the term is known but the end is contingent upon a certain event. The best example is life insurance. Payment of the premium starts at the time of purchase and continues while the insured is alive, and the end occurs at the time of the insured's death.

3. *Perpetuity:* an annuity for which the beginning of the term is known but the end is infinite. An example is a certain principal fund that is kept invested indefinitely and thus continues to generate interest income indefinitely.

The following annuities are based on the timing of the compounding process:

1. *Simple annuity:* an annuity for which the compounding occurs at a time matching the time of payment. For example, the interest on an annuity that is due at the end of each quarter would be compounded quarterly at the end of each quarter.

2. *General annuity:* an annuity for which the compounding process does not match the payment intervals, so interest is assessed either more or less often than the payments. An example is a quarterly annuity where the interest is compounded monthly, called a **complex annuity**.

The most common annuity in the financial world, an ordinary and certain annuity, is our focus hereafter.

4.2. FUTURE VALUE OF AN ORDINARY ANNUITY

The regular compounding formula that we learned in the preceding chapter $[FV = CV(1 + r)^n]$ is designed to work on a single fund invested at one point in time. It will not work when we deal with a set of funds or payments deposited into an account periodically, as in the case of annuities. Next, we describe how the appropriate formula for this type of financial transactions can be obtained.

Let a payment of a quarterly annuity be A and assume a term of annuity of 1 year. The first payment would be due at the end of the first quarter and the last would be due at the end of the last quarter (see Figure 4.1). We can arrange the compounded FVs of these payments as follows:

$$\text{1st:} \quad FV = A(1 + r)^3$$

$$\text{2nd:} \quad = A(1 + r)^2$$

$$\text{3rd:} \quad = A(1 + r)^1 = A(1 + r)$$

$$\text{4th:} \quad = A(1 + r)^0 = A$$

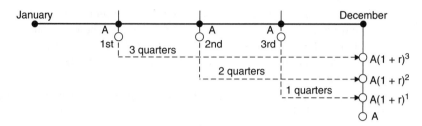

FIGURE 4.1

Now, let's replace the four quarters by any other number of intervals, such as n; then we get

$$1\text{st}: \quad FV = A(1+r)^{n-1}$$

$$2\text{nd}: \quad = A(1+r)^{n-2}$$

$$\text{the second before the last}: \quad = A(1+r)^2$$

$$\text{the first before the last}: \quad = A(1+r)^1$$

$$\text{the last}: \quad = A$$

The total FV would be

$$FV = A + A(1+r) + A(1+r)^2 + \cdots + A(1+r)^{n-2} + A(1+r)^{n-1} \quad (1)$$

Multiplying by $1+r$, we get

$$FV(1+r) = A(1+r) + A(1+r)^2 + A(1+r)^3$$
$$+ \cdots + A(1+r)^{n-2+1} + A(1+r)^{n-1+1} \quad (2)$$

If we subtract equation (2) from (1), we get

$$FV - FV(1+r) = A - A(1+r)^n$$

$$FV[1 - (1+r)] = A[1 - (1+r)^n]$$

$$FV(-r) = A[1 - (1+r)^n]$$

$$FV = \frac{A[1 - (1+r)^n]}{-r}$$

Multiplying by $(-1/-1)$, we get

$$\boxed{FV = \frac{A[1+r)^n - 1]}{r}}$$

This is the formula for a future value of an ordinary annuity.

Example 4.2.1   At age 25 Adam started to contribute to his retirement account by making monthly contributions of \$100. If his IRA pays $6\frac{1}{4}\%$ interest compounded semiannually, how much will he collect when he retires at age 65, and how much will he make on his investment?

His annuity would be semiannual: that is, $600(100 \times 6)$. His maturity would also be counted by semiannual terms; that is, $(65-25) \times 2 = 80$. His interest rate would be on a semiannual rate: that is, $.0625/2 = .03125$.

$$\text{FV} = \frac{A[(1+r)^n - 1]}{r}$$

$$= \frac{600[(1+.03125)^{80} - 1]}{.03125}$$

$$= 205{,}922$$

$$\text{total contributions} = 1{,}200 \times 40 = 48{,}000$$

$$\text{total interest} = 205{,}922 - 48{,}000$$

$$= 157{,}922$$

Example 4.2.2   Lori wanted to deposit \$150 a month in an account bearing 4% interest compound quarterly. It is for her 6-year-old son, who would cash it when he starts his college education at age 18. How large will the son's education fund be?

A is quarterly and is equal to $150 \times 3 = 450$; r is on a quarterly rate: $.04/4 = .01$; n is measured by quarters $= (18-6) \times 4 = 48$.

$$\text{FV} = \frac{A[(1+r)^n - 1]}{r}$$

$$= \frac{450[(1+.01)^{48} - 1]}{.01}$$

$$= 27{,}550$$

The future value formula above can be rewritten based on the table value (Table 7 in the Appendix), which calculates the term $[(1+r)^n - 1]/r$ based on an annuity amount of \$1.00, called $S_{\overline{n}|r}$. The future value formula can therefore be rewritten as

$$\boxed{\text{FV} = A \cdot S_{\overline{n}|r}}$$

Example 4.2.3   Use the table value to obtain the future value in Example 4.2.2.

Looking up across the interest rate .01 and the term 48, we read the $S_{\overline{48}|01}$ value as 61.222608.

$$FV = A \cdot S_{\overline{n}|r}$$
$$= 450(61.222608)$$
$$= 27{,}550$$

4.3. CURRENT VALUE OF AN ORDINARY ANNUITY

We can obtain the formula for the discounted future value of an ordinary annuity by reversing the amounts of the quarterly payments in the example to derive the current value of annuity (see Figure 4.2). The discounted values of each annuity payment, according to the formula $CV = FV/(1+r)^n$, would be

$$CV_4 = \frac{A}{(1+r)^4} + \frac{A}{(1+r)^3} + \frac{A}{(1+r)^2} + \frac{A}{(1+r)^1}$$
$$CV_4 = A(1+r)^{-4} + A(1+r)^{-3} + A(1+r)^{-2} + A(1+r)^{-1}$$

Multiply both sides by $(1+r)^4$:

$$CV_4(1+r)^4 = A(1+r)^{-4}(1+r)^4 + A(1+r)^{-3}(1+r)^4$$
$$+ A(1+r)^{-2}(1+r)^4 + A(1+r)^{-1}(1+r)^4$$
$$CV_4(1+r)^4 = A + A(1+r) + A(1+r)^2 + A(1+r)^3$$

Considering all payments as future values, we can replace A's with FV

$$CV_4(1+r)^4 = FV + FV(1+r) + FV(1+r)^2 + FV(1+r)^3 = FV_{all}$$

For any number of payments, such as n, we can write the equation as

$$CV_n(1+r)^n = FV$$
$$CV_n = \frac{FV}{(1+r)^n}$$
$$CV_n = \frac{A[(1+r)^n - 1]/r}{(1+r)^n}$$

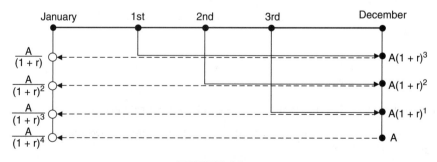

FIGURE 4.2

Rearranging

$$\text{CV}_n \text{ or } \text{CV}_{\text{all}} \text{ or just CV} = \frac{A[(1+r)^n - 1](1+r)^{-n}}{r}$$

The final formula for the current value of all payments (or the discounted FV) is

$$\boxed{\text{CV} = \frac{A[1 - (1+r)^{-n}]}{r}}$$

Example 4.3.1 A man wants to cash in his trust fund, which pays him $500 a month for the next 10 years. The interest on the fund is $6\frac{1}{2}\%$ compounded monthly. How much would he receive?

The monthly interest rate is $.065/12 = .0054$; the monthly term is $10 \times 12 = 120$.

$$\text{CV} = \frac{A[1 - (1+r)^{-n}]}{r}$$

$$= \frac{500[1 - (1 + .0054)^{-120}]}{.0054}$$

$$= 48,518.85$$

Example 4.3.2 What is the current value of an annuity involving $3,750 payable at the end of each quarter for 7 years at an interest rate of 8% compounded quarterly?

The quarterly rate is $.08/4 = .02$; the term in quarters: $7 \times 4 = 28$.

$$\text{CV} = \frac{A[1 - (1+r)^{-n}]}{r}$$

$$= \frac{3,750[1 - (1 + .02)^{-28}]}{.02}$$

$$= 79,804.77$$

The current value formula above can be rewritten in terms of the table value (Table 8 in the Appendix). This value is based on calculating the term $[1 - (1+r)^{-n}]/r$ for the periodic payment of $1.00, denoted $a_{\overline{n}|r}$. The current value formula can therefore be rewritten as

$$\boxed{\text{CV} = A \cdot a_{\overline{n}|r}}$$

Example 4.3.3 Find the CV in Example 4.3.2 using the table value. Looking at Table 8 for an interest rate of 2% and a maturity of 28, we read the value (21.281272):

$$CV = A \cdot a_{\overline{n}|r}$$
$$= 3,750(21.281272)$$
$$= 79,804.77$$

4.4. FINDING THE PAYMENT OF AN ORDINARY ANNUITY

We can rearrange the formulas of both the future value and the current value of the annuity and rewrite them in terms of A.

- Annuity payment (A) when FV is given:

$$FV = \frac{A[(1+r)^n - 1]}{r}$$

$$\boxed{A = \frac{FV \cdot r}{(1+r)^n - 1}}$$

- Annuity payment (A) when the current value is given:

$$\boxed{A = \frac{CV \cdot r}{1 - (1+r)^{-n}}}$$

Example 4.4.1 Samantha is planning to buy a new car for $17,000 upon her graduation 5 years from now. She is investing in an ordinary annuity account paying 7% interest compounded monthly. How much should she be depositing each month?

Since the future value is given ($17,000), we use the first formula. The monthly $r = .07/12 = .00583$; the monthly terms are $5 \times 12 = 60$ months.

$$A = \frac{FV \cdot r}{(1+r)^n - 1} = \frac{17,000(.00583)}{(1 + .00583)^{60} - 1}$$
$$= 237.48 \text{ per month}$$

Example 4.4.2 Jim deposited $18,000 in his savings account. He planned to set an automatic monthly payment to his son for the next 6 years. If the account bears an $8\frac{1}{2}\%$ interest compounded quarterly, how much would his son receive at the end of each month?

Since the current value is available ($18,000), we use the second formula. The quarterly rate of interest would be $.085/4 = .02125$; the terms in quarters $= 6 \times 4 = 24$.

$$A = \frac{CV \cdot r}{1 - (1 + r)^{-n}}$$

or

$$A = \frac{CV}{[1 - (1 + r)^{-n}]/r}$$

$$= \frac{18{,}000}{[1 - (1 + .02125)^{-24}/.02125]}$$

$$= 965.21 \text{ per quarter}$$

$$\frac{965.21}{3} = 321.73 \text{ per month}$$

Using the table method, we can rewrite the annuity payment formulas as

$$A = FV \cdot \frac{1}{S_{\overline{n}|r}}$$

and since $1/S_{\overline{n}|r} = 1/a_{\overline{n}|r} - r$,

$$A = FV \left(\frac{1}{a_{\overline{n}|r}} - r \right)$$

and the annuity when the current value is available is

$$A = CV \cdot \frac{1}{a_{\overline{n}|r}}$$

where

$$\frac{1}{a_{\overline{n}|r}} = \frac{r}{1 - (1 + r)^{-n}}$$

Example 4.4.3   Use the table method to determine how much Samantha's quarterly deposit would be if the interest rate is 8% compounded quarterly for 5 years.

$r = .08/4 = .02; n = 5 \times 4 = 20$; the table value of $1/a_{\overline{20}|.02} = 1/16.3514 = .061157$.

$$A = FV \left(\frac{1}{a_{\overline{n}|r}} - r \right)$$

$$- 17,000 \left(\frac{1}{16.3514} - .02 \right)$$

$$= 17,000(.061157 - .02)$$

$$= 699.67 \qquad \text{the quarterly deposit}$$

Example 4.4.4   Determine how much Jim's son would receive as a semiannual payment if the interest rate is 12% compounded semiannually for 6 years. Use the table method.

$CV = 18,000; r = .12/2 = .06; n = 6 \times 2 = 12;$ the   table   value   of $1/a_{\overline{12}|.06} = .119277$.

$$A = CV \cdot \frac{1}{a_{\overline{n}|r}}$$

$$= 18,000 \cdot \frac{1}{a_{\overline{12}|.06}}$$

$$= 18,000(.119277)$$

$$= 2,147$$

4.5. FINDING THE TERM OF AN ORDINARY ANNUITY

The term of annuity (n) can be found by rearranging the future value of an annuity formula:

$$FV = \frac{A[(1+r)^n - 1]}{r}$$

$$FV \cdot r = A[(1+r)^n - 1]$$

$$\frac{FV \cdot r}{A} = (1+r)^n - 1$$

$$\frac{FV \cdot r}{A} + 1 = (1+r)^n$$

$$\ln \left(\frac{FV \cdot r}{A} + 1 \right) = n \ln(1+r)$$

$$\boxed{n = \frac{\ln[(FV \cdot r/A) + 1]}{\ln(1+r)}}$$

It can also be found by rearranging the current value of an annuity formula:

$$CV = \frac{A[1 - (1+r)^{-n}]}{r}$$

$$FV \cdot r = CV \cdot r = A[1 - (1+r)^{-n}]$$

$$\frac{CV \cdot r}{A} = 1 - (1+r)^{-n}$$

$$(1+r)^{-n} = 1 - \frac{CV \cdot r}{A}$$

$$-n \ln(1+r) = \ln\left(1 - \frac{CV \cdot r}{A}\right)$$

$$\boxed{n = -\frac{\ln\left[1 - \left(\frac{CV \cdot r}{A}\right)\right]}{\ln(1+r)}}$$

Example 4.5.1 A small business owner wants to buy equipment for $60,000. He can save $500 a week for his future purchase. Determine how long it will take him if his account is paying $7\frac{3}{4}\%$ interest compounded quarterly.

The quarterly rate $= .0775/4 = .019375$. The deposit (A) is $500 weekly $= $6,500 quarterly.

$$n = \frac{\ln\left[(FV \cdot r/A) + 1\right]}{\ln(1+r)}$$

$$= \frac{\ln[(60,000)(.019375)/6,500 + 1]}{\ln 1.019375}$$

$$= \frac{\ln 1.1789}{\ln 1.0194}$$

$$= \frac{.1646}{.0192}$$

$$= 8.6 \text{ quarters}$$

Example 4.5.2 Sev deposited a prize of $230,000 in an ordinary annuity account that pays $9\frac{1}{4}\%$ interest compounded monthly. She wants to withdraw $5,000 a month as cash. Find the number of months she would be able to withdraw.

The monthly rate is $.0925/12 = .0077$; the monthly withdrawal $= $5,000.

$$n = -\frac{\ln[1 - (CV \cdot r/A)]}{\ln(1+r)}$$

$$= -\frac{\ln[1 - (230,000)(.0077)/5,000]}{\ln(1 + .0077)}$$

$$= -\frac{\ln .64542}{\ln 1.0077}$$

$$= 57 \text{ months}$$

4.6. FINDING THE INTEREST RATE OF AN ORDINARY ANNUITY

Both the future and current value formulas of annuity cannot be solved mathematically for a practical value of r. Therefore, finding r can be done depending on table value and by interpolation. **Linear interpolation** is a method to calculate an unknown value positioned between two known values by using comparative proportions. Interpolation can help deduce an unknown annuity interest rate by using the closest table rates above and under that rate. But before that we need to arrange the information given for a problem and make it compatible with the information in the table, which shows n and r for an annuity payment of \$1. The formula for such an arrangement is

$$\boxed{\text{FV} = A_{\overline{n}|r}}$$

FV is a future value of an annuity that is obtained by multiplying a certain annuity payment by the appropriate table value, as we have seen before. This table value is read across a certain set of n and r. The symbol $\overline{n}|r$ refers to the table value.

Example 4.6.1   Wayne sets up an interest-bearing fund for his son in which the son receives \$200 four times a year: at the end of March, June, September, and December for 8 consecutive years. The son will have received a total of \$8,800 at the end of the 8 years. What interest rate would make this possible given that the fund compounds its interest quarterly?

Using the formula above, we can find the targeted table value:

$$\text{FV} = A_{\overline{n}|r} \qquad n = 8 \times 4 = 32$$

$$8,800 = 200_{\overline{32}|r}$$

$$\frac{8,800}{200} = 3_{\overline{2}|r}$$

$$44 = 3_{\overline{2}|r}$$

A table reading of 44 or close to 44 can be traced down at $n = 32$. Looking at Table 7 in the Appendix, we can see that there are two values close to 44. They are: 44.22702961, corresponding to an interest rate of 2%, and 32 terms; and 43.30793563, corresponding to an interest rate of $1\frac{7}{8}\%$, and 32 terms. By comparing these sets of values through the linear interpolation, we can find the interest rate corresponding to the value of 44 fairly accurately.

$$44 - 43.30 \left\{ \begin{array}{l} \left[\begin{array}{l} 43.30793563 \\ 44 \\ 44.22702961 \end{array} \right. \begin{array}{c} 1\frac{7}{8} \\ r \\ 2 \end{array} \right\} \begin{array}{c} r - 1\frac{7}{8} \\ \\ 2 - 1\frac{7}{8} \end{array} \right.$$

(Table Value / Interest Rate (%))

By symmetric property and cross-multiplication, we can find the r value.

$$\frac{44 - 43.30793563}{44.22702961 - 43.30793563} = \frac{r - 1.875}{2 - 1.875}$$

$$r = \frac{1.809809259}{.91909398}$$

$$r = 1.969\%$$

4.7. ANNUITY DUE: FUTURE AND CURRENT VALUES

Annuity due is just like an ordinary annuity except that the payments occur at the beginning of each payment term instead of at the end. Therefore, the entire term of an annuity due starts at the date of the first payment and ends at the end of the interval of the last payment. In other words, it would end one payment interval after the last payment has been paid. Insurance premiums and property rentals are typical examples of annuities due. As a result of the change in the timing of payments, the future and current values would also change. The future value of an annuity due formula would reflect that change:

$$\boxed{FV_d = \left[A \frac{(1+r)^n - 1}{r} \right] (1+r)}$$

Example 4.7.1 Jack and his wife will deposit \$550 in their savings account at the first day of each month for the next 5 years in order to have a large enough down payment to buy a larger house. If their savings account pays 7% interest compounded monthly, how much down payment will they collect?

The monthly rate $= .07/12 = .00583$; $n = 5 \times 12 = 60$.

$$FV_d = 550 \left[\frac{(1 + .00583)^{60} - 1}{.00583} \right] (1 + .00583)$$

$$= 39,606$$

Example 4.7.2 If Simon deposits \$200 at the beginning of each week for 2 years in an account that pays 2.5% compounded weekly, how much would he have at the end of the year?

The weekly rate $= .025/52 = .00096$; $n = 52 \times 2 = 104$.

$$FV_d = 200 \left[\frac{(1 + .00096)^{104} - 1}{.00096} \right] (1 + .00096)$$

$$= 21,920$$

The current value of an annuity due can be obtained using the following formula:

$$CV_d = \left[A \frac{1 - (1 + r)^{-n}}{r} \right] (1 + r)$$

Example 4.7.3   What is the current value of an annuity due for an investment fund paying $5\frac{3}{4}\%$ interest compounded monthly if \$750 is deposited at the beginning of each month for 3 years?

The monthly rate $= .0575/12 = .00479$; $n = 3 \times 12 = 36$.

$$CV_d = 750 \left[\frac{1 - (1 + .00479)^{-36}}{.00479} \right] (1 + .00479)$$

$$= 24,864$$

The formulas for the future and current values of an annuity due can also be rewritten in terms of the table values of $S_{\overline{n}|r}$ and $a_{\overline{n}|r}$:

$$FV_d = A \cdot S_{\overline{n}|r} \cdot (1 + r)$$

$$CV_d = A \cdot a_{\overline{n}|r} \cdot (1 + r)$$

Example 4.7.4   Using the table method, find the down payment for Jack and his wife in Example 4.7.1 if they deposit \$6,600 at the beginning of each year for 5 years at an interest rate of 7% annually.

We look up the $S_{\overline{n}|r}$ Table 7 in the Appendix, across 7% interest and 5 years maturity to get the value $S_{\overline{5}|.07} = 5.750739$.

$$FV_d = A \cdot S_{\overline{n}|r} \cdot (1 + r)$$

$$= A \cdot S_{\overline{5}|.07} \cdot (1 + .07)$$

$$= 6,600(5.750739)(1.07)$$

$$= 40,612$$

Example 4.7.5 Using Table 8 in the Appendix, find the current value of the annuity due in Example 4.7.3 if the interest rate is 6% compounded semiannually and the deposit is $4,500 twice a year for 5 years.

The interest rate $r = .06/2 = .03$; $A = 4,500$; $n = 2 \times 5 = 10$.

$$CV_d = A \cdot a_{\overline{n}|r} \cdot (1 + r)$$

$$= A \cdot a_{\overline{10}|.03} \cdot (1.03)$$

$$= 4,500(8.530203)(1.03)$$

$$= 39,537$$

4.8. FINDING THE PAYMENT OF AN ANNUITY DUE

Just like the ordinary annuity payment, the annuity due payment can be obtained by a rearrangement of either the future value, or the current value formula, depending on which one is available. If the future value is given, the payment formula would be

$$A_d = \frac{FV \cdot r}{(1 + r)^{n+1} - (1 + r)}$$

If the current value is given, the payment formula would be

$$A_d = \frac{CV \cdot r}{(1 + r) - (1 + r)^{1-n}}$$

Example 4.8.1 Sam purchased a stereo set on his appliance store card. He was told that with the store rate of 18% compounded monthly, he would end up paying $1,744 on a 2-year payment plan. How much would he pay at the start of each month?

Since we know the future value ($1,744), we apply the A_d with the future value formula:

$$A_d = \frac{FV \cdot r}{(1 + r)^{n+1} - (1 + r)}$$

The monthly rate $= .18/12 = .015$; $n = 2 \times 12 = 24$.

$$A_d = \frac{1,744(.015)}{(1 + .015)^{25} - (1 + .015)}$$

$$= 60$$

Example 4.8.2 What will be the cash price of the stereo set in Example 4.8.1? How much total interest will Sam pay? What will be the annual percentage rate?

$$CV_d = A\left[\frac{1-(1+r)^{-n}}{r}\right](1+r)$$

$$= 60\left[\frac{1-(1+.015)^{-24}}{.015}\right](1+.015)$$

$$= \$1,220 \text{ cash price}$$

$$\text{total interest} = 1,744 - 1,220 = 524 \text{ for 2 years}$$

$$524 \div 2 = 262 \text{ interest per year}$$

$$\frac{262}{1,220} = 21.47\% \text{ annual percentage rate.}$$

Example 4.8.3 What would the monthly payment be if Sam chooses another brand, whose price is \$1,500?

Since we know the current value (\$1,500), we would use the A_d with current value formula:

$$A_d = \frac{CV \cdot r}{(1+r) - (1+r)^{1-n}}$$

$$= \frac{1,500(.015)}{(1+.015) - (1+.015)^{1-24}}$$

$$= 74$$

Sam pays \$74 at the beginning of each month for 2 years.

4.9. FINDING THE TERM OF AN ANNUITY DUE

The term (n) can also be found depending on whether the future value or the current value is given. If the future value is given, the n formula would be

$$n = \frac{\ln\left[1 + \dfrac{FV \cdot r}{A(1+r)}\right]}{\ln(1+r)}$$

and if the current value is given, the n formula would be

$$n = -\frac{\ln\left[1 - \dfrac{CV \cdot r}{A(1+r)}\right]}{\ln(1+r)}$$

Example 4.9.1   Tim and Cathy will need a down payment of $15,000 for their first home. If they can save $600 at the beginning of each month in an account bearing $8\frac{1}{2}\%$ compounded monthly, determine how long it would take them to buy a home.

The monthly rate $= .085/12 = .007083$.

$$n = \frac{\ln\left[1 + \dfrac{FV \cdot r}{A(1+r)}\right]}{\ln(1+r)}$$

$$n = \frac{\ln\left[1 + \dfrac{(15,000)(.007083)}{(600)(1 + .007083)}\right]}{\ln(1 + .007083)}$$

$$n = \frac{\ln(1.176)}{\ln(1.007083)}$$

$$n = 22.95 \quad \text{or} \quad 23 \text{ months}$$

Example 4.9.2   If Tim and Cathy want to purchase a home for $180,000 at an interest rate of 7% compounded annually, and if they can pay a $1,500 mortgage at the beginning of each month, find how many years it would take them to pay it off.

The monthly rate $= .07/12 = .00583$.

$$n = \frac{\ln\left[1 - \dfrac{CV \cdot r}{A(1+r)}\right]}{\ln(1+r)}$$

$$n = -\frac{\ln\left[1 - \dfrac{(180,000)(.00583)}{(1,500)(1 + 00583)}\right]}{\ln(1 + 00583)}$$

$$n = -\frac{\ln(.3045)}{\ln(1.00583)}$$

$$n = 204.6 \text{ months or about 17 years}$$

4.10. DEFERRED ANNUITY

A deferred annuity is an ordinary annuity except that the payoff starts at a later time. In this case, the first payment is made after a certain time has been passed according to the financial contract. This time is called the **deferment period**. The important note to make here is that because of the fact that the payments are made at the end of each payment term as an ordinary annuity, calculation of the deferment period needs attention. For example, if the first payment is made at the fifth payment period, the annuity would be deferred four payment periods, and by the same logic, we conclude that if an annuity is deferred for seven payment periods, its first payment has to be made at the end of the eighth payment. The time line in Figure 4.3 shows a 10-payment annuity that won't be paid until the fifth payment period, resulting in a four-period deferment. If we are to calculate the current value of this annuity, we normally have to bring back the future value for 15 periods, and because this would not be correct, we would count for the deferred period by subtracting the current value when $n = 5$ from the current value when $n = 15$. In this case, we would have the correct current value 10 periods back.

Example 4.10.1 Determine the current value of a deferred annuity of $250 a month for 18 months from January 2008 to June 2009 if the first payment is due at the end of April 2008, given that the interest rate is 7% compounded monthly (see Figure E4.10.1).

Since the first payment is made at the end of April 2008, there would be three months of deferment: January, February, and March. The future value would occur (or the annuity will be paid off) on September 2009, three months after the original date of June 2009, because of the three months' skip in advance. As for the current value calculation, it can be done by bringing back the future value from September 2009 to January 2008; but that is 21 periods, and that is why we also need to calculate the CV of the deferment period and subtract it from the entire CV. Therefore, we calculate two current values; the first where n is equal to the regular annuity term of 18 months plus the deferred period of 3 months ($n = 18 + 3 = 21$). The second current value is for the deferment only where n is equal to 3. Then we subtract the second CV from the first to get the correct CV, where n is 18.

The monthly rate $= .07/12 = .00583$.

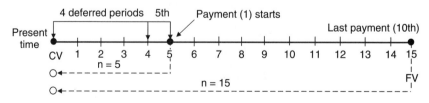

FIGURE 4.3

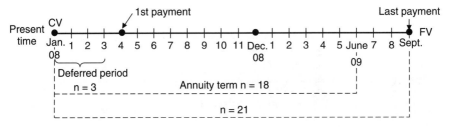

FIGURE E4.10.1

$$CV_1 = \frac{A[1 - (1 + r)^{-n}]}{r}$$

$$= \frac{250[1 - (1 + .00583)^{-21}]}{.00583}$$

$$= 4,927.85$$

$$CV_2 = \frac{250[1 - (1 + .00583)^{-3}]}{.00583}$$

$$= 741.34$$

$$CV = CV_1 - CV_2$$

$$= 4,927.85 - 741.34$$

$$= 4,186.50$$

4.11. FUTURE AND CURRENT VALUES OF A DEFERRED ANNUITY

Using the table value, we can obtain both the future value of the deferred annuity (FV_{def}) and the current value (CV_{def}). The future value is the same as the future value for an ordinary annuity:

$$\boxed{FV_{def} = A \cdot S_{\overline{n}|r}}$$

Example 4.11.1   Find the future value of an annuity of $500 payable quarterly for 2 years where the first payment is made at the end of 6 months given that the interest rate is 8%.

$A = \$500; r = .08/4 = .02; n = 8$ (see Figure E4.11.1).

$$FV_{def} = A \cdot S_{\overline{n}|r}$$

$$= 500 \cdot S_{\overline{8}|.02}$$

$$= 500(8.582969)$$

$$= 4,291$$

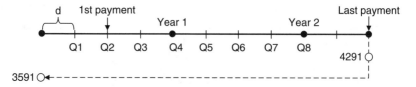

FIGURE E4.11.1

The current value formula for a deferred annuity (CV_{def}) can be written

$$CV_{def} = A \cdot a_{\overline{n}|r}(1+r)^{-d}$$

where d is the deferment period.

Example 4.11.2   Calculate the current value of the deferred annuity of Example 4.11.1.

$A = 500; r = .02; n = 8; d = 1; a_{\overline{8}|.02} = 7.325481.$

$$CV_{def} = A \cdot a_{\overline{n}|r}(1+r)^{-d}$$
$$= 500(7.325481)(1 + .02)^{-1}$$
$$= 3,590.92$$

4.12. PERPETUITIES

A **perpetuity** is an annuity with a payment that goes on forever. The perpetuity term is infinite and therefore only the perpetuity's current value is of concern. Most perpetuities are streams of interest paid on a fixed principal that is held in an account indefinitely. Endowment accounts, philanthropic funds, and cancelable preferred stock dividends are all examples of perpetuities. The initial and original principal of perpetuity would be the current value (CV). The interest earned on the principal would be the perpetuity payment (A), and the rate of interest (r) would be a compound rate, as in all annuities. The term of perpetuity (n) would be equal to ∞. Calculation of the periodic interest as payment would therefore be determined by

$$A = r \cdot CV_{\infty}$$

From this equation we can also obtain the equations for the rate of interest (r) and the current value (CV):

$$r = \frac{A}{CV_{\infty}}$$

$$\boxed{CV_\infty = \frac{A}{r}}$$

Example 4.12.1   An amount of \$750,000 was given by a former professor's wife to his department for setting up a new scholarship in his name. The department invested the gift at $10\frac{3}{4}\%$ interest compounded annually. Find the size of the annual scholarship.

$CV = 750,000; r = .1075.$

$$A = r \cdot CV_\infty$$
$$= .1075(750,000)$$
$$= 80,625 \quad \text{annual scholarship fund}$$

Example 4.12.2   A family left a portion of its fortune to a local church, which invested the fund at $8\frac{2}{3}\%$ interest compounded semiannually. If the church receives \$2,650 a month as payment from that fund, Find the original family gift to the church.

$A = 2,650 \times 6 = 15,900$ semiannual payment; $r = .08662/2 = .0433$ semiannual rate.

$$CV_\infty = \frac{A}{r}$$
$$= \frac{15,900}{.0433} = 367,205 \quad \text{initial family gift}$$

Example 4.12.3   A man wishes to give his choice of charities \$150 each quarter for an indefinite time. He has \$5,000 to invest in a perpetuity for this purpose. What should the interest rate be for him to make his wish come true?

$$r = \frac{A}{CV}$$
$$= \frac{150}{5,000} = .03 \quad \text{quarterly rate}$$

$.03 \times 4 = .12$ or 12% annual interest rate

Unit II Summary

For the centrality of the time value of money as a concept in all financial literature and calculations, this is the first unit of the core material of this book. It is the foundation of the theory of finance and therefore is treated as a service unit for the remainder of the book. Most, if not all, of the original concepts and calculation methods and formulas used later in the book are detailed here. Four major topics constitute this units material: simple interest, bank discounts, compound interest, and annuities. In the first chapter we dissected the simple interest formula and showed how any one of its four elements can be found. As one of those four elements, the current value was singled out to explain the method of simple discount as a process reverse to that of the simple interest accumulation. The term of maturity was calculated in days, and the difference between ordinary and exact interest was shown as well as the way to obtain one in terms of the other. Equations of value and time were detailed with examples and time line diagrams. Focal and due dates were calculated, and partial payments were explained in both the merchants and U.S. methods. Finally, the dollar-weighted method to obtain the simple interest rate was explained with examples. The time line diagram was presented as a visual aid to simplify the examples and make their comprehension easier.

Different from the simple discount method discussed in the first chapter as the process to get the current value of a future amount, another discount method is explained in the second chapter. In the bank discount procedure the lending bank assesses the interest on a loan but deducts it immediately at the time of dispensing the loan amount to the borrower. The amount dispersed is called the proceeds, and the rate at which the total amount of the loan was discounted to the proceeds is called the discount rate. The discount formula is analogous to the simple interest formula. Finding all four elements of the formula was demonstrated and a comparison of the discount rate to the interest rate was made. Two of the most popular applications of the bank discount are that of discounting a promissory note and purchase of a U.S Treasury bill. They were explained with several examples. Finally, the difference between a simple discount and the bank discount was addressed and explained with practical examples.

In the third chapter we moved from the simple interest technique to compound interest. It is more involved and has a wider use and higher applications. The popularity of compound interest stems from its "magical" ability to accumulate

Mathematical Finance, First Edition. M. J. Alhabeeb.
© 2012 John Wiley & Sons, Inc. Published 2012 by John Wiley & Sons, Inc.

gains. The standard mathematical formula for compound interest is considered the cornerstone of the theory of finance. We learned how to apply this formula and how to manipulate it to get all four elements: the future value, the current value, the rate of compound interest, and the term of maturity. Reversing the future value back in time produced a discounted current value just as in the simple interest case, but there is an obvious technical difference this time. We also learned how important it was to change the frequency of compounding according to several conversion periods. This produced several types of compounding, which we calculated and applied in real-life situations. We also explained how nominal rates of interest may differ from an actual effective rate of interest, and learned how to go back and forth between the two. Just as with simple interest, we learned the equations of value and time when we have compound interest, which are very important and popular topics in finance. Last but not least was the topic of knowing the maturity period required to get certain multiples of the initial principal through some simple formulas for magic numbers.

The most frequently asked question by students regards when one would use the regular compounding formula and when one would use the annuity formula. The answer is obtained through knowing the frequency of such financial transactions as making deposits or payments. The regular compounding formula is designed to handle the financial accumulation of a single amount left in an account one time. The annuity formulas are designed to handle more than one transaction over a period of time, such as making periodic deposits or regular interval payments. Here the compounding would work as many times as in the designated conversion periods, whereas in the regular compounding formula, the compounding process would be applied once considering the length of the time from the initiation of transaction to the end of maturity time. In this sense, annuities are sets of periodic transactions involving equal amounts at equal intervals of time. We discussed three major types of annuities: the ordinary, for which the transaction is made at the end of the interval period, the annuity due, for which the transaction is made at the start of the interval period, and the deferred annuity, which is an ordinary annuity in nature but whose transaction is deferred to other times. Annuities were also discussed on the basis of their terms, such as the annuity certain, the contingent annuity, and the perpetuities. For the three types of annuity, ordinary, due, and deferred, the major formulas were presented and applied. Future value, current value, payment, interest rate, and term of maturity were all obtained in terms of other variables. In addition to the formula method, a table method was used to get most of these values.

As with other types of mathematical problems in previous chapters, we used time line diagrams to illustrate the details of transactions. We recommend this way to students as a helpful method to lay out the information of the problem and solving it with clarity. It would make a lot of difference in the basic comprehension to sketch a problem diagram, especially in problems of annuity when the only difference sometimes between one type and another is the timing of the payments.

List of Formulas

Simple interest

Total interest:

$$I = \text{CV} \cdot r \cdot n$$

Rate of interest:

$$r = \frac{I}{\text{CV} \cdot n}$$

Term of maturity:

$$n = \frac{I}{\text{CV} \cdot r}$$

Current value:

$$\text{CV} = \frac{I}{r \cdot n}$$

Future value:

$$\text{FV} = \text{CV}(1 + rn)$$

Rate of interest when FV is known:

$$r = \frac{\text{FV}/\text{CV} - 1}{n}$$

Term of maturity when FV is known:

$$n = \frac{\text{FV}/\text{CV} - 1}{r}$$

Mathematical Finance, First Edition. M. J. Alhabeeb.
© 2012 John Wiley & Sons, Inc. Published 2012 by John Wiley & Sons, Inc.

Ordinary interest:

$$I_0 = I_e \left(1 + \frac{1}{72}\right)$$

$$\text{or } I_0 = 1.014 I_e$$

Exact interest:

$$I_e = I_0 \left(1 + \frac{1}{73}\right)$$

$$\text{or } I_e = \frac{I_0}{1.014}$$

Equivalent time:

$$\bar{n} = \frac{\sum P_i n_i}{\sum P_i}$$

Interest rate by the dollar-weighted method:

$$r = \frac{E - [(B + D) - W]}{Bt + D(t - t_1) - W(t - t_2)}$$

Bank discount

Discounted proceeds:

$$C = FV(1 - dn)$$
$$C = FV - D$$

Future value:

$$FV = \frac{C}{1 - dn}$$

Discounting term:

$$n = \frac{1 - (C/FV)}{d}$$

Discounting rate:

$$d = \frac{1 - (C/FV)}{n}$$

Interest rate:

$$r = \frac{d}{1 - dn}$$

Discount rate in terms of interest rate:

$$d = \frac{r}{1 + rn}$$

Discount rate in terms of a bid:

$$d = \frac{360 - 3.6B}{n}$$

Compound interest

Future value:

$$FV = CV(1 + r)^n$$

Current value:

$$CV = \frac{FV}{(1 + r)^n}$$

Discount factor:

$$DF = \frac{1}{(1 + r)^n}$$

Interest rate:

$$r = \sqrt[n]{\frac{FV}{CV}} - 1$$

Term of maturity:

$$n = \frac{\ln(FV/CV)}{\ln(1 + r)}$$

Effective interest rate:

$$R = \left(1 + \frac{r}{m}\right)^m - 1$$

Continuous compounding—future value:

$$FV = CV \cdot e^{rn}$$

Continuous compounding—current value:

$$CV = FV \cdot e^{-rn}$$

Rule of 72:

$$n = \frac{72}{r}$$

Rule of 114:

$$n = \frac{114}{r}$$

Rule of 167:

$$n = \frac{167}{r}$$

Annuities

Future value of an ordinary annuity:

$$FV = \frac{A[(1+r)^n - 1]}{r}$$

$$FV = A \cdot S_{\overline{n}|r}$$

Current value of an ordinary annuity:

$$CV = \frac{A[1 - (1+r)^{-n}]}{r}$$

$$CV = A \cdot a_{\overline{n}|r}$$

Payment of an ordinary annuity (FV is given):

$$A = \frac{FV \cdot r}{(1+r)^n - 1]}$$

$$A = FV \cdot \frac{1}{S_{\overline{n}|r}}$$

$$A = FV \left(\frac{1}{a_{\overline{n}|r}} - r \right)$$

Payment of an ordinary annuity (CV is given):

$$A = \frac{CV \cdot r}{1 - (1+r)^{-n}}$$

$$A = CV \cdot \frac{1}{a_{\overline{n}|r}}$$

Term of an ordinary annuity:

$$n = \frac{\ln\left[(FV \cdot r/A) + 1\right]}{\ln(1+r)}$$

Future value of an annuity due:

$$FV_d = A\left[\frac{(1+r)^n - 1}{r}\right](1+r)$$

$$FV_d = A \cdot S_{\overline{n}|r} \cdot (1+r)$$

Current value of an annuity due:

$$CV_d = A\left[\frac{1 - (1+r)^{-n}}{r}\right](1+r)$$

$$CV_d = A \cdot a_{\overline{n}|r} \cdot (1+r)$$

Payment of an annuity due (FV is given):

$$A_d = \frac{FV \cdot r}{(1+r)^{n+1} - (1+r)}$$

Payment of an annuity due (CV is given):

$$A_d = \frac{CV \cdot r}{(1+r) - (1+r)^{1-n}}$$

Term of annuity due (FV is given):

$$n = \frac{\ln\{1 + [FV \cdot r/A(1+r)]\}}{\ln(1+r)}$$

Term of an annuity due (CV is given):

$$n = -\frac{\ln\{1 - [CV \cdot r/A(1+r)]\}}{\ln(1+r)}$$

Future value of a deferred annuity:

$$FV_{def} = A \cdot S_{\overline{n}|\,r}$$

Current value of a deferred annuity:

$$CV_{def} = A \cdot a_{\overline{n}|\,r}(1+r)^{-d}$$

Perpetuity:

$$A = r \cdot CV_{\infty}$$

Rate of a perpetuity:

$$r = \frac{A}{CV_{\infty}}$$

Current value of a perpetuity:

$$CV_{\infty} = \frac{A}{r}$$

Exercises for Unit II

Simple Interest

1. How much will the interest be on a loan of $2,350 at a simple interest of 7% for 6 months?

2. If a loan of $5,750 is taken out at a simple interest of 5.25% for a year and a half, how much interest will the borrower pay?

3. At what rate of simple interest would a loan of $3,000 be taken for 2 years if the total interest is $375?

4. At what rate of interest would Janet take out a loan of $13,000 for 3.5 years if she pays $4,322 in interest?

5. How long will it take for $900 to earn $486 at a 12% rate of simple interest?

6. How many years will it take Jack to pay off his debt of $5,700 at a simple rate of 8.5% and pay a total interest of $1,090?

7. Find the principal of a loan taken at 8% for 3 years when the total interest is $516.

8. What is the current value of a loan taken at $3\frac{3}{4}$% for 4 years if the interest paid is $1,350?

9. What is the current value of a loan taken at $7\frac{1}{4}$% for 5 years if the borrower has to pay a total of $11,500?

10. If you deposit $600 at a bank paying $4\frac{1}{2}$% simple interest, how much will you collect after 11 months?

11. If you want your balance to be $8,000 in 4 years and if you deposit your money at a bank paying $6\frac{1}{4}$%, how much is the initial deposit?

12. Find the amount of money that $700 would grow to if you leave it for 6 years at a bank paying $11\frac{1}{2}$% simple interest?

13. What is the rate of interest for Linda, who saved $2,500 in her havings account and collected $2,590 after 20 months?

Mathematical Finance, First Edition. M. J. Alhabeeb.
© 2012 John Wiley & Sons, Inc. Published 2012 by John Wiley & Sons, Inc.

14. Michael wants his $3,200 to grow to $4,500 in 3 years. He has a savings account paying simple interest on savings. What rate of interest would help him achieve his goal?

15. If Michael can only get $9\frac{1}{2}$ % interest on his savings, how long will he have to wait to get his needed $4,500?

16. Suppose that Michael was able to add $300 to his deposit of $3,200 and was willing to wait 6 months past the 3 years to reach his goal of getting $4,500. What interest rate would he need to make his goal come true?

17. Find the exact and approximate time between October 2, 2010 and June 15, 2011.

18. Find the amount of simple interest on $2,100 from September 2, 2010 to June 15, 2011 using the banker's rule.

19. Find the amount of interest on $3,000 at 7% for 50 days using both ordinary and exact interest.

20. Jennifer owes $600 due in 9 months and $1,500 plus 6% interest due in 3 months. She wants to pay off both debts in a single payment in 11 months. How much should she pay if the money is worth 5%?

21. If the rate of interest is $7\frac{1}{2}$%, when will one single payment of $3,550 discharge the following three debts: $550 due in 30 days, $1,300 due in 45 days, and $1,700 due in 70 days?

22. What is the ordinary interest if the exact interest is $320.59?

23. What is the exact interest if the ordinary interest is $117.50?

24. What is the simple interest rate for a fund that starts with $1,550 on July 15, a deposit of $730 37 days later, a withdrawal of $250 100 days later, and a balance of $2,211?

25. Find the simple rate of interest on the following transactions using the dollar-weighted method.

   (a) $312.50 initial deposit on March 15
   (b) $617.70 another deposit on May 25
   (c) $115.20 withdrawal on June 17
   (d) $250.00 deposit on July 1
   (e) $1,229 balance on October 8

Bank Discount

1. Find the proceeds from a loan of $2,300 for 2 years at a discount rate of 4.9%.

2. What will the proceeds be for a loan of $900 discounted by $75?

3. Discount $1,250 for 5 months at a simple interest of 9% and find the simple discount.

4. What are the bank discount and the proceeds if Gen settles her debt of $3,700 at the end of 90 days at a discount rate of 13%?

5. If you receive $820.25 as the proceeds of a 7-month loan at 6%, what is the loan amount?

6. Find the discount rate if you get only $1,803.75 when you borrow $1,850 for 120 days.

7. Find the discount rate for Jimmy, who received proceeds of $3,308 for a $3\frac{1}{2}$-month loan of $3,400.

8. How long would it take to pay off a loan of $2,900 at 11% discount when the bank charges $638?

9. What is the term of discounting for your friend who borrowed $5,600 at 13% and received only $5,054?

10. Find the equivalent rate of interest for a discount rate of $7\frac{1}{2}$% for 60 days.

11. If a note is discounted at 14% for 4 years, what is the equivalent interest rate?

12. An interest rate of $15\frac{1}{2}$% was offered to Tim on a loan for 6 months, but he was thinking of a comparable discount rate. What would that discount rate be?

13. Bryan received a promissory note for lending $3,000 for 90 days at 13% simple interest. If this note is sold to a bank charging 11% interest, would Bryan make any money, and how much would the bank make?

14. A friend buys merchandise for $3,500 and signs a 120-day non-interest-bearing promissory note. Find the proceeds if the vendor sells the note to his bank, which charges 12% interest, and if the merchandise costs the vendor $1,600, how much would he make?

15. A local company has a bid of $89.56 on a 180-day U.S. Treasury bill of $20,000. What is the purchase price, and what are the total discount and rate of return expressed in interest and discount terms?

Compound Interest
1. $2,500 is left in an account that pays $7\frac{1}{2}$% interest rate for 4 years. Calculate how much it would grow to (a) if the rate of interest is simple and (b) if the rate is compound. (c) What is the difference between the two future values?

2. Calculate the future value of a savings account of $10,000 for 6 years at $9\frac{1}{4}$% compound interest.

3. Find the current value of $7,950 due in 5 years if the money is worth $8\frac{1}{2}$ %.

4. When Brook was born, his parents put down a deposit for him in an investment account earning $12\frac{3}{4}\%$. When he graduated from college at age 22, he cashed in that account, receiving \$35,036. How much was the initial deposit?

5. On her fifth birthday, Corenza received a gift from her grandparents. It was an opening of an investment account with an initial deposit of \$5,000 that she can cash when she is 25. How much will she get if the compound interest rate is 14%?

6. Gen is interested in seeing how the money grows. Her mom suggests that she take the money in her piggy bank and deposit it in a local bank paying $8\frac{1}{2}\%$ interest compounded quarterly. If she finds \$700 in the piggy bank, how much will it grow to in 10 years?

7. What rate of compound interest will turn \$3,500 into \$10,870 in 10 years?

8. If Maria wants to collect \$16,750 in 5 years and has an option to deposit an initial amount of \$5,000 in an account bearing interest compounded monthly, what rate of interest will help her achieve her goal?

9. If Sev wants to compare the accumulations of a fund with a starting value of \$2,800 deposited for 3 years in an account bearing $13\frac{3}{4}\%$ interest, what will be the accumulated values if the compounding is (a) annual; (b) semiannual; (c) quarterly; (d) monthly; (e) weekly; (f) daily; (g) continuous?

10. How long will it take (in years) for \$11,450 to accumulate to \$29,545.88 if the money is worth 18% compounded semiannually?

11. Find the term of maturity in months for a fund with $CV = \$7,000, FV = \$10,866.80$, and $r = 14\frac{3}{4}\%$ compounded monthly.

12. Kevin was told that his loan bears 12% interest. Knowing that the compounding term will be monthly, what is the effective rate of interest?

13. Find the effective interest rate if the stated nominal rate is $5\frac{1}{2}\%$ compounded weekly.

14. What sum of money due at the end of 6 years is equivalent to \$2,100 due at the end of 10 years if the compound interest rate is $10\frac{1}{2}\%$?

15. If \$920 is due in 2 years, find an equivalent amount of debt at the end of 10 months, and at the end of 30 months. Consider a compound interest rate of $7\frac{1}{4}\%$.

16. Your friend wants to pay off her two debts in a single payment. The first debt is \$570 due in 8 months, and the second is \$1,380 due $1\frac{1}{2}$ in years. What will that single payment be if she wants to make it at the end of 1 year given a compound interest rate of 4.9%?

17. Charles had a debt of \$9,000 which is due in 1 year with 6% interest compounded quarterly. He has already made 2 payments: \$1,200 three months

after receiving his loan, and $2,570 four months after that payment. How much would he still have to pay to settle all debt at the original maturity?

18. Marty plans to discharge his debt of $3,500 in two payments, $1,500 in 10 months, and $2,000 in 15 months. If he changes his mind and wants to pay his debt off in one payment, when would that be if the interest rate is 24%?

19. Find n after a loan of three installments to be paid off in a single payment. The installments are $7,000 in 2 years, $9,000 in 3 years, and $4,000 in 4 years. Consider a rate of $5\frac{1}{2}$ %.

20. How long will it take an investment fund of $4,000 to be (a) $8,000, (b) $12,000, and (c) $20,000 if the interest rate is 10%?

21. How many years will it take to have an initial investment grow 36-fold if the interest rate is 4%?

22. If an investor is planning to multiply his initial investment 50-fold when would that be possible if the interest rate is $15\frac{1}{2}$ %?

Annuities
1. Calculate the accumulated value of an ordinary annuity of $4,200 a year for 6 years if the money is worth $7\frac{1}{2}$ %.

2. Find the future value of the cash flow of $600 a month for 5 years at 9% interest compounded monthly.

3. If Gabe makes a $450 deposit into his savings fund at the end of each quarter for 6 years, how much will he be able to collect at the end of the sixth year if the money is worth $6\frac{3}{4}$ %?

4. If Brenda contributes $630 at the end of each month to her retirement account that pays $8\frac{3}{4}$ % compounded semiannually, how much will she have when she retires 20 years from the start of contributions?

5. If $970 must be paid to an organization at the end of each month for the next 10 years, how much money is needed now if the interest is 12%?

6. Find the discounted value of an ordinary annuity of $1,490 a month for 3 years if the interest is $14\frac{1}{4}$ % compounded quarterly.

7. What is the current value of an annuity of $7,500 paid at the end of each half-year for 10 years in an account bearing $11\frac{1}{2}$ % compounded annually?

8. A couple wants to renovate their house in 3 years. They need $27,000 which they plan to save for in monthly payments in an account that pays $8\frac{1}{2}$ % compounded monthly. How much would their monthly savings be?

9. Wayne wants to set up a monthly payment for his daughter, who plans to live in another state for the next four years. He deposited $40,000 for her automatic bank payments. If the bank pays $7\frac{1}{4}$ % interest compounded semiannually, how much will she receive each month?

10. Rosemary would like to buy a chalet overlooking the Alps for $300,000. She can save up to $10,000 a month in an account paying 15% interest compounded monthly. How long will it take her to wait?

11. A local college receives a $500,000 gift from an alumnus's widow to establish a scholarship of $10,000 a year. If the college invests the gift in an account bearing $11\frac{1}{2}$ % interest compounded quarterly, how many years will the gift last?

12. A restaurant owner wants to buy new kitchen equipment for $25,000. He would like to pay for it through saving up $2,000 a week in a fund that pays 10% interest compounded monthly. How long should he wait to save the entire amount?

13. If the future value of an annuity is $35,507.50 and the quarterly payment is $1,750 for 9 years, how much will the annuity interest rate be?

14. If you deposit $220 in your savings account at the beginning of each month for 2 years and if your account bears $6\frac{1}{4}$ % compounded monthly, how much will you save at the end of the 2 years?

15. Find the current value of an annuity due of $900 each week for $1\frac{1}{2}$ years at 8% interest compounded weekly.

16. Find the future fund for Kelly, who is saving $350 at the beginning of each month for the next 4 years, if her savings account bears $7\frac{1}{2}$ % interest compounded quarterly.

17. A $75,000 mortgage is obtained at 9%. It should be paid in 20 years. Find the payment at the start of each month.

18. Herb is a self-employed agent who is setting his own retirement fund. He is depositing $17,000 a year for the next 20 years. How much will he be able to collect for his retirement given that his retirement fund bears 11% interest compounded semiannually?

19. Jen needs $20,000 in the near future. She is saving up $875 at the beginning of each month in an account bearing 15% interest compounded annually. When will the $20,000 be available?

20. Find the future value of an annuity of $2,615 payable monthly for $3\frac{1}{2}$ years that starts after 3 months. The interest rate would be 7%.

UNIT III

Mathematics of Debt and Leasing

1 Credit and Loans

Credit for both businesses and consumers is an important element in the health of the economy, as long as it can be well managed. Credit understanding and management must be part of a long-term spending and saving plan, and for that it can become a determinant of economic growth. Unless access to credit acts to create a false sense of prosperity and leads to harmful overextension, it can actually reap great benefits by expanding current consumption, stimulating saving, and bringing about economic growth and robustness. Borrowing can basically increase today's income and expand one's current capacity to buy. However, all borrowing is made against future income, which will be reduced by the repayment of debt and its interest. This means that a person's future income will not be able to maintain the level of consumption unless it is increased by what can compensate for all debt repayments and other variables, such as inflation and changes in one's needs and wants. The bottom line is that borrowing and saving involve the choice of spending more or less today versus less or more tomorrow. Making the right choice requires a good understanding of what credit is all about and how debt can best be managed. Our focus here is on understanding the types of debt and loans, the process and calculation of repayment, how debt is amortized, and how the cost of credit is determined.

1.1. TYPES OF DEBT

Generally, and according to the structure of repayment, debt is categorized into two major categories: noninstallment debt and installment debt.

Noninstallment Debt

Noninstallment debt is also called **open-ended debt**. This category includes:

1. Single-payment debt, where the full balance of both principal and interest is paid at a certain agreed-upon time.
2. A family of revolving and charge account credits, such as credit cards and certain business charge cards, such as department store or gas company charge accounts. It also includes personal lines of credit, home equity lines

Mathematical Finance, First Edition. M. J. Alhabeeb.
© 2012 John Wiley & Sons, Inc. Published 2012 by John Wiley & Sons, Inc.

of credit, and service credit. The payment for this type of debt is either a minimum payment such as 2 to 3% of the full balance, or any other optional payment exceeding the usual minimum payment.

Installment Debt

Installment debt is also called **close-ended debt**. This is the most common type of debt and is also known by its association with amortization. An amortized debt is any interest-bearing debt on certain borrowed money that, by contract, is to be paid off in a series of equal payments at regular intervals, usually months, for a certain period of maturity. This can be a short term (up to one year), as in some personal loans and consumer debt; an intermediate term (up to five years), as in auto loans, home improvement, and loans for consumer durables; or a long term, as in mortgages and business loans.

We focus here on the nature and dynamics of the amortization process and later on its "cousin," the sinking fund process, but before that, let's discuss the issues of generating and breaking down the interest and knowing the actual annual percentage rate (APR). One of the most distinct features of an amortized loan is how the payment is broken down between the portion that pays off the principal and the portion that pays off the interest.

1.2. DYNAMICS OF INTEREST–PRINCIPAL PROPORTIONS

The interest portion in the repayment of a loan is usually determined according to the interest rate (r) and time of maturity (n) and whether the interest is simple or compound. The principal portion would be determined simply by the difference between the payment amount and the interest portion for any interval. The question is whether those portions remain the same throughout the entire series of intervals and also, how they relate to each other. Generally, there are two ways to deal with the dynamics of the interest and principal proportions as they relate to themselves and to each other during the entire series of payments.

The Level Method

In the level method, both interest and principal proportions remain the same throughout the entire maturity, and as a result, their relationship to each other remains consistent for the entire period. The following example illustrates this constant and consistent relationship.

Example 1.2.1 Let's consider a loan of $1,600 at 12% interest for 1 year as it breaks down into 12 monthly payments (PYTs):

$$P = \$1,600$$
$$I = P \cdot r \cdot t$$
$$= \$1,600(.12)(1)$$

$$= \$192$$

$$n = 12 \qquad \text{number of payments}$$

$$\text{PYT} = \frac{P + I}{n}$$

$$= \frac{\$1,600 + \$192}{12}$$

$$= \$149.33$$

This monthly payment (MP) is broken down into the monthly interest portion (MIP) and the monthly principal portion (MPP).

$$\text{MIP} = \frac{I}{n}$$

$$= \frac{\$192}{12}$$

$$= \$16$$

$$\text{MPP} = \text{PYT} - \text{MIP}$$

$$= \$149.33 - \$16$$

$$= \$133.33$$

Table E1.2.1 shows how the MIP of $16 and the MPP of $133.33 remain the same throughout the entire 12 months.

TABLE E1.2.1

PYT No.	PYT Fraction	MIP	MPP	MP
1	$\frac{1}{12}$	16	133.33	149.33
2	$\frac{1}{12}$	16	133.33	149.33
3	$\frac{1}{12}$	16	133.33	149.33
4	$\frac{1}{12}$	16	133.33	149.33
5	$\frac{1}{12}$	16	133.33	149.33
6	$\frac{1}{12}$	16	133.33	149.33
7	$\frac{1}{12}$	16	133.33	149.33
8	$\frac{1}{12}$	16	133.33	149.33
9	$\frac{1}{12}$	16	133.33	149.33
10	$\frac{1}{12}$	16	133.33	149.33
11	$\frac{1}{12}$	16	133.33	149.33
12	$\frac{1}{12}$	16	133.37	149.37
	$\frac{12}{12}$	192 +	1600 =	1792

The Rule of 78 Method

Under this method, interest and principal portions are in a reverse relationship. While the interest portion decreases, the principal portion increases throughout the maturity period. Interest for each payment is determined by a certain fraction calculated according to the rule of 78. This rule obtained its name based on the total number of monthly payments within a year $(1 + 2 + 3 + \cdots + 12 = 78)$. The number (78) serves as a denominator of a fraction used to determine the interest portion for any month. This number would also change based on the maturity term for a total number of payments. For example, 24 payments would total \$300 $(1 + 2 + 3 + \cdots + 24 = \$300)$, 36 payments would total \$666 $(1 + 2 + 3 + \cdots + 36 = \$666)$, and so on.

The fraction to determine the amount of interest for any payment is

$$K_i = \frac{j}{D} \qquad j = 1, 2, \ldots, n$$

The numerator (j) is the number of payments placed in reverse order. For example, in a 12-payment loan, j for the first payment is 12, and the second payment is 11, and so on, until the last payment is 1. The denominator would be 78 for 12 payments, \$300 for 24 payments, \$666 for 36 payments, and so on. A general formula to get the total number of digits to any number of payments, quickly and more conveniently is:

$$D = \frac{n(n + 1)}{2}$$

where D is the denominator and n is the total number of payments in the entire maturity time, such as 12 for a year, 24 for two years, 36 for three years, and so on. For a 5-year term, the denominator in the rule of 78 method would be

$$D = \frac{60(60 + 1)}{2}$$
$$= 1,830$$

The interest is obtained by multiplying the total interest (I) by the appropriate fraction for that payment. For example, the monthly portion for the fourth payment in a 24-payment loan is

$$\text{MIP}_4 = \frac{21}{\$300}(I)$$

The numerator (21) is obtained as the number corresponding to payment 4 in reverse, as in:

1	2	4	5	6	7	8	9	10	11	12	13	14	15	16	17	18	19	20	21	22	23	24
24	23	21	20	19	18	17	16	15	14	13	12	11	10	9	8	7	6	5	4	3	2	1

and the denominator 300 is obtained by

$$D = \frac{n(n+1)}{2} = \frac{24(24+1)}{2} = 300$$

Example 1.2.2 Let's apply the rule of 78 method to break down the monthly payment (PYT) in Example 1.2.1 between interest and principal.

$$PYT = \$149.33$$

For example, the sixth payment would be broken down into

$$MIP_6 = \frac{7}{78}(192)$$

$$= \$17.23$$

$$MPP_6 = \$149.33 - \$17.23$$

$$= \$132.10$$

and so on, for all payments and their breakdown, as shown in Table E1.2.2.
    Note that the 6th payment took the order of 7 in a one-year loan.

The Declining Balance Method

In the declining balance method, interest on a loan is applied on the outstanding balance as it decreases month by month due to the principal portions being deducted. This is why the balance and interest show decreasing figures, while the

TABLE E1.2.2

PYT No. 5	PYT Fraction	MIP	MPP	MP
1	12/78	29.54	119.79	149.33
2	11/78	27.08	122.25	149.33
3	10/78	24.62	124.71	149.33
4	9/78	22.15	127.18	149.33
5	8/78	19.69	129.64	149.33
6	**7/78**	**17.23**	**132.10**	**149.33**
7	6/78	14.77	134.56	149.33
8	5/78	12.31	137.02	149.33
9	4/78	9.85	139.48	149.33
10	3/78	7.38	141.95	149.33
11	2/78	4.92	144.41	149.33
12	1/78	2.46	146.91	149.37
	78/78	192	1600	1792

principal portions show increasing figures throughout the payoff schedule. At the end and in the last payment, the principal portion would have reached a maximum, the interest portion would have reached a minimum, and the outstanding balance would have reached zero, which is the total payoff point of the loan.

1.3. PREMATURE PAYOFF

One might ask: What difference would it make to have one method or another to determine interest if the interest paid is the same, such as the $192 of Example 1.2.2 and in both methods? The answer is that it would make no difference whatsoever if the loan is paid off at full maturity. However, if for some reason the borrower decides at some point to pay off the loan prematurely, the rule of 78 method would make him pay more interest than would the level method. This is because the rule of 78 method charges interest in a descending manner, with higher interest paid in the early payments than in later payments.

Example 1.3.1 Let's suppose that the borrower of the loan in Example 1.2.2 decided to pay off the entire loan after making only four payments. What would be:

(a) the balance due
(b) the total interest to be paid in both methods and
(c) the difference in interest paid between the two methods?

Under the level method:

(a) The balance due would be the remaining principal only, since there will be no interest assessed after the balance of the principal has been paid. In this case, eight principal proportions would make the balance due:

$$\text{balance due} = (\$133.33 \times 7) + \$133.37$$
$$= \$1,064.37$$

(b) The interest paid would be what has been paid for the first four payments:

$$\text{interest paid} = \$16 \times 4$$
$$= \$64$$

Under the rule of 78 method: Since the principal proportions are different from each other, the balance due would be either the sum of all of the remaining eight principal proportions or the total of the principal proportion minus the first four paid proportions, which is easier:

(a)   Balance due $= \$1,600 - (\$119.79 + \$122.25 + \$124.71 + \$127.18)$

$$= \$1,106.07$$

(b)   Interest paid $= \$29.54 + \$27.08 + \$24.62 + \$22.15$

$\qquad\qquad\qquad = 103.39$

(c)   The difference in interest $= \$103.39 - \$64$

$\qquad\qquad\qquad\qquad\qquad = \$39.39$

which renders the rule of 78 method more costly to borrowers but more rewarding to lenders, especially in the case of the premature payoff of a loan.

There is another way to find the balance due when a loan is paid prematurely under the rule of 78 method. We divide the total number of payments (n) into two parts: the number of payments already paid (N_p) and the remaining unpaid number of payments (N_u).

$$N_u = n - N_p$$

So, in Example 1.3.1, N_u would be eight payments ($8 = 12 - 4$). Since the rule of 78 denominator was determined by

$$D_n = \frac{n(n+1)}{2}$$

D_u can also be determined by

$$D_u = \frac{N_u(N_u + 1)}{2}$$

and the fraction of the interest on the remaining payments, which is called the **rebate factor** (RF), would be obtained by

$$RF = \frac{D_u}{D_n} = \frac{[N_u(N_u + 1)]/2}{[n(n+1)]/2}$$

$$RF = \frac{N_u(N_u + 1)}{n(n+1)}$$

The amount of remaining interest that should not be paid, called the rebate (Rb), is calculated as part of the entire interest (I) as determined by RF.

$$Rb = RF(I)$$

This portion of interest is excluded from the total of all remaining payments (TN_u) to get what should be paid or the balance due (B_d) when a loan is paid prematurely.

$$B_d = TN_u - Rb$$

Let's apply this method to Example 1.2.2.

$$RF = \frac{N_u(N_u + 1)}{n(n + 1)}$$

$$= \frac{8(8 + 1)}{12(12 + 1)}$$

$$= .46154$$

$$Rb = RF(I)$$

$$= .46154(192)$$

$$= \$88.61$$

$$TN_u = PYT(N_u)$$

$$= \$149.33(8)$$

$$= \$1,194.64$$

$$B_d = TN_u - Rb$$

$$= \$1,194.64 - \$88.61$$

$$= \$1,106.03$$

$$I = \$103.39 + \$88.61$$

$$= \$192$$

Some financial institutions impose a fee called a **prepayment penalty** on paying off a loan prematurely. This fee is actually calculated as the difference between the balance due according to the rule of 78 and what the borrower might assume to pay, which is most likely to be the remaining portion of the principal (as in the level method).

In Example 1.3.1 the balance due was \$1,106.03 according to the rule of 78. The borrower may assume that what is due is just the remaining portion of the principal, which in the level method was calculated as \$1,064.37. The difference of \$41.66 ($\$1,106.03 - \$1,064.37$) would be the prepayment penalty.

1.4. ASSESSING INTEREST AND STRUCTURING PAYMENTS

According to different types of loans, interests are assessed and payments are structured in different ways. Next we discuss the various types of loans and how interest and payments are calculated.

Single-Payment Loans

Repayment for a single-payment loan must be in full, including the principal and total interest, and be made only once, at a certain agreed-upon time. It is simple and straightforward and carries none of the complications usually associated with breaking down payments and figuring out the correct fraction of interest, or dealing with calculating the right due date, and so on. As a result of simplicity and plainness, the actual annual percentage interest (APR) would exactly match the stated APR in this type of loan. Total interest (I) on the borrowed amount (P) would be determined in the standard way:

$$I = P \cdot r \cdot t$$

where r is the interest rate or stated APR and t is the time of maturity. The payoff amount would simply be one payment:

$$\boxed{\text{PYT} = P + I}$$

Example 1.4.1   Joe borrowed \$5,000 from his cousin to buy new equipment for his computer lab. They agreed that the amount would be paid off in full plus 6.5% interest after 2 years on the day the fund was received by Joe. How much will Joe be paying back?

$$
\begin{aligned}
I = P \cdot r \cdot t \\
= \$5,000(.065)(2) \\
= \$650 \\
\text{PYT} = P + I \\
= \$5,000 + \$650 \\
= \$5,650
\end{aligned}
$$

If the APR_s stated was 6.5%, what was the actual APR_a in this example?

$$\text{APR}_a = \frac{AI}{P}$$

where AI is the annual interest and P is the principal or the original amount borrowed (\$5,000).

$$
\begin{aligned}
\text{AI} &= \frac{I}{t} \\
&= \frac{\$650}{2} \\
&= \$325 \\
\text{APR}_a &= \frac{\$325}{\$5,000} \\
&= .065 \quad \text{or} \quad 6.5\%
\end{aligned}
$$

This shows that the stated APR is the same as the actual APR ($APR_s = APR_a$) for the single-payment loan. However, it might not be the same with other types of loans or even with this one when the original amount received by the borrower is different from the stated principal, such as in the case of a discounted loan.

Example 1.4.2   If in Example 1.4.1 Joe was borrowing the $5,000 from a bank that follows the discounted interest method, the total interest of $650 would be taken by the lender up front, so that Joe would receive $4,350 ($5,000 − $650). Then the actual APR would be

$$APR_a = \frac{\$325}{\$4,350}$$

$$= .075 \text{ or } 7.5\%$$

and that confirms that in the discounted loan, the actual annual percentage rate would be larger than the rate stated.

$$APR_a > APR_s$$

Add-On Interest Loans

Add-on interest loans use add-on as a method to calculate interest. They are popular for most commercial banks and savings and loans firms, and they constitute most consumer finance installment loans. In this type of loan, interest is assessed and charged up front but is not actually received immediately. It is calculated based on the whole principal and for the entire length of the maturity time. The whole interest is added to the principal and the total is divided by the maturity time to determine the size of the periodic payment, which is usually a monthly payment.

$$PYT = \frac{P + I}{n} \tag{1}$$

where PYT is the periodic payment, P is the principal or the amount borrowed, n is the maturity time measured in equal intervals such as months, and I is the total interest, which is determined by

$$I = P \cdot r \cdot t \tag{2}$$

which would allow the modification of equation (1) into equation (3), using n instead of t:

$$PYT = \frac{P + P \cdot r \cdot n}{n}$$

$$\boxed{PYT = \frac{P(1 + rn)}{n}} \tag{3}$$

Example 1.4.3   Based on an add-on interest, Emily borrowed $6,750 at 8.25% interest for 3 years. Determine her quarterly payment.

$P = \$6{,}750; r = .0825/4 = .0206; n = 3 \times 4 = 12.$

$$\text{PYT} = \frac{P(1+rn)}{n}$$

$$= \frac{\$6{,}750[1 + .0206(12)]}{12}$$

$$= \$701.72 \quad \text{per quarter.}$$

Discount Loans

Just as in the add-on loan, interest is assessed up front in the discount loan, but opposite to the add-on loan, the entire interest is deducted immediately from the amount of loan sought and the borrower is given the rest of the principal. However, as an installment loan, the full principal would be broken down into monthly payments by dividing it by the maturity time. The interest rate is called the discount rate here and is denoted d, the total interest is called the total discount and denoted D, and the actual amount received by the borrower is denoted P_0 or the principal obtained:

$$D = P \cdot d \cdot n$$

$$P_0 = P - D$$

$$P_0 = P - P \cdot d \cdot n$$

$$\boxed{P_0 = P(1 - dn)}$$

$$\text{PYT} = \frac{P}{n}$$

So this is the only type of installment loan where the monthly payment consists entirely of the principal proportion since the full interest was deducted in advance. It is noteworthy here to mention that both the add-on and discount loan structures assume that the entire principal is owed for the entire maturity time, whereas reality shows that the principal is reduced gradually over time as payments are made, plus the fact that in a discount loan, less than the full principal is used by the borrower to begin with. This discrepancy between assumption and reality leads to the fact that the actual APR in these methods would be different from the APR stated. Below we present two formulas used to estimate the actual annual percentage rate (APR_a).

$$\boxed{\text{APR}_a^1 = \frac{2KI}{P(n+1)}} \quad \text{first formula}$$

where K is the number of payments within a year, I is the total interest, P is the principal or the original amount borrowed, and n is the total number of payments throughout the maturity period.

Example 1.4.4   To calculate the actual annual percentage rate in Example 1.4.3, we use four quarterly payments for K, the combined total of $6,750 for P, 12 quarterly payments within 3 years for n, and a total interest of $1,670.62 for I, as obtained by

$$I = P \cdot r \cdot t$$
$$= \$6,750(.0825)(3)$$
$$= \$1,670.62$$
$$\text{APR}_a^1 = \frac{2KI}{P(n+1)}$$
$$\text{APR}_a^1 = \frac{2(4)(\$1,670.62)}{\$6,750(12+1)}$$
$$= .152 \text{ or } 15.2\%$$

The second formula used to calculate the actual annual percentage rate (APR_a^2) for the installment loans with add-on interest is called the **n-ratio formula**:

$$\boxed{\text{APR}_a^2 = \frac{K(95n+9)I}{12n(n+1)(4P+I)}} \qquad \text{second formula}$$

where K is the number of payments within a year, n is the total number of payments throughout the entire maturity period, P is the principal or the amount borrowed, and I is the total interest.

Example 1.4.5   Applying the second formula to Example 1.4.4 yields

$$\text{APR}_a^2 = \frac{4[95(12)+9]\$1,670.62}{12(12)(12+1)[4(\$67,750)+\$1,670.62]}$$
$$= .143 \text{ or } 14.3\%$$

The two formulas produce different estimations of APR, but both yielded an actual APR that is larger than the one stated for this loan:

$$\text{APR}_a < \text{APR}_s$$

which confirms that the add-on interest method does overestimate the interest charged.

1.5. COST OF CREDIT

Consumers and businesses pay billions of dollars every year for loans and for the charges they run on credit cards and other charge accounts. Two major factors actually determine the power to increase the cost of credit: the interest rate and the time to keep an outstanding balance. Both of these factors have a positive relationship to the amount of interest charged on the balance. The higher the interest rate and the longer the time, the higher the interest paid by the borrower.

A significantly costly action that borrowers could take is to keep paying only the minimum payments on their revolving accounts for a long period. This action can increase the time required to pay off debt to an unfathomable length and, as a result, would push the amount of interest paid substantially, especially if the minimum payment is set up at the lower level of 2% or less of the outstanding balance. The industry standard for credit card and charge accounts is 2 to 3%. Let's work through an example.

Example 1.5.1   Jennifer fell in love with a complete home entertainment system that was for sale at $3,000. The department store is offering a monthly payment of only $60 for customers who buy the system on the store credit card, which made it sound very easy to handle. Given that the store charges 20% APR on its charge card, how long will it take Jennifer to pay off this purchase, and how much total interest will she end up paying?

The $60 is a minimum payment set up at 2% of the outstanding balance, according to the store policy.

$P = \$3,000$; APR $= 20\%$; monthly rate $= .20/12$.

The monthly payment (PYT) of $60 is going to be split between a monthly interest portion (MIP) and a monthly principal portion (MPP).

$$\text{MIP} = \$3000 \times \frac{.20}{12} = \$50$$

$$\text{PYT} = \$60$$

$$\text{MPP} = \$60 - \$50 = \$10$$

$$\text{principal payoff time (PPT)} = \frac{P}{\text{MPP}} = \frac{\$3,000}{\$10} = 300 \text{ months}$$

$$= \frac{\$300}{12} = 25 \text{ years}$$

$$\text{total interest paid} = \text{MIP} \cdot \text{PPT}$$

$$= \$50(300)$$

$$= \$15,000$$

Using this method of charging interest, it would take Jennifer 25 years and a total interest of $15,000 to pay off the $3,000 for the entertainment system if she chooses to pay the minimum payment only. In fact, the reality would be even worse! The reason is that in our simple calculation above, we assumed that the split between interest $50 and principal $10 stays consistent throughout the maturity time. In reality, the split changes, which leads to adding more time for the payoff. Running the payoff schedule in the computer financial calculator shows that the actual annuity split system would make the payoff time 524 months or more than 43 years, and the total interest would actually be less, $12,126. Table E1.5.1 shows how long the $3,000 charge would be paid off in years and how

TABLE E1.5.1   Cost of $3000 in Installments by Total Interest Paid (I) and Time to Pay Off in Years (t)

Minimum Payment (%)	APR							
	5%		10%		15%		20%	
	t(years)	I($)	t (years)	I($)	t (years)	I($)	t (years)	I($)
2	12	685	15.3	1,831	22	4,185	43.6	12,126
3	8.75	443	10.2	1,050	12	1,937	15.25	3,361
4	7	327	7.75	738	8.75	1,271	10	1,989
5	5.9	260	6.4	570	7	948	7.75	1,418

much it would cost in interest under four possible APR levels (5%, 10%, 15%, 20%) and four possible minimum payment settings (2%, 3%, 4%, and 5%). In our example, if Jennifer makes her payment 5% (3% more than the minimum payment she chose), she can cut the payoff time to less than 8 years (a reduction of 82%) and cut down the interest to only $1,418 (a reduction of 88%). Note how large a difference a lower interest rate makes in both the time and money paid as a cost of credit. The best scenario in Table E1.5.1 is taking this money as a loan at 5% and choosing 5% or more as a minimum payment percentage. At 5% minimum payment, the total interest would be only $260, a far cry from the $12,126 (a reduction of 98%), and the payoff time would be less than 6 years (a reduction of 86.5% compared to the 43.6 years).

1.6. FINANCE CHARGE AND AVERAGE DAILY BALANCE

Most credit statements, such as a consumer's credit card statement, show a finance charge, which is the interest cost, calculated by applying the monthly rate of the APR on the average daily balance of the account. There are some varieties of the application according to specific lenders and depending on what they include in, or exclude out of, that balance. Generally, the **average daily balance** (ADB) is calculated by

$$\boxed{\text{ADB} = \frac{\sum_{i=1}^{k} b_i t_i}{Cy} \qquad I = 1, 2, \dots, K}$$

where b_i is the daily balance of the account, t_i is the time for which balance b remains outstanding, and CY is the billing cycle.

Example 1.6.1   For the month of October, Linda's credit card statement shows the following:

$190	carried over from September
$50	payment on October 5
$72.95	health club charge on October 11

TABLE E1.6.1

Date	Days Balance per	Purchase/Cash Advance ($)	Payments ($)	Balance ($)
10/1	4			190.00
10/5	6	72.95	50.00	140.00
10/11	6	210.85		212.95
10/17	5	26.90		423.80
10/22	7	18.75		450.70
10/29	3			469.45
10/31				469.45

$210.85   J.C. Penney charge on October 17

$26.90   Friendly's charge on October 22

$18.75   T.J. Max charge on October 29

Calculate her finance charge if the APR is 18% and the billing cycle is 31 days.

$$\text{ADB} = \frac{\sum_{i=1}^{K} b_i t_i}{Cy}$$

$$= \frac{\begin{array}{c}(\$190 \times 4) + (\$140 \times 6) \\ +(\$212.95 \times 6) + (\$423.80 \times 5) \\ +(\$450.70 \times 7) + (\$469.45 \times 3)\end{array}}{31}$$

$$= \$308.39$$

$$\text{monthly rate (MR)} = \frac{.18}{12} = .015$$

$$\text{monthly finance charge (MFC)} = \text{ADB} \cdot \text{MR}$$

$$= \$308.39(.015)$$

$$= \$4.63$$

The varieties of the average daily balance include:

1. ADBs that exclude new purchases, where the interest is assessed only on the balance carried over from the preceding month.
2. ADBs that include new purchases, where interest is assessed on both the balance carried over and the new balance. This variety is of two types, one with a grace period that allows new purchases to be excluded only when there is no previous balance, and a second with no grace period that allows the inclusion of both previous and current balances.
3. ADBs with two cycles which have no grace period for the current cycle or for the preceding cycle. It basically doubles the finance charge by assessing interest on both the preceding and current cycles.

1.7. CREDIT LIMIT VS. DEBT LIMIT

The credit limit and the debt limit are not interchangeable terms as they may seem! The **credit limit** is the maximum level of credit that would be offered by a lender to borrowers on an open-ended credit account based on the lender's criteria. The **debt limit** is the maximum level of debt that a borrower would allow for himself or herself based on affordability and the person's future capacity to meet the debt repayment obligations. It is a subjective assessment based on both, understanding one's financial situation and on some objective affordability criteria. Rationality dictates that the debt limit should be lower than the potential credit limit offered because the borrower's self-interest is better judged personally rather than depending on the lender's commercial offers to extend credit.

While lenders determine credit limit based on their own considerations and depending on the borrower's credit score, a borrower's debt limit can generally be assessed based on three indicators:

1. *Debt payment/disposable income ratio.* The ratio of all debt payments (excluding mortgage payment) to disposable income should not exceed 20%. A ratio between 20 and 30% would mean that a borrower is perhaps "excessively over-indebted."

$$\boxed{\frac{DP}{DI} \leq .20}$$

where DP is debt payments and DI is disposable income.

2. *Debt/equity ratio.* Debt in this ratio would also exclude mortgage debt. It is an aggregate measure of one's state of solvency. Debt (D) in this measure should not be more than one-third of equity (E).

$$\boxed{\frac{D}{E} \leq .33}$$

3. *Continuous debt measure.* This measure is a more qualitative assessment of the state of indebtedness that one could be experiencing. Financial experts believe that if a person is unable to clear his debt within a four- to five-year period, it could be a strong indicator that this person is somehow dependent on debt and may not get out of it completely.

Example 1.7.1   Determine the debt limit status for Peter when his gross income is $85,000, his taxes are $21,250, and his monthly debt payments include:

Mortgage: $1,920

Auto loan: $230

Personal loan: $275

Credit card 1: $318

Credit card 2: $165

First, we have to unify, whether we calculate annually or monthly, and second, we have to exclude the mortgage payment. We may get the monthly disposable income and keep all other debt monthly as given.

$$\text{annual disposable income} = \text{gross income} - \text{taxes}$$

$$= \$85,000 - \$21,250 = \$63,750$$

$$\text{monthly disposable income} = \frac{\$63,750}{12} = \$5,312.50$$

$$\text{all debt} = \$230 + \$275 + \$318 + \$165 = \$988$$

$$\frac{\text{DP}}{\text{DI}} = \frac{\$988}{\$5,312.50} = 18.6\%$$

Peter is still okay but is almost on the margin. His situation would not be comfortable if he incurs more debt.

Example 1.7.2   Jill and Tom have total monetary assets of $6,700, investment assets of $77,000, and tangible assets of $305,000, including the value of their home, estimated at $250,000. They also have short-term liabilities totaling $2,320 and long-term liabilities totaling $209,000, including $187,000 in outstanding mortgage loan balance. How would this couple assess their debt limit?

Here we have enough information on the couple's assets and liabilities. The best criterion would be the debt/equity ratio. First we get their equity or net worth. We have to exclude the mortgage from both assets (home value) and from liabilities (mortgage balance).

$$\text{total assets} = \$6,700 + \$77,000 + (\$305,000 - \$250,000)$$

$$= \$138,700$$

$$\text{total liabilities} = \$2,320 + (\$209,000 - \$187,000)$$

$$= \$24,320$$

$$\text{equity} = \$138,700 - \$24,320 = \$114,380$$

$$\frac{D}{E} = \frac{\$24,320}{\$114,380} = 21.3\%$$

Jill and Tom are in good shape, and their debt limit has not reached an alarming level.

2 Mortgage Debt

Mortgage debt is long-term debt incurred by obtaining a loan granted specifically to purchase real estate property, with the debt collateralized by the property itself. This means that the lender would have the legal right to foreclose on the property should the borrower default on the loan contract. Since the amount of such a loan is large and gets much larger with the added interest, paying off such a loan is made gradual and extended over a long period of time, often extending to 30 years. The process of structuring the graduated payments and breaking them down between principal and interest portions, called **amortization**, is described next as a way of dealing with mortgage loans specifically, although it is a general method that can be applied to other types of loans.

2.1. ANALYSIS OF AMORTIZATION

According to Guthrie and Lemon (2004), the invention of the amortization process was a breakthrough in the borrowing world that occurred in the first third of the nineteenth century. Until that time, all loans were paid off with interest in a single whole payment at the end of a certain maturity period or by self-determined installments. The first home mortgage loan to be formally amortized was obtained by Comly Rich in 1831 and financed by the Oxford Provident Building Association of Frankford, Pennsylvania. It was to purchase a three-room house in Frankford for a total of $375. That was the leap forward that forever changed the structure of long-term installment loans and ushered in their popularity and wide prevalence. Since then, amortization of debt has become the most important application of annuities in the financial world and one of the most vital business transactions.

Under the amortization method, the long term of maturity, usually anywhere between 15 and 30 years, determines the number of payments, which are usually monthly. As a certain interest rate is applied, a sequence of equal payments is set to show how the entire loan and its interest are to be paid off by the end of maturity. This sequence of payments, their breakdown between principal and interest, and their reduction in the outstanding balance is called the **amortization schedule**.

Table 2.1 shows a partial amortization schedule for a $100,000 loan at 8% interest for a 30-year term. The entire 360 payments throughout the life of the

Mathematical Finance, First Edition. M. J. Alhabeeb.
© 2012 John Wiley & Sons, Inc. Published 2012 by John Wiley & Sons, Inc.

TABLE 2.1 Amortization Schedule Loan:
$100,000; Term: 30 Years; Interest: 8% Fixed

Term Years	PYT No.	PYT Amount ($)	MPP ($)	MIP ($)	Total PYT to Date ($)	Loan Balance ($)
1st	1	733.76	67.09	666.67	733.76	99,932.91
	2	733.76	67.55	666.21	1,467.52	99,865.36
	3	733.76	68.00	665.76	2,201.28	99,797.36
	4	733.76	68.44	665.32	2,935.04	99,728.91
	5	733.76	68.91	664.85	3,668.80	99,660.00
	6	733.76	69.36	664.40	4,402.56	99,590.64
	7	733.76	69.83	663.93	5,136.32	99,520.81
	8	733.76	70.29	663.47	5,870.08	99,450.52
	9	733.76	70.76	663.80	6,603.84	99,379.76
	10	733.76	71.23	662.53	7,337.60	99,308.53
	11	733.76	71.71	662.05	8,071.36	99,236.82
	12	733.76	72.19	661.57	8,805.12	99,164.63
2nd	24	733.76	78.18	655.58	17,610.24	98,259.94
5th	60	733.76	99.30	634.46	44,025.60	95,069.85
10th	120	733.76	147.95	585.81	88,051.20	87,724.70
15th	180	733.76	220.42	513.34	132,076.80	76,781.55
20th	240	733.76	328.38	405.38	176,102.40	60,477.95
25th	300	733.76	489.25	244.51	220,128.00	36,188.09
30th	360	733.76	728.90	4.86	264,153.60	0

loan are not shown, but the first year is shown in its monthly details and then the figures jump to the 2nd, 5th, 10th, 15th, 20th, 25th, and 30th years, only to give a general sense of what the sequence of payments would look like later.

Observations   Looking at the amortization table, we observe the following:

- Due to a fixed interest rate that stays the same at 8% throughout the full term of 30 years, mortgage payment stays the same at $833.76. It would have not stayed the same had this loan been taken as an adjustable rate loan.
- The monthly principal portion (MPP) starts at as little as $67.09, which is only around 9% of the payment. It increases slowly and consistently throughout the term until, on the last payment, it reaches its maximum, $728.90, which is more than 99% of the payment.
- Exactly contrary to the monthly principal portion, the monthly interest portion (MIP) starts at the maximum, $666.67, which is about 91% of the payment. It decreases slowly and consistently throughout the term until it reaches only $4.86 on the last payment, which is about. (.006) of the payment. This structure is designed so that lenders would get as much of their interest as early as possible.
- The "total payment to date" column shows all payments made accumulatively so that the marginal change is always equal to one payment. In other

words, any figure in this column is obtained by multiplying the payment by its number. For example, after 10 years of assuming this loan, the borrower would have paid $88,051.20 ($733.76 × 120).

- The loan balance starts at $99,932.91, which is the loan amount minus the first principal payment ($100,000 − $67.09), and it gets decreased each month by the amount of the principal portion of the month before. In other words, any month's balance is the balance of the preceding month reduced by the principal portion of the current month. For example, the balance at the 8th month ($99,450.52) is ($99,520.81 − $70.29).

- The interest portion (MIP) for each month is determined by multiplying the loan balance from the preceding month by the monthly interest. For example, the interest portion for the 10th payment ($662.53) is obtained by multiplying the loan balance of the 9th month ($99,379.76) by the monthly interest rate .08/12 [$99,379.76 × (.08/12) = $662.53].

- The principal portion (MPP) for each month is obtained by subtracting the interest portion from the monthly payment: PYT − MIP = MPP. For example, the principal portion of $71.23 is obtained by subtracting $662.53 from the payment of $733.76 ($733.76 − $662.53 = $71.23).

- The monthly payment is normally calculated as an ordinary annuity payment of a current value, the current value being the amount of the loan. The following formula is applied to get the payment (A) when the current value of the loan (CV) is $100,000, the interest rate (r) is 8% compounded monthly, and the maturity (n) is 30 years broken down into 360 payments:

$$A = \frac{CV \cdot r}{1 - (1 + r)^{-n}}$$

$$= \frac{\$100,000(.08/12)}{1 - [1 + (.08/12)]^{-360}}$$

$$= \$733.76$$

We can also obtain the payment by the table method, using the following formula:

$$A = CV \cdot \frac{1}{a_{\overline{n}|r}}$$

$$A = (100,000)(.0073376)$$

$$A = 733.76$$

Example 2.1.1   Jimmy is interested in a two-bedroom townhouse, the asking price for which is $98,000. His mortgage officer told him that the best interest rate he could get would be 5% if he chooses the 15-year term. What would his monthly payment be? Use both the math formula and table methods, given that the table value of $1/a_{\overline{180}|(.05/12)}$ is .00790794.

$$r = \tfrac{.05}{12}; \; n = 12 \times 15 = 180.$$

By the formula method:

$$
\begin{aligned}
A &= \frac{CV \cdot r}{1 - (1+r)^{-n}} \\
&= \frac{\$98,000(.05/12)}{1 - (1+.05/12)^{-180}} \\
&= \$774.98
\end{aligned}
$$

By the table method:

$$
\begin{aligned}
A &= CV \cdot \frac{1}{a_{\overline{180}|(.05/12)}} \\
&= \$98,000(.00790794) \\
&= \$774.98
\end{aligned}
$$

Table 2.2 shows precalculated factors that can quickly assist in calculating installment loan payments based on $1,000 at different APRs and different terms. For example, the monthly payment for a $125,000 loan obtained at 9% APR for 5 years would be $295.50 (125 × 20.76).

TABLE 2.2 Monthly Payment for a $1,000 Installment Loan

APR (%)	1 / 12	2 / 24	3 / 36	4 / 48	5 / 60	6 / 72	7 / 84	8 / 96	9 / 108	10 / 120
				Maturity (year/month)						
1	83.79	42.10	28.21	21.26	17.09	14.32	12.33	10.84	9.69	87.60
2	84.24	42.54	28.64	21.70	19.53	14.75	12.77	11.28	10.13	9.20
3	84.69	42.98	29.08	22.13	17.97	15.19	13.21	11.73	10.58	9.66
4	85.15	43.42	29.52	22.58	18.42	15.65	13.67	12.19	11.04	10.12
5	85.61	43.87	29.97	23.03	18.87	16.10	14.13	12.66	11.52	10.61
6	86.07	44.32	30.42	23.49	19.33	16.57	14.61	13.14	12.01	11.10
7	86.53	44.77	30.88	23.95	19.80	17.05	15.09	13.63	12.51	11.61
8	86.99	45.23	31.34	24.41	20.28	17.53	15.59	14.14	13.02	12.113
9	87.45	45.68	31.80	24.88	**20.76**	18.03	16.09	14.65	13.54	12.67
10	87.90	46.14	32.27	25.36	21.25	18.53	16.60	15.17	14.08	13.22
11	88.38	46.61	32.74	25.85	21.74	19.03	17.12	15.71	14.63	13.78
12	88.85	47.07	33.21	26.33	22.24	19.55	17.65	16.25	15.18	14.35
13	89.32	47.54	33.69	26.83	55.75	20.07	18.19	16.81	15.75	14.93
14	98.79	48.01	34.18	27.33	23.27	20.61	18.74	17.37	16.33	15.53
15	90.26	48.49	34.67	27.83	23.79	21.14	19.27	17.95	16.92	16.13
16	90.73	48.96	35.16	28.34	24.32	21.69	19.86	18.53	17.53	16.75
17	91.20	49.44	35.65	28.85	24.85	22.25	20.44	19.12	18.14	17.38
18	91.68	49.92	36.15	29.37	25.39	22.81	21.02	19.72	18.76	18.02
19	92.16	50.41	36.66	29.90	25.94	23.38	21.61	20.33	19.39	18.67
20	92.63	50.90	37.16	30.43	26.49	23.95	22.21	20.95	20.13	19.33

Knowing the relevant formula would allow us not only to calculate the payment under the amortization system but also to calculate the other variables in terms of the other known values. Therefore, we can calculate, for example, the unpaid balance of any loan at any time given knowledge of the payment, interest rate, and maturity. The unpaid balance would, in this case, be the current value of the annuity, or in other words, the value of what is left of the loan discounted at the given interest rate of the loan, and its specific conversion.

Example 2.1.2 What is the unpaid balance of the loan in Table 2.2 after the borrower has paid for a full year?

The borrower has made 12 payments. The remaining are $360 - 12 = 348$, which would be the value for our n. We apply the following formula:

$$CV = \frac{A[1 - (1+r)^{-n}]}{r}$$

$$= \frac{\$733.76\left[1 - (1 + .08/12)^{-348}\right]}{.08/12}$$

$$= \$99,164.63$$

which is exactly the balance at the end of the first year on the amortization schedule table.

Example 2.1.3 What would be the balance of a mortgage loan of $112,000 at 10.5% interest for a 25-year term if the borrower has already paid his monthly payment of $1,057.48 for 12 years?

In 12 years 144 payments are made. Therefore, the unpaid payments would be $(25 \times 12) - 144 = 156$.

$$CV = \frac{\$1,057.48\left[1 - (1 + .105/12)^{-156}\right]}{.105/12}$$

$$= \$89,807.07$$

which would be the balance at the 12th-year line (144th payment line) of the amortization schedule for this specific loan. We can also reach the same answer if we solve by the table value of $a_{\overline{n}|r} = a_{\overline{156}|7/8}$, which is 84.92554867, given that the monthly interest rate is $(.105/12 = .00875)$ which is 7/8 in the table.

$$CV = A \cdot a_{\overline{n}|r}$$

$$= A \cdot a_{\overline{156}|7/8}$$

$$= \$1,057.48(84.92554867)$$

$$= \$89,807.07$$

We can also use the CV formula to calculate the value of a loan in the market at specific conditions, especially at times of interest rate changes. Accordingly, we can find the market value of a loan by calculating the value of the remaining balance as discounted at the current market interest rate instead of the rate at which the loan was originated.

Example 2.1.4   Suppose that the loan in the preceding examples was to be obtained by another bank after 15 years of its origination. Suppose also that the current interest rate in the market is 11.25%. What price should the bank pay?

The answer lies in finding the current value of the remaining balance [the remainder of the 180 payments; $360 - (15 \times 12)$], as discounted at 11.25% interest.

$$CV = \frac{A[1 - (1 + r)^{-n}]}{r}$$
$$= \frac{\$733.76\left[1 - (1 + .1125/12)^{-180}\right]}{.1125/12}$$
$$= \$63,675.40$$

This is the value of the loan at the new market interest rate compared to its value ($76,781.55) as it is shown on the balance at the 15th-year line of the amortization schedule table.

Example 2.1.5   Amortization is not exclusive to mortgage loans, as illustrated here by the following example of an amortized auto loan:

Nathalie purchased a car from a local dealer for $10,000, financed at a promotional rate of 9.5% over a 3-year term. Her monthly payment was $304. But she had to go abroad, so she asked her bank to take over the loan. If the market interest rate is 12%, how much will the bank pay for this loan? Use the formula and table methods.

Formula method:

$$CV = \frac{A[1 - (1 + r)^{-n}]}{r}$$
$$= \frac{\$304\left[1 - (1 + .12/12)^{-36}\right]}{.12/12}$$
$$= 9,152.68$$

Table method:

$$CV = A \cdot a_{\overline{36}|.01}$$
$$= \$304(30.10750504)$$
$$= 9,152.68$$

If for any reason the number of payments remaining is not known, it can be calculated using the other information, such as the remaining balance, interest rate, and monthly payment. The remainder of the maturity (n) can be obtained by the following formula:

$$n = \frac{\ln\left[1 - (\text{CV} \cdot r)/A\right]}{\ln(1 + r)}$$

Example 2.1.6   Suppose that Caitlin received information on the mortgage loan that her late father signed. The information revealed a remaining balance of $36,188.09 on the loan which was obtained at 8%, and monthly payments of $733.76. She wants to know how many payments are left.

$$n = \frac{\ln\left[1 - \dfrac{\$36,188.09(.08/12)}{\$733.76}\right]}{\ln[1 + (.08/12)]}$$

$$= 60 \text{ payments}$$

A total of 60 payments remain, which will take Caitlin 5 years to pay. It means that her father has made 25 years' worth of payments on that loan. This is confirmed by looking at the balance of the 25th year on Table 2.1, which is $36,188.09.

2.2. EFFECTS OF INTEREST RATE, TERM, AND DOWN PAYMENT ON THE MONTHLY PAYMENT

The monthly payment of a loan is, of course, affected by how high or low the interest rate is, and how long the term of maturity is. A classic manipulation has been known that makes borrowers extend the term if they cannot lower the interest rate in order to have more affordable monthly payments which they can live with for the long haul. Table 2.3 gives general estimates of the monthly payments for each $1,000 of any mortgage loan at a choice of a certain interest rate and term of maturity.

For example, we want to compare how the monthly payment differs if we choose a term of 15 years vs. 30 years for a $140,000 mortgage at 5.5% interest. The reading for 5.5 and 15 is 8.17, which makes the monthly payment $1,143.80 (140 × 8.17) and the reading for 5.5 and 30 is 5.68, which makes the monthly payment $795.20. We can conclude immediately that choosing a shorter term (15 years) would save $348.60 ($1,143.80 − $795.20) in monthly payments alone, in addition to a huge saving in total interest paid over the life of the loan. That is over 30% in monthly savings. As to finding a better rate, we can compare, for example, how a 1% or 1.5% difference in the interest rate can affect our monthly payment. Suppose that the $140,000 loan is obtained at 4% instead of 5.5% for the 15-year term. The monthly payment would be $1,036 (140 × 7.40)

TABLE 2.3   Monthly Payments Including Principal and Interest for Each $1,000 of a Mortgage Loan Financed at Four Different Interest Rates

Interest Rate (%)	Term (years)			
	15	20	25	30
1	5.98	4.60	3.77	3.22
1.5	6.21	4.83	4.00	3.45
2	6.44	5.06	4.24	3.70
2.5	6.67	5.30	4.49	3.95
3	6.91	5.55	4.74	4.22
3.5	7.15	5.80	5.01	4.49
4	**7.40**	6.06	5.28	**4.77**
4.5	7.65	6.33	5.56	5.07
5	7.91	6.60	5.84	5.37
5.5	**8.17**	6.88	6.14	**5.68**
6	8.44	7.16	6.44	5.99
6.5	8.71	7.45	6.75	6.32
7	8.99	7.75	7.07	6.65
7.5	9.27	8.06	7.39	6.99
8	9.55	8.36	7.72	7.34
8.5	9.85	8.68	8.05	7.69
9	10.14	8.99	8.39	8.05
9.5	10.44	9.32	8.74	8.41
10	10.75	9.65	9.09	8.77

compared to $1,143.80, offering a savings of $107.80 ($1,143.80 − $1,036) a month (9.4%) and would be $667.80 (140 × 4.77) for the 30-year term, offering even higher savings of $127.40 ($795.20 − $667.80), which is 16%.

Table 2.4 shows both the monthly payment and the total interest paid over the life of a $70,000 mortgage loan at many combinations of four possible interest rates and four terms. A borrower would be able to see the benefit of paying more monthly to avoid a lot of interest over time. Of course, choosing a shorter term and a lower rate would be the best choice, if it at all possible. For example, at 6% interest, choosing a 15-year term instead of 30 years would result in having a monthly payment of $591 instead of $420, but paying $36,000 instead of $81,000 in total interest. That is, an increase of $171 (40.7%) in the monthly payment would cause a decrease of $45,000 (55.5%) in total interest.

$$\$591 - \$420 = \$171 \qquad \$81,000 - \$36,000 = \$45,000$$

$$\frac{171}{420} \times 100 = 40.7\% \qquad \frac{45,000}{81,000} \times 100 = 55.5\%$$

This gain would become more dramatic at a higher interest rate, such as 12%. For a 15-year term, the monthly payment would be $840 instead of $720 for the 30-year term. An increase of $120 ($840 − $720), 16.6% in the monthly

TABLE 2.4   Total Interest Paid and Monthly Payments of a $70,000 Loan Financed at Different Interest Rates and Terms

	Interest Rates							
	6%		8%		10%		12%	
Term (years)	PYT ($)	Total Interest ($)	PYT ($)	Total Interest ($)	PYT ($)	Total Interest ($)	PYT ($)	Total Interest ($)
30	**420**	**81,000**	524	115,000	614	151,000	**720**	**189,000**
25	451	65,000	540	92,000	636	121,000	737	151,000
20	502	50,000	586	71,000	676	92,000	771	115,000
15	**591**	**36,000**	669	50,000	752	65,000	**840**	**81,000**

TABLE 2.5   Monthly Payment of a $100,000 Mortgage Loan by Size of Down Payment at a 7% Interest Rate for 20 Years

Down Payment ($)	Amount Financed ($)	Monthly Payment ($)
5,000	95,000	736.53
10,000	90,000	**697.77**
15,000	85,000	659.00
20,000	80,000	**620.24**
25,000	75,000	581.48

payment, would yield a saving of $108,000 (57%) in total interest. So, at higher interest rates, accepting a little higher payment, if affordable, would yield a huge saving in the interest paid over time.

$$\$840 - \$720 = \$120 \qquad \$189,000 - \$81,000 = \$108,000$$

$$\frac{\$120}{\$840} \times 100 = 16.6\% \qquad \frac{\$108,000}{\$189,000} = 57.1\%$$

The down payment is another factor that affects the monthly payment and the interest paid over the life of the loan. Table 2.5 shows how the monthly payment drops as the down payment gets larger. For example, as the borrower doubles his down payment from $10,000 to $20,000, an increase by 100%, the monthly payment drops from $697.77 to $620.24, only $77.53, or 11%. Here it becomes a matter of the borrower's subjective assessment of the opportunity cost of her cash. The major question for the decision making becomes: Would the saving of $77.53 a month be worth the sacrifice of an additional $10,000 in cash?

2.3. GRADUATED PAYMENT MORTGAGE

The mortgage payments we have seen in the amortization section were set according to the traditional method of breaking the loan into equal payments. However, this is not the only way to set up mortgage loans. In graduated payment mortgages, the payment starts low, increases over a certain period of time, and then stays fixed to the end of the loan's maturity. This method can be very helpful for borrowers who predict that within a certain period they will be able to afford more than they do at the time of taking out the loan. To make this work, the lender decides on a particular growth rate (g) by which the payment increases in k years. Before we set up the formula for this type of mortgage, let's address the formula for the current value of future payments, which grow at a constant rate. It is the basis on which we can build up the graduated payment formula.

Let's assume that we have a loan, whose first payment is A, that grows at a constant rate (g) throughout the maturity time n, and that r is the loan interest rate. Therefore, we can write the current value for those increasing future payments as

$$CV = A(1+g)\left[\frac{1 - \left(\frac{1+g}{1+r}\right)^n}{r-g}\right]$$

Example 2.3.1 Calculate the present value of a stream of cash flows, the first of which is \$2,000, that grow at a rate of 5% annually for 10 years given that the interest rate is $8\frac{1}{2}\%$.

$$CV = A(1+g)\left[\frac{1 - \left(\frac{1+g}{1+r}\right)^n}{r-g}\right]$$

$$CV = 200(1+.05)\left[\frac{1 - \left(\frac{1+.05}{1+.085}\right)^{10}}{.085 - .05}\right]$$

$$CV = 16,774$$

If we rewrite the formula in monthly terms, we get $CV = 16,774$

$$CV = A(1+g)\left[\frac{1 - \left[\frac{1+(g/12)}{1+(r/12)}\right]^{12n}}{(r/12)-(g/12)}\right]$$

Based on the premise of this formula, we can write the graduated payment formula as

$$CV = A \sum_{i=0}^{k} (1+g)^i \left[\frac{1-(1+r/12)^{-12}}{r/12} \right] \left(1+\frac{r}{12}\right)^{-12i}$$
$$+ (1+g)^k \left[\frac{1-(1+r/12)^{12(n-k)}}{r/12} \right] \left(1+\frac{r}{12}\right)^k$$

where

        CV: is the present value of the increasing future payments up to k years.
        A: is the first payment.
        k: is the number of years in which the payment continues to increase.
        g: is the constant rate at which the payments increase.
        i: is the number of each year of k period.
        r: is the interest rate for the loan.

Because this formula is tedious, we would use ready values (TVs) from Table 2.6 to simplify the calculations. Therefore, the formula above would be rewritten as

$$CV = A(TV_{r,g})$$

and therefore the initial payment (A) would be obtained by

$$A = \frac{CV}{TV_{r,g}}$$

where CV is the amount of the loan and TV is the table value obtained across the loan interest rate (r) and the growth rate (g).

Example 2.3.2   Consider a graduated payment type of mortgage of $150,000 at $9\frac{1}{2}\%$ interest for 30 years. During the first 7 years of the term, the payment increases at the end of each year by a rate of 5%, at which point it will remain fixed until the end of maturity. What would the payment be during the first year?

$$A = \frac{CV}{TV_{r,g}}$$
$$= \frac{150,000}{TV_{.095,.05}}$$
$$= \frac{150,000}{143.151659}$$
$$= 1,047.84$$

TABLE 2.6 TY Values Based on Selected Interest Rates and Growth Rates

Growth Rate, g	Loan Interest Rate, r (%)									
	9.0	9.5	10.0	10.5	11.0	11.5	12.0	12.5	13.0	13.5
.5	126.644463	121.160698	116.065745	111.325244	106.908263	102.786929	98.936099	95.333086	91.657377	88.790416
1.0	129.049925	123.435060	118.218683	113.365578	108.844034	104.625483	100.684146	96.996771	93.542332	90.301805
1.5	131.498844	125.750320	120.410161	115.442265	110.814132	106.496478	102.462889	98.689526	95.154840	91.839335
2.0	133.991838	128.107067	122.640725	117.555818	112.819039	108.400377	104.272759	100.411759	96.795292	93.403375
2.5	136.529518	130.505869	124.910914	119.706746	114.859235	110.337628	106.114182	102.163872	98.464063	94.994282
3.0	139.112518	132.947321	127.221288	121.895576	116.935214	112.308697	107.987601	103.946280	100.161548	96.612430
3.5	141.741480	135.432030	129.572412	124.122838	119.047479	114.314059	109.893459	105.759406	101.888147	98.258193
4.0	144.417028	137.960579	131.964840	126.389053	121.196520	116.354176	111.832193	107.603662	103.644249	99.931943
4.5	147.139814	140.533581	134.399147	128.694763	123.382846	118.429526	113.804257	109.479475	105.430258	101.634062
5.0	149.910501	143.151659	136.875916	131.040518	125.606975	120.540601	115.810112	111.387281	107.246587	103.364939
5.5	152.729729	145.815411	139.395714	133.426847	127.869407	122.687870	117.850204	113.327501	109.093631	105.124950
6.0	155.598169	148.525473	141.959131	135.854315	130.170668	124.871831	119.924999	115.300577	110.971811	106.914493

2.4. MORTGAGE POINTS AND THE EFFECTIVE RATE

Mortgage loan points are additional charges imposed by lenders and paid in advance by the borrower. It is just a way to increase the lender's profitability. The point would increase the cost of borrowing for the borrower by pushing up the effective interest rate. The points are usually set as 1% of the amount borrowed. For example, if the mortgage loan of $200,000 at 7% comes with 3 points, it means that $6,000 ($200,000 × .01 × 3) is deducted from the proceeds, giving the borrower $194,000 at 7% on $200,000. To assess the impact of points on the effective interest rate, let's consider an example.

Example 2.4.1   Consider a mortgage loan of $85,000 at 9% and 2 points, with a maturity of 20 years.

First, let's calculate the monthly payment. $r = .09/12 = .0075$; $n = 20 \times 12 = 240$.

$$A = CV \left[\frac{r}{1 - (1+r)^{-n}} \right]$$
$$= 85,000 \left[\frac{.0075}{1 - (1 + .0075)^{-240}} \right]$$
$$= \$764.77$$

So the borrower would pay $764.77 for 20 years, but he, in fact, does not get $85,000 but $85,000 minus 2%. That is $85,000 $(1 - .02) = \$83,300$. If we plug in the payment of $764.77, the term of 20 years (240 months), and the loan proceeds of $83,300 in a mortgage calculator, we get the rate (r) as 9.28%. This means that the 2 points raised the interest rate from 9% to 9.28%, a total increase of 3% in the rate or 1.5% per point.

2.5. ASSUMING A MORTGAGE LOAN

An existing mortgage loan can be assumed by a buyer at some point of its maturity, when the buyer would assume responsibility to pay for the loan and go by its original contract until the end of maturity. This is particularly valuable when a buyer wants to get a loan with an interest rate from the past, especially when it was lower than what is available currently. The loan that would be subject to such a transfer has to be an "assumable" loan according to the original contract. To calculate the gain to the buyer, let's take this example.

Example 2.5.1   Suppose that at a time when the interest rate on mortgage loans is 9%, you find a homeowner who is willing to let you assume his existing mortgage, obtained at $6\frac{1}{2}$% interest 10 years ago, with a balance remaining for the next 20 years of $95,000. How would you assess the gain or loss if you assume such a loan?

First, let's calculate the monthly payment on the unpaid balance of $95,000 at the loan's original interest rate of $6\frac{1}{2}\%$. $r = .065/12 = .00542$; $n = 20 \times 12 = 240$.

$$A = CV \left[\frac{r}{1 - (1 + r)^{-n}} \right]$$

$$= \$95,000 \left[\frac{.00542}{1 - (1 + .00542)^{-240}} \right]$$

$$= \$708.73$$

and the payment at the current market rate of 9%, with $r = .09/12 = .0075$ and $n = 20 \times 12 = 240$, would be

$$A_2 = \$95,000 \left[\frac{.0075}{1 - (1 + .0075)^{-240}} \right]$$

$$= \$95,000 \left[\frac{.0075}{a - (1 + .0075)^{-240}} \right]$$

$$= \$854.73$$

The difference between the two payments would be the monthly savings for the buyer:

$$A_2 - A_1 = \$854.73 - \$708.73 = \$146$$

We should consider the cash flow of these monthly savings of $146 a month for the next 20 years by calculating the present value of this cash flow, which would be the total saving for the buyer throughout the life of the loan, with $r = .0075$ and $n = 240$.

$$CV = \frac{A[1 - (1 + r)^{-n}]}{r}$$

$$= \frac{\$146[1 - (1 + .0075)^{-240}]}{.0075}$$

$$= \$16,227.16$$

So this is the total gain from assuming the loan. Determining the net gain or loss is just a matter of comparing this total gain to the total cost of assumption.

2.6. PREPAYMENT PENALTY ON A MORTGAGE LOAN

Lenders basically get their earnings from the interest paid on what they lend. Of course, the longer they collect interest, the better. This is why they protect themselves by clauses such as the **prepayment penalty**, which is a charge they impose on the borrower if the loan is paid off before its normal maturity. Lenders set their prepayment penalties differently, and they also vary them according to types of loans. Typically, we can say that most prepayment penalties involve imposing a charge equal to 3 to 6 months of interest on the unpaid balance of the loan. If a borrower ends up paying the prepayment penalty, the cost of borrowing would be higher, and that can be translated into getting an interest rate that is actually higher than is stated in the contract.

Example 2.6.1   Suppose that a mortgage loan of $250,000 at 5% for 25 years is paid off after only 3 years. The mortgage contract has a prepayment penalty of 5 months of interest on the remaining balance. Calculate the prepayment penalty.

First, we calculate the original monthly payment of the loan. $r = .05/12 = .0042$; $n = 25 \times 12 = 300$.

$$A = CV \left[\frac{r}{1 - (1 + r)^{-n}} \right]$$

$$= \$250,000 \left[\frac{.0042}{1 - (1 + .0042)^{-300}} \right]$$

$$= \$1,467.31$$

Second, we use the monthly payment to get the unpaid balance of the loan for the remaining 22 years $(25 - 3)$. $r = .0042$; $n = 22 \times 12 = 264$.

$$CV = \frac{A[1 - (1 + r)^{-n}]}{r}$$

$$= \frac{\$1,467.31[1 - (1 + .0042)^{-264}]}{.0042}$$

$$= \$233,818.75$$

The prepayment penalty is five monthly interest payments of the remaining balance at 5%.

$$\$233,818.75(.05) \left(\frac{5}{12} \right) = \$4,871.22$$

2.7. REFINANCING A MORTGAGE LOAN

If the terms of getting a new mortgage loan become more favorable to homeowners than they have been, most people prefer to refinance their existing mortgage loans. Usually, a better interest rate and lower financing cost are the major attractions in refinancing. A major incentive for borrowers is to get a lower monthly payment and other benefits, such as moving from a variable interest loan to a fixed interest loan. There is, of course, a cost to refinance. Although lenders vary in charging such a cost, it would probably include a certain number of points plus a lump sum to cover the cost of the transactions of originating and finalizing a new loan and paying off the old. The worthiness for borrowers would be summarized as getting more benefits than is offset by the costs. To assess the gain from refinancing, let's consider an example.

Example 2.7.1   Joyce is paying $1,184.87 in monthly mortgage payments. She wants to refinance her existing mortgage loan of $100,000 at 14% interest for 30 years that she obtained 4 years ago. Her mortgage officer informed her that he could get her a rate of 10% but that the finance cost would include paying a prepayment penalty on the existing loan, which is equal to 6 months' interest on the balance of the loan, plus a closing cost of $2,000. Would it be worth it for Joyce to refinance, and if it is worth it, how long should Joyce stay in this house to justify paying for the refinance?

First, we calculate the remaining balance of the existing loan. $r = .14/12 = .01167$; $n = 26 \times 12 = 312$.

$$CV = \frac{A[1 - (1+r)^{-n}]}{r}$$

$$= \frac{\$1,184.87[1 - (1+.01167)^{-312}]}{.01167}$$

$$= \$98,837.32$$

Second, we calculate the prepayment penalty of 6 months' interest at 14%.

$$\$98,837.32(.14)\left(\frac{1}{2}\right) = \$6,918.61$$

Next, we calculate the monthly payment on the new loan of $98,837.21 at 10% interest for the remaining maturity of 26 years. $r = .10/12$; $n = 26 \times 12 = 312$.

$$A = CV\left[\frac{r}{1 - (1+r)^{-n}}\right]$$

$$= \$98,837.32\left[\frac{.10/12}{1 - (1+.10/12)^{-312}}\right]$$

$$= \$890.50$$

Then we can calculate the monthly saving for the borrower as the difference between the old and new payments:

$$\$1,184.87 - \$890.50 = \$294.37$$

Next, we calculate the present value of the stream of savings for the remaining 26 years at the new interest rate of 10%.

$$CV = \frac{A[1 - (1+r)^{-n}]}{r}$$

$$CV = \frac{\$294.37\left[1 - (1 + .10/12)^{-312}\right]}{.10/12}$$

$$= \$32,672.31$$

This is the present value of all the gain that Joyce would enjoy from the refinance. It has to be assessed against the refinancing cost, which includes both the prepayment penalty and the closing cost:

$$\text{total refinancing cost} = \$6,918.61 + \$2,000 = \$8,918.61$$

Now, the net gain for Joyce would be the difference between all the gain and all the cost:

$$\$32,672.31 - \$8,918.61 = \$23,753.70$$

That would be what Joyce gains out of going through the refinance of her current mortgage loan. However, this gain is throughout the remaining life of the loan of 26 years. If we divide the total refinance cost of $8,918.61 that she would pay by the monthly saving she gets, we obtain what is called the **refinance recovery time** (RRT):

$$RRT = \frac{\text{total refinance cost}}{\text{monthly saving of refinance}}$$

$$= \frac{\$8,918.61}{\$294.37} = 30.3$$

which means that for Joyce, going through the refinancing process would not be worthwhile unless she stays in the house at least 30 months in order to recover the cost paid.

2.8. WRAPAROUND AND BALLOON PAYMENT LOANS

At a time of high interest rates, lenders would earn more by collecting higher interest from borrowers. But in these times, fewer and fewer borrowers can afford to have loans. So lenders would face lower demand on their products. One strategy to avoid this situation and boost earnings is for lenders to replace old low-interest loans with new high-interest loans. They would only be able to do that if they make this replacement a condition in granting a loan related to the old one. So **wraparound loans** are loans at higher interest that are wrapped around the package of a new loan as a stipulation imposed on the borrower in order to be granted the new loan. This package may also include the imposition of a **balloon payment**, which is a particularly large final payment. To understand how this package is paid for, let's work an example.

Example 2.8.1   Owners of a small business needed a loan of $500,000 at 11% for 20 years (see Figure E2.8.1). The bank loan officer informed them that to get this loan they must agree to:

1. Rewrite their old loan of 9% interest and an unpaid balance of $350,000 into a new loan at 11% and pay it off over the maturity of the new loan (20 years), although only 10 years is remaining to pay it off according to the current contract.
2. Pay a balloon payment at the end of the 15th year.

How would this arrangement be paid?

First, we calculate the monthly payment on the old loan. $r = .09/12$; $n = 10 \times 12 = 120$.

$$A_1 = CV \left[\frac{r}{1 - (1+r)^{-n}} \right]$$

$$= \$350,000 \left[\frac{.09/12}{1 - (1 + .09/12)^{-120}} \right]$$

$$= \$4,433.65$$

Then we calculate the monthly payment of the new loan, which includes the new amount $500,000 and the old balance wrapped around it ($500,000 + $350,000 = $850,000) at the new rate of 11%.

$$A_2 = \$850,000 \left[\frac{.11/12}{1 - (1 + .11/12)^{-240}} \right]$$

$$= \$8,773.60$$

Finally, we calculate the balloon payment that would cover the last five years of maturity, which would be the present value of the last five years of payments of

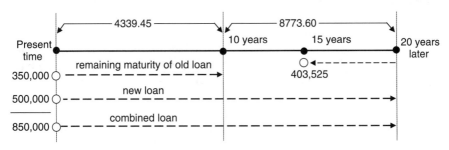

FIGURE E2.8.1

$8,873.60. $n = 5 \times 12 = 60$.

$$CV_B = \frac{A[1 - (1+r)^{-n}]}{r}$$

$$= \frac{8,773.60\left[1 - (1+.11/12)^{-60}\right]}{.11/12}$$

$$= \$403,524.48$$

This is the last payment that would be made at the end of the 15th year, to cover the next 5 years until the end of maturity at the end of the 20th year.

- For the first 10 years, the payment would equal the difference between the new and old payments: $8,773.60 - \$4,433.65 = \$4,339.95$.
- For the years 11 to 15, the new payment of $8,773.60 is valid.

2.9. SINKING FUNDS

The sinking fund method is not related particularly to mortgage debt, but it is comparable to the amortization method. The sinking funds are set up specifically to accumulate money in a systematic way for a specific purpose. Typically, they are used to satisfy a future financial need, such as to retire certain debt, redeem bond issues, purchase new equipment, or replace worn-out machines. Knowing the amount of the money needed in the future, the time it is needed for, and the interest rate necessary to help it grow makes this type of fund an ordinary annuity with its payment to be determined. The payment would serve as a periodic and equal deposit into the fund. The sequence of deposits would earn interest and accumulate toward the targeted future value. In the case of debt retirement, the sinking fund would have two separate functions:

1. To continue to pay the interest on the debt as it comes due.
2. To build up an amount sufficient to pay off the principal as it comes due.

This would imply that the two functions may have to deal with different interest rates. Since it is always more expensive to borrow than to save, the rate to pay the interest would probably be higher than the rate to be earned on the deposits to build up the principal payoff.

Similar to the amortization method, the sinking fund method would structure the principal payoff accumulation into what is called a **sinking fund schedule**. It typically shows the sequence of deposits made, the interest they earn, and how the fund gets accumulated toward building up the targeted amount by the end of the term. That amount will exactly equal the principal.

Example 2.9.1   A small business has a loan of $6,500 that has to be paid off in 3 years at 14.5% interest payable quarterly. The business opens an account for a sinking fund to pay off the principal of its debt at the end of the 3 years. This account pays 13% interest compounded quarterly.

(a) What would be the quarterly interest payment?

(b) What would the sinking fund schedule look like?

(a) The total interest on the loan would be obtained by

$$I = P \cdot r \cdot t$$
$$= \$6,500(.145)(3)$$
$$= \$2,827.50$$

The quarterly payment of interest (QI) would be

$$QI = \frac{I}{n} = \frac{\$2,827.50}{3 \times 4} = \$235.62$$

(b) The key in the sinking fund schedule is the quarterly deposit that this borrower has to make to accumulate (Table E2.9.1) enough to pay off the $6,500 at the end of 3 years. We treat this as an ordinary annuity payment that has to be obtained when the future value is known ($6,500): r is quarterly ($.13/4 = .0325$) and $n = 3 \times 4 = 12$.

$$A = \frac{FV \cdot r}{[(1 + r)^n - 1]}$$
$$= \frac{(\$6,500)(.0325)}{[(1 + .0325)^{12} - 1]}$$
$$= \$451.54$$

We can also obtain this payment by the table method where the annuity payment (A) is:

$$A = FV \cdot \frac{1}{S_{\overline{n}|r}}$$
$$= FV \left(\frac{1}{a_{\overline{n}|r}} - r \right)$$
$$= \$6,500 \left(\frac{1}{a_{\overline{12}|.0325}} - .0325 \right)$$
$$= \$6,500(.10196719 - .0325)$$
$$= \$451.54$$

Observations

• The first values to be calculated are the values in the sinking fund accumulations column. Each value would be a future value of an ordinary annuity

TABLE E2.9.1   Sinking Fund Schedule

Year 1 Quarter 1		Interest on Fund ($)	Deposit ($)	Deposit Growth ($)	Sinking Fund Accumulations ($)	Book Value ($)
		—	451.54	451.54	451.54	6,048.45
	2	14.68	451.54	466.22	917.76	5,582.24
	3	29.83	451.54	481.37	1,399.12	5,100.88
	4	45.47	451.54	497.01	1,896.13	4,603.87
2	5	61.62	451.54	513.16	2,409.30	4,090.70
	6	78.30	451.54	529.84	2,939.14	3,560.86
	7	95.52	451.54	547.06	3,486.20	3,013.80
	8	113.30	451.54	564.84	4,051.04	2,448.96
3	9	131.66	451.54	583.20	4,634.24	1,865.76
	10	150.61	451.54	602.15	5,236.40	1,263.60
	11	170.18	451.54	621.72	5,858.12	641.88
	12	190.39	451.50	641.89	6,500.00	0
		1081.56	5418.44	6500.00		

that can be obtained by

$$FV = \frac{A[(1+r)^n - 1]}{r}$$

For example, the value for the second quarter would be

$$FV = \frac{\$451.54[(1 + .0325)^2 - 1]}{.0325}$$
$$= \$917.76$$

Also, any value in this column can be calculated using the table method. For example, to find how much the fund has accumulated at the end of the ninth quarter, we use

$$FV = A \cdot S_{\overline{n}|r}$$
$$= \$451.54 S_{\overline{9}|.0325}$$
$$= \$451.54(10.26319401)$$
$$= \$4,634.24$$

The values in the sinking fund accumulation column can also be obtained from within the sinking fund schedule. If each value in the column is added to the deposit growth value of the following quarter, the next accumulation value can be obtained. For example, we add $451.54 to $466.22 to get $917.76, we add $917.66 to $481.37 to get $1399.12, and so on.

- Applying the quarterly rate of. 0325 on the sinking fund accumulated balances would give us the quarterly interests that the fund earns as recorded in the second column. For example, the interest on the first deposit is credited to the account in the second quarter, $14.68 = $451.54 \times .0325$, the interest on the second balance is credited in the third quarter, $29.83 = $917.76 \times .0325$, and so on.

- The values in the column of the deposit growth are obtained by adding the deposits to their respective interests. For example, in the eighth quarter, the growth is $546.84, which is $451.54 + $113.30.
- The final column shows the book value, which is the principal ($6,500) minus the accumulated values per quarter. For example, the book value for the seventh quarter is $3,013.80, which is $6,500 − $3,486.20. When the accumulated fund reaches the goal of $6,500 at the end of the 12th quarter, the book value would be zero, meaning that the debt has been paid off and its balance has been cleared ($6,500 − $6,500 = 0$).
- The total of the deposit growth column is the target growth figure, $6,500, which has grown out of the total of the 12 deposits of $5,418.44 by earning interest of $1,081.56.

$$12 \times \$451.54 = \$5,418.44 + \$1,081.56 = \$6,500$$

Example 2.9.2 To build a new gym, a local sport club borrows $500,000 at 7% interest to be paid semiannually for 15 years. The club board set up a sinking fund paying 6% compounded monthly. This fund is to pay both the interest on the loan on time and to retire the original principal in 15 years.

(a) What would the interest be on the loan?
(b) How much would the monthly deposit be on the sinking fund?
(c) How much would the club pay a month for both interest and principal?

(a) Interest payment:

$$\text{semiannual rate} = \frac{.07}{2} = .035$$

$$\$500,000 \times .035 = \$17,500 \text{ paid every 6 months}$$

(b) The monthly deposit into the sinking fund:

monthly rate $= .06/12 = .005$ or $\frac{1}{2}\%$; $n = 15 \times 12 = 180$.

$$\begin{aligned}
A &= \frac{FV \cdot r}{(1+r)^n - 1} \\
&= \frac{\$500,000(.005)}{(1+.005)^{180} - 1} \\
&= \$1,719.29 \quad \text{monthly deposit into the sinking fund}
\end{aligned}$$

and by the table method, where the value of $1/a_{\overline{180}|\frac{1}{2}\%}$ is given as .00843857, A would be

$$A = \text{FV}\left(\frac{1}{a_{\overline{180}|\frac{1}{2}\%}} - .005\right)$$

$$= \$500,000[(.00843857) - .005]$$

$$= \$1,719.29$$

(c) Total payment: We obtain the monthly payment for the interest fund in the same way. Monthly rate $= .06/12 = .005$; $n = 6$ months.

$$A = \frac{\text{FV} \cdot r}{[(1+r)^n - 1]}$$

$$= \frac{\$17,500(.005)}{[(1+.005)^6 - 1]}$$

$$= \$2,880.42$$

and by the table method,

$$A = \text{FV}\left(\frac{1}{a_{\overline{n}|r}} - r\right)$$

$$= \$17,500\left(\frac{1}{a_{\overline{6}|\frac{1}{2}\%}} - .005\right)$$

$$= \$17,500[(.16959546) - .005]$$

$$= \$2,880.42$$

$$\text{monthly total payment} = \text{monthly interest payment}$$

$$+ \text{ monthly principal payment}$$

$$= \$2,880.42 + \$1,719.29$$

$$= \$4,599.71$$

Example 2.9.3 A city hall administration issued its local bonds totaling \$2,000,000 and carrying a quarterly interest of 8%. The city needed to set up a sinking fund to redeem the bonds in 10 years. If the fund pays 7.5% compounded quarterly, what size deposit would the city have to make into the fund, what size interest payment would have to be made, and what would the total be?

Interest on the bonds will be $(r = .08/4 = .02)$

$$I = P \cdot r \cdot t$$
$$= \$2,000,000(.08)(\tfrac{1}{4})$$
$$= \$40,000 \qquad \text{each quarter}$$

The deposit into the fund will be $(r = .07/4 = .01875 = 1\tfrac{7}{8}; \ n = 10 \times 4 = 40)$

$$A = \frac{FV \cdot r}{(1 + r)^n - 1}$$
$$= \frac{\$2,000,000(.01875)}{(1.01875)^{40} - 1}$$
$$= \$34,018.26$$

By the table method,

$$A = FV \left(\frac{1}{a_{\overline{n}|r}} - r \right)$$
$$= \$2,000,000 \left(\frac{1}{a_{\overline{40}|1\tfrac{7}{8}\%}} - .01875 \right)$$
$$= \$2,000,000(.03575913 - .01875)$$
$$= \$34,018.26$$

$$\text{total payment per quarter} = \text{quarterly interest} + \text{quarterly deposit}$$
$$= \$40,000 + \$34,018.26$$
$$= \$74,018.26$$

2.10. COMPARING AMORTIZATION TO SINKING FUND METHODS

In the amortization method, a loan gets paid off through a sequence of equal payments, each of which includes a portion of interest and a portion of principal. When all payments are made, the entire loan (principal and interest) would be paid off. In the sinking fund method, there would be two accounts for two separate purposes: one to pay the interest as it comes due, and the other to build up the fund and wait for it until it is equal to the principal. At that time the principal would be paid off in a single payment.

This important difference leads to another major difference, having to deal with two different interest rates in the case of the sinking fund, compared to one interest rate in the case of amortization. This might lead to having a different debt cost between the two methods unless all interest rates are equal. If the sinking fund receives a rate of interest on its deposits lower than the rate of interest paid on debt, the periodic cost for the sinking fund method would be higher than that of the amortization method. Let's consider the following case. Suppose that r_1 is the interest rate per period on debt in both methods, r_2 is the interest rate earned on the sinking fund deposit for the same period, n is the term of maturity, P is the principal of a loan, and A_1 and A_2 are the interest and principal payments in the two methods. Then:

$$\text{Amortization: } A_1 = \frac{P}{a_{\overline{n}|r_1}} = \text{P}r + \frac{P}{S_{\overline{n}|r_1}}$$

$$\text{Sinking fund: } A_2 = \text{P}r_1 + \frac{P}{S_{\overline{n}|r_2}}$$

Then it becomes clear that:

- If $r_1 > r_2 \rightarrow S_{\overline{n}|r_1} > S_{\overline{n}|r_2} \rightarrow A_1 < A_2$, the amortization is less costly than the sinking fund.
- If $r_1 = r_2$, then $A_1 = A_2$, and there is no difference.
- If $r_1 < r_2 \rightarrow S_{\overline{n}|r_1} < S_{\overline{n}|r_2} \rightarrow A_1 > A_2$, the sinking fund is less costly.

This would hold only if the interest rate paid in the two methods is equal and the question becomes whether or not the interest rate earned on the sinking fund deposit can compensate.

3 Leasing

Leasing is defined as having the right to use a product and reap all the benefits from its use while leaving the ownership title in the hands of the owner, who would be the lease grantor. The party who uses the product under the lease contract is called the **lessee** and the party who owns the product, originates, and offers the lease contract is called the **lessor**. Both parties must estimate the lease costs and benefits to them, and before signing and carrying out the contract, expect to have the benefits exceed the costs. For the lessor, it is a form of investment in which she expects to get an investment rate of return equal to her own required cost of capital. The crucial question for the lessee is: Which would be less expensive: to buy the product or to lease it? Since leasing is involved in a series of periodic payments for a certain amount of time, it is assumed to be a form of debt financing, and since the alternative to leasing is to buy on credit, making the right choice between the two becomes a matter of debt management. In this analysis we compare the costs of the two alternatives in terms of:

1. The present value of the future payments involved in both transactions
2. The after-tax form of transactions

We analyze the lessee's side first and then the lessor's side.

3.1. FOR THE LESSEE

The Cost of Buying on Credit

The cost to buy on credit is estimated by the present value of all the payments to be made toward paying off the purchase price of the product. Not surprisingly, the present value turns out to be the purchase price itself. This is simply because we are switching the direction of the money value. So a product is purchased today for x amount of money. It would be paid off in many payments and with interest r, it would accumulate at the end as a future value of $x + m$. Now, if this future value is discounted back to the present using the same rate of r, it would return to x. This is the situation without considering the tax effect, but interestingly, when we calculate the present value of the after-tax payments using the after-tax discount rate, we end up at the purchase price back in the present.

Mathematical Finance, First Edition. M. J. Alhabeeb.
© 2012 John Wiley & Sons, Inc. Published 2012 by John Wiley & Sons, Inc.

The following example illustrates how the present value of all payments made for a purchase on credit equals the purchase price for both before- and after-tax considerations.

Example 3.1.1 A machine is purchased on credit for $2,990.60 at 20% interest and is paid off in 5 years at $1,000 per year (see Figure E3.1.1). What would be the present value of all five payments?

$$CV = \frac{A[1 - (1 + r)^{-n}]}{r}$$

$$= \frac{\$1,000[1 - (1 + .20)^{-5}]}{.20}$$

$$= \$2,990.60$$

As we did before, if we bring those five payments of $1,000 each from their own time in the future back to the present time, we discount them individually as current values, where adding them up in the present forms the present value of all (Table E3.1.1a).

Now, let's find the after-tax present value of those payments, where the cost of borrowing would be the after-tax effective cost of debt. Let's assume that

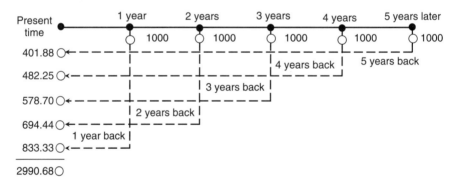

FIGURE E3.1.1

TABLE E3.1.1a

Year	Payments ($)	Current Value Factor $(1 + r)^{-n}$	Current Value ($)
1	1,000	$(1 + .20)^{-1}$	833.33
2	1,000	$(1 + .20)^{-2}$	694.44
3	1,000	$(1 + .20)^{-3}$	578.70
4	1,000	$(1 + .20)^{-4}$	482.25
5	1,000	$(1 + .20)^{-5}$	401.88
	5,000		2,990.60 present value

the tax rate is 40%. We should know that since the payment is split between principal and interest portions, only the interest part would be subject to taxes. The after-tax interest would therefore be

$$I_{at} = I_{bt}(1 - t)$$

where I_{at} is the after-tax interest, I_{bt} is the before-tax interest, and t is the tax rate.

Table E3.1.1b shows how each payment of $1,000 is split between the principal and interest portion, where only the interest portion is subject to taxes, but both the principal portion and the tax-adjusted interest would constitute the total after-tax burden. The present value is calculated for all those annual after-tax burdens, to be at the end equal to the original purchase price of $2,990.60, as we have seen with the before-tax case.

As an example, let's follow the first payment of $1,000: It is split between the interest and principal portions in this way:

- The interest portion is obtained by applying the interest rate (20%) on the original principal (the purchase price):

$$\$2,990.60 \times .20 = \$598.12$$

- The rest of the payment would be the principal portion:

$$\$1,000 - \$598.12 = \$401.88$$

- This payment of the principal portion is taken from the original principal of $2,990.60.

$$\$2,990.60 - \$401.88 = \$2588.72$$

- We apply the tax rate of 40% on the interest portion:

$$\$598.12(1 - .40) = \$358.87$$

- We add this tax-adjusted interest to the principal portion to get the total after-tax burden for year 1, which comes out of payment 1.

$$\$358.87 + \$401.88 = \$760.75$$

- We discount this total after-tax burden using an after-tax rate:

$$\text{after-tax rate} = .20(1 - .40) = .12$$

$$CV_1 = \frac{760.75}{(1 + .12)^1} = \$679.24$$

TABLE E3.1.1b   Present Value of After-Tax Payments

(1) Year	(2) Payment	(3) Interest Portion [(5) × .20]	(4) Principal Portion [(2) − (3)]	(5) Balance of Principal $2990.60 [(5) − (4)]	(6) After-Tax Burden of Interest [(3)(1 − .4)]	(7) Total After-Tax Burden [(4) + (6)]	(8) PV of After-Tax Burden [(7) × αd.f.]
1	1,000	598.12	401.88	2,588.72	358.87	760.75	679.24
2	1,000	517.74	482.26	2,106.46	310.64	792.90	632.09
3	1,000	421.29	578.71	1,527.75	252.78	831.48	591.84
4	1,000	305.55	694.45	833.30	183.33	877.78	557.85
5	1,000	166.66	833.30	0	100.00	933.30	529.58
	5,000	2,009.40	2,990.60				2,990.60

- We add up all the current values of the after-tax burdens in column (8) to get the present value, which is $2,990.60, exactly equal to the original purchase price.
- Note that all the interest portions add up to $2,009.40, which if added to the principal of $2,990.60 gives us $5,000, which is equal to five yearly payments of $1,000 each.

The Cost of Leasing

Four major components comprise the cost of leasing:

1. The direct cost of the lease payments
2. The opportunity cost of depreciation
3. The opportunity cost of the residual value
4. The costs of service and maintenance.

The cost of payments is plain and direct; the opportunity cost of depreciation reflects the fact that the person who leases assets would forgo the tax benefits of the depreciation of that asset which can be claimed by the owner of the asset. The same logic goes for the opportunity cost of the residual or salvage value of the asset that can be claimed by the owner only.

So when the asset is leased, the lessee forgoes such a benefit. The costs of service and maintenance is the only component that can be claimed by the lessee because most, if not all, leased assets come with a built-in service and maintenance clause. But if the asset is purchased, the service and maintenance costs have to be carried by the buyer. Therefore, the first three components are added elements, but the fourth, the costs of service and maintenance, is the only one to be deducted from the leasing cost. All the elements are to be considered in terms of their present value, and all are to be considered in after-tax status.

$$PV(LC) = [PV(pyt) + PV(D) + PV(slvg)] - PV(svc)$$

where LC is the leasing cost, pyt is the lease payment, D is the annual depreciation of the leased asset, slvg is the salvage or residual value of the leased asset, and svc is the maintenance service of the leased asset.

- The present value of payments would be obtained by

$$PV(pyt) = (1 - t)pyt(a_{\overline{mn}|mr})$$

where t is the tax rate, mr is the monthly after-tax interest rate, and mn is the monthly term of leasing. It can also be obtained by the regular present value of an annuity formula:

$$PV(pyt) = (1 - t)pyt \cdot \frac{1 - (1 + mr)^{-mn}}{mr}$$

- The present value of depreciation would be obtained by

$$PV(D) = (t)D(a_{\overline{n}|r})$$

where $(t)D$ is the tax on the annual depreciation of the leased asset, r is the annual after-tax interest rate, and n is the annual term of leasing. This can also be obtained by the regular formula

$$PV(D) = (t)D\left[\frac{1 - (1 + r)^{-n}}{r}\right]$$

- The present value of salvage would be obtained by

$$PV(\text{slvg}) = CV(\text{slvg}) = (1 - t)\text{slvg}(v^n)$$

where slvg is the salvage or residual value of the leased asset adjusted for taxes. The current value is taken because it is a one-time process. It can also be obtained by the regular formula

$$CV(\text{slvg}) = (1 - t)\text{slvg}\left[\frac{1}{(1 + r)^n}\right]$$

where r is the annual after-tax rate and n is the annual term of leasing.
- The present value of services would be obtained by

$$PV(\text{svc}) = (1 - t)\text{svc}(a_{\overline{mn}|mr})$$

where svc is the monthly maintenance service charge on the leased asset being adjusted for taxes, mn is the monthly term of leasing, and mr is the monthly after-tax interest rate. The regular formula would be

$$PV(\text{svc}) = (1 - t)\text{svc}\left[\frac{1 - (1 + mr)^{-mn}}{mr}\right]$$

The following example illustrates how all these four elements are put together to constitute the present value of leasing.

Example 3.1.2 Our department needs a new and more capable copying machine. The budget committee is debating whether to buy a machine or lease one. The purchase price of $40,000 can be financed at 9% interest. It is estimated that this machine will have a productive lifetime of 8 years, after which it will be salvaged at 10% of its original price. The local vender is offering to lease this machine for 8 years at a monthly payment of $850 a month, which includes free regular service, whose value is $90 a month. Given that the tax rate is 36% and the asset depreciation method is a straight line, would it be better to buy or lease this machine?

The present value of the cost of financing the purchase at 9% is the purchase price itself, $40,000. The present value of the cost of leasing is the sum of three components minus the present value of the service:

$$PV(L) = [PV(\text{pyt}) + PV(D) + PV(\text{slvg})] - PV(\text{svc})$$

PV(pyt):

$$\text{monthly payment} = \$850$$

$$\text{after-tax payment} = \$850(1 - .36) = \$544$$

$$n = 8 \times 12 = 96 \text{ months}$$

$$\text{adjusted interest rate} : .09(1 - .36) = .0576$$

$$\text{monthly adjusted rate} = \frac{.0576}{12} = .0048$$

$$PV(\text{pyt}) = \frac{A[1 - (1 + r)^{-n}]}{r}$$

$$= \frac{\$544[1 - (1 + .0048)^{-96}]}{.0048}$$

$$= \boxed{\$41,766}$$

PV(D):

$$\text{depreciation } (D) = \$40,000 - \$40,000(.10)$$

$$= \$36,000 \text{ for 8 years}$$

$$= \frac{\$36,000}{8}$$

$$= \$4,500 \text{ each year}$$

$$\text{tax-adjusted depreciation } (Dt) = \$4,500(.36) = \$1,620$$

$$PV(D) = \frac{A[1 - (1 + r)^{-n}]}{r}$$

$$= \frac{\$1,620[1 - (1 + .0576)^{-8}]}{.0576}$$

$$= \boxed{\$10,156}$$

PV(slvg): The salvage value is 10% of the original price.

$$\$40,000 \times .10 = \$4,000$$

$$PV = CV = \frac{\$4,000}{(1+.0576)^8}$$

$$= \boxed{\$2,555.60}$$

PV(svc):

adjusted service $= \$200(1 - .36) = \$128$

$$PV = \frac{A[1 - (1+r)^{-n}]}{r}$$

$$= \frac{\$128[1 - (1+.0048)^{-96}]}{.0048}$$

$$= \boxed{\$9,827.34}$$

$$PV(L) = [(\$41,766 + \$10,156 + \$2,555.60)] - \$9,827.34$$

$$= \boxed{\$44,650}$$

The present value of the cost of leasing is larger than the present value of the cost of buying, so it would be better to buy the machine.

3.2. FOR THE LESSOR

For the lessor, leasing is an investment transaction. He would want to earn the highest possible return. So he would set the contract on his terms and decide on the payments according to his own required cost of capital. The present value of the lease payments plus the present value of depreciation and salvage value would equal the price he pays for the asset. The tax adjustment is still considered here except that the tax rate on the lessor may be different from that on the lessee part, and the interest rate is his required rate of return, but it would not be adjusted for taxes.

Example 3.2.1   A local rental shop is leasing a piece of heavy equipment whose value is \$80,000, which would depreciate by the straight-line method to 10% of its value in 5 years. That value would be considered its salvage value at the end of the fifth year. If the shop's required cost of capital is 18% and their taxes are 40%, what will the monthly lease payment be?

$$PV(\text{original value}) = PV(\text{pyt}) + PV(D) + PV(\text{slvg})$$

PV(pyt): Here, the lease payment is the unknown but we know that it should be tax-adjusted by being multiplied by $1 - t$.

$$\text{monthly interest rate} = \frac{.18}{12} = .015 \qquad \text{monthly term} = 5 \times 12 = 60$$

$$PV(\text{pyt}) = \frac{(1 - .4)L[1 - (1 + r)^{-n}]}{r}$$

$$= \frac{.6L[1 - (1 + .015)^{-60}]}{.015}$$

$$= \boxed{.6L(39.380)}$$

PV(D):

$$\text{depreciation} = \$80,000 - \$80,000(.10) = \$72,000$$

$$\text{annual depreciation} = \frac{\$72,000}{5} = \$14,400$$

$$\text{tax-adjusted depreciation} = Dt = \$14,400(.4) = \$5,760$$

$$PV(D) = \frac{\$5,760[1 - (1 + .18)^5]}{.18}$$

$$= \boxed{18.012}$$

PV(slvg):

$$\text{salvage value} = .10(\$80,000) = \$8,000$$

$$CV(\text{slvg}) = \frac{\$8,000}{(1 + .18)^5}$$

$$= \boxed{\$3,497}$$

$$PV(\text{original value}) = PV(\text{pyt}) + PV(D) + PV(\text{slvg})$$

$$\$80,000 = .6L(39.380) + 18.012 + \$3,497$$

$$23.628L = \$58,491$$

$$L = \frac{\$58,491}{23.628}$$

$$= \boxed{\$2,475.49}$$

For the rental shop to achieve the desired rate of return, the lease payment should be set at $2,475.49 a month.

Unit III Summary

The debt incurred by consumers and businesses is not necessarily a bad thing unless it gets out of control. If understood and managed well, debt can work as an economic stimulant and effective factor in economic growth. Debt can literally boost the purchasing power and practically expand current consumption. But the critical point remains that expanding current consumption may mean reducing future consumption if income cannot be increased to compensate. Therefore, accepting and enjoying the access to credit becomes a choice of having more consumption for today or for tomorrow. That is why understanding how debt works is an essential part of successful financial management.

Based on the repayment structure, debt can be either installment debt or non-installment debt. Installment debt is most common and is associated with the process of amortization, which can handle large debt by breaking down the repayment process into periodic payments. Periodic payments are most likely to be uniform throughout the maturity period, but they can be different in how they are broken down between the interest and principal portions. Under the level method, the interest and principal portions both remain the same throughout the length of maturity. In the rule of 78 and the declining balance methods, the interest and principal portions both change throughout the maturity period. How interest is charged and collected is another issue where loans can be either add-on, discount, or single payment. When the reduction in the principal throughout the subsequent payments is not considered to reduce interest, a higher actual and effective interest rate can be calculated—hence the distinction between nominal and real APR.

The cost of credit to borrowers can increase dramatically with a higher interest rate and a long repayment period. Paying only the minimum payment on revolving credit can be really costly if it continues for a long time. There is an essential difference between the credit limit, which is assessed by the creditor, and the debt limit, which should be assessed by the borrower. It is much better for any borrower for the debt limit to be lower than the credit limit.

Mortgages and sinking funds are structured by the process of amortization and are dealt with as ordinary annuities. Their payments and other variables can be calculated by both the formula method and the table method.

Mathematical Finance, First Edition. M. J. Alhabeeb.

List of Formulas

The rule of 78

Fraction to determine interest portion:

$$K_i = \frac{j}{D}$$

Proper denominator for all payments:

$$D_n = \frac{n(n+1)}{2}$$

Proper denominator for the remaining payment:

$$D_u = \frac{N_u(N_u+1)}{2}$$

Rebate factor:

$$\text{RF} = \frac{D_u}{D_n}$$

Rebate:

$$\text{Rb} = \text{RF(I)}$$

Add-on interest

Add-on payment:

$$\text{PYT} = \frac{P(1+rn)}{n}$$

Actual APR formula 1:

$$\text{APR}_a^1 = \frac{2KI}{P(n+1)}$$

Actual APR formula 2 (n-ratio formula):

$$\text{APR}_a^2 = \frac{K(95n + 9)I}{12n(n + 1)(4P + I)}$$

Average daily balance

Average daily balance:

$$\text{ADB} = \frac{\sum_{i=1}^{K} b_i t_t}{c_y}$$

Monthly finance charge:

$$\text{MFC} = \text{ADB} \cdot \text{MR}$$

Debt limit

Debt payment/disposable income ratio:

$$\frac{\text{DP}}{\text{DI}} \le .20$$

Debt/equity ratio:

$$\frac{D}{E} \le .33$$

Amortization

Monthly payment of an amortized loan:

$$A = \frac{\text{CV} \cdot r}{1 - (1 + r)^{-n}}$$

Monthly payment of an amortized loan using table value:

$$A = \text{CV} \cdot \frac{1}{a_{\overline{n}|r}}$$

Balance of an amortized loan:

$$\text{CV} = \frac{A[(1 + r)^{-n}]}{r}$$

Balance of an amortized loan using table value:

$$\text{CV} = A \cdot a_{\overline{n}|r}$$

Maturity of an amortized loan:

$$n = \frac{\ln[1 - (CV \cdot r)/A]}{\ln(1 + r)}$$

Sinking fund

Deposit to a sinking fund:

$$A = \frac{FV \cdot r}{(1 + r)^n - 1}$$

Deposit to a sinking fund using table value:

$$A = FV \cdot \frac{1}{S_{\overline{n}|r}}$$

Sinking fund accumulation:

$$FV = \frac{A[(1 + r)^n - 1]}{r}$$

Sinking fund accumulation using table value:

$$FV = A \cdot S_{\overline{n}|r}$$

Exercises for Unit III

1. For a loan of $1,300 at an annual interest of 11% for 18 months, using the level method, calculate (**a**) the monthly payment, (**b**) the monthly interest portion (MIP), and (**c**) the monthly principal portion (MPP).

2. Calculate the actual annual percentage rate of interest for the loan in Exercise 1 using both formulas APR_a^1 and APR_a^2.

3. How is a monthly payment of $900 broken down between interest and principal if the annual interest is 10% and maturity is for $1\frac{1}{2}$ years? What would be the actual annual percentage for the loan if you use formula APR_a^1 ?

4. Margaret obtains a personal loan of $1,200 at $8\frac{3}{4}$% interest for 1 year. Given that the bank follows the rule of 78, construct the payment schedule showing five columns: payment number, payment fraction, monthly interest portion, monthly principal portion, and monthly payment.

5. Suppose that Margaret is given another choice: increasing her loan to $1,500 but at 10% interest and she can spread the payments over 2 years. If the method is still the rule of 78, what will be (**a**) her monthly payment and (**b**) the breakdown between interest and principal for (1) the 1st payment, (2) the 10th payment, and (3) for the 23rd payment?

6. If you consider monthly payments for a loan, what will be the determining denominator (D) in the rule of 78 method if the maturity of such a loan is $3\frac{1}{2}$ years?

7. Calculate D for a loan maturity of 52 months if the bank follows the rule of 78 when calculating interest.

8. Glenn had an auto loan of $4,500 at 7% annual interest for 3 years. He decides to pay off the remaining balance after he has made 29 payments. What will be the balance due be under the rule of 78 method?

9. What will Glenn's balance be if he has made 20 payments and decides to pay the rest in one lump sum at the 21st payment? Suppose that his lender is following the level method.

10. Lynn had a $2,000 personal loan at $9\frac{1}{2}$% for 18 months. Suppose that she is given the choice between the level method and the rule of 78 method and that she wants to pay off the remaining balance after a year of making

Mathematical Finance, First Edition. M. J. Alhabeeb.
© 2012 John Wiley & Sons, Inc. Published 2012 by John Wiley & Sons, Inc.

monthly payments. Which method would she choose, and what difference in interest would she pay?

11. What is the rebate (Rb) on a loan of $2,000 at 5% annual interest if the borrower decides to pay it all after 10 payments of the 24-payment maturity time?

12. Find the monthly payment of a $4,000 loan for 48 months at an add-on 9% annual interest. Also, find the final balance if this loan is to be paid off at the 30th payment (assuming that 29 payments have been made).

13. Based on an add-on interest, Sally obtains a loan for $3,400 at $6\frac{1}{2}$% annual interest for 4 years. What will be her semiannual payment and the total interest paid on the loan?

14. How long will it take Kate to pay for her furniture set if she chooses to pay only the minimum monthly payment of 2% of the outstanding balance? Her furniture set cost $3,750 financed at $19\frac{3}{4}$% annual interest.

15. How long will it take Kate to pay for the same purchase if the minimum payment is changed to 3% of the outstanding balance of $3,750 but the interest rate is raised to 23%? How much interest will she pay in total?

16. Calculate the average daily balance (ADB) and the finance charge on Drew's account that has 9.5% APR and a 30-day billing cycle. His January statement shows:

    $425.89 balance carried from December

    $133.15 K-Mart charge on January 5

    $76.95 Greek Cuisine restaurant charge on January 8

    $150.00 payment paid on January 10

    $25.25 Texaco gas charge on January 11

    $33.15 Target charge on January 21

    $80.00 health club charge on January 26

    $17.85 Wal-Mart charge on January 28

17. The Smith family has a gross income of $85,000 a year and they pay the following bills:

    annual taxes: $19,550

    monthly mortgage: $1,850

    credit card 1: $320 a month

    credit card 2: $190 a month

    personal loan: $113 twice a month

    department store charge account: $65.00 a month

Calculate their debt limit.

18. Calculate the debt limit for the Mitchell family if their financial statement reads as follows:

total assets, including their home value of $380,000: $478,700

short-term liabilities: $6,700

long-term liabilities, including a mortgage balance of $298,000: $320,000

19. Jane purchased her first home for $95,000. She made a down payment of 20% and financed the rest at 8% for 20 years. Find her monthly payment by both the formula method and the table method.

20. For a mortgage loan of $110,000 and a down payment of $20,000 financed at 10% for 25 years, construct the amortization schedule for the first year. Show five columns: payment number, monthly interest portion, monthly principal portion, monthly payment, and the balance.

21. A condominium is purchased for $80,000 with a down payment of $12,000 at an annual interest rate of 9% for 15 years. Calculate the unpaid balance after 10 years of making payments.

22. What would be the market value of the condominium in Exercise 21 after 12 years of making payments if the interest rate goes up to 11%? Use the formula method and the table method.

23. How many mortgage payments were left on Harry's house if the remaining balance is $72,375.15 given that he purchased it at 14% interest and has been making a monthly payment of $900?

24. What is the semiannual payment to finance $200,000 in a sinking fund that pays 12% annual interest? Also, calculate the total deposits of the sinking fund and the interest earned by the semiannual payment.

25. The owner of a local pizzeria needs $250,000 in 5 years to open another branch. How much should be deposited in the company's sinking fund that earns 8% compounded monthly?

26. Jack borrowed $25,000 at 8% for 5 years. He also opened a sinking fund account that pays 6%. How much should he pay for the interest on his loan? What size deposit should he make to his sinking fund to pay off the principal of his loan on time?

27. Use both the formula method and the table method to calculate the quarterly deposit into a sinking fund paying 7% to pay off a debt of $400,000 in 12 years. Also, calculate the interest payment on this $8\frac{1}{2}$% loan.

UNIT IV

Mathematics of Capital Budgeting and Depreciation

1 Capital Budgeting

One of the most crucial decisions made in the business world is an investment decision, where investors have to choose the most worthy project to fund. It refers to the highest responsibility of the decision makers to see if a certain capital fund should be allocated in the next budget as a specific investment project. Given that alternative investment opportunities differ in many aspects, such as the level of risk associated with them, and their capacities to yield future returns, the criteria for choice would be to assess carefully the proposed alternatives, and select the potentially most profitable. Capital expenditure is an outlay of funds which a firm would rely on to bring enough returns to cover and exceed the initial investment. Therefore, capital budgeting is a process to review, analyze, and select those projects that promise to be the most rewarding in the medium and long runs.

Cash flow analysis is most helpful in determining the profitability of capital. It involves the evaluation and comparison of two flows: the cash outflows, which consist mainly of the initial capital funds allocated to an investment project as well as the capital expenditures throughout the life of assets in their productive process; and the cash inflows, which consist primarily of estimates of the returns expected on an investment project. They also include certain allowances for the depreciation of the productive assets and their residual values. While cash outflows are normally measured at their current values, the value of cash inflows has to be brought back from future maturity to the present by discounting them at the firm's cost of capital. Vital to the analysis of cash flow are major concepts such as the net present value, internal rate of return, profitability index, and capitalized cost. We focus on these concepts in this chapter, and in the second chapter we address the issues of depreciation and depletion of assets.

1.1. NET PRESENT VALUE

Net present value (NPV) is an analytical method, which uses the discounted cash flow to provide a tool to determine how profitable an investment is. It compares the current value of all capital expenditures in an investment project with the value of the returns expected from that project as discounted at an interest rate equal to the firm's marginal cost of capital. Recall that the current value (CV) of

Mathematical Finance, First Edition. M. J. Alhabeeb.
© 2012 John Wiley & Sons, Inc. Published 2012 by John Wiley & Sons, Inc.

a future return (FV) is

$$CV = \frac{FV}{(1+r)^n}$$

or is the future value (FV) multiplied by the discount factor:

$$CV = FV(1+r)^{-n}$$

Also recall that for a stream of future returns, the present value is

$$PV = \sum FV(1+r)^{-n}$$

Therefore, if the capital expenditure for initial investment (I_0) at its current value is compared to the present value of its future returns, the result would be the net present value (NPV):

$$NPV = [FV_1(1+r)^{-1} + FV_2(1+r)^{-2} + \ldots + FV_n(1+r)^{-n}] - I_0$$

$$\boxed{NPV = \sum_{n=1}^{N} FV(1+r)^{-n} - I_0}$$

that is, for multiple returns at various maturities. The objective is to have the future return cover or exceed the capital spent initially on an investment. Therefore, a positive net present value (NPV $\geq$ 0) would indicate a promisingly profitable investment and probably lead to project approval and capital allocation. On the other hand, a negative net present value (NPV $<$ 0) would indicate a loss of the capital spent and may very well lead to the rejection of that investment proposal and block any attempt to allocate funds to it in the next budget.

Example 1.1.1   A proposal to expand a fast-food restaurant calls for an initial capital investment of \$42,000 and promises that the return on investment would be at least \$14,000 for each of the next 5 years. Would the franchise company approve if it has a 10% cost of capital?

$$PV = \sum_{n=1}^{5} \frac{FV_n}{(1+r)^n}$$

$$= \frac{FV_1}{(1+r)^1} + \frac{FV_2}{(1+r)^2} + \frac{FV_3}{(1+r)^3} + \frac{FV_4}{(1+r)^4} + \frac{FV_5}{(1+r)^5}$$

$$= \frac{\$14,000}{(1+.10)^1} + \frac{\$14,000}{(1+.10)^2} + \frac{\$14,000}{(1+.10)^3} + \frac{\$14,000}{(1+.10)^4} + \frac{\$14,000}{(1+.10)^5}$$

$$= \$12,727.27 + \$11,570.25 + \$10,518.41 + \$9,562.19 + \$8,692.90$$

$$= \$53,071.02$$

We can also obtain the present value by the table method as

$$PV = FV \cdot a_{\overline{n}|r}$$

where the table value across an interest rate of 10% and a maturity of 5 years is 3.791.

$$PV = \$14,000a_{\overline{5}|.10}$$

$$= \$14,000(3.791)$$

$$= \$53,074$$

$$NPV = PV_{in} - I_0$$

$$= \$53,072 - \$42,000$$

$$= \$11,072$$

The franchise administration would approve the expansion project as a potentially successful investment.

Example 1.1.2   The development committee in a construction company is studying two investment proposals whose cash inflows are projected for the next four years (see Table E1.1.2). Both proposals ask for a capital allocation of \$200,000, but the cost of capital for the first project is 8% and for the second is $7\frac{1}{2}$%. Which of the two proposals would be approved?

For project I:

$$PV = \frac{\$35,000}{(1+.08)^1} + \frac{\$40,000}{(1+.08)^2} + \frac{\$50,000}{(1+.08)^3} + \frac{\$120,000}{(1+.08)^4}$$

$$= \$32,407.41 + \$34,293.55 + \$39,691.61 + \$88,203.58$$

$$= \$194,596.15$$

TABLE E1.1.2

	Cash Inflows	
Year	Project I: $r = 8\%$	Project II: $r = 7\frac{1}{2}\%$
1	35,000	40,000
2	40,000	40,000
3	50,000	95,000
4	120,000	100,000

By the table value:

$$PV = (\$35,000 \times .92592593) + (\$40,000 \times .85733882)$$
$$+ (\$50,000 \times .79383224) + (\$120,000 \times .73502985)$$
$$= \$194,596.13$$
$$NPV_I = PV_{in} - I_0$$
$$= \$194,596.15 - \$200,000 = -\$5,403.85$$

For project II:

$$PV = \frac{\$40,000}{(1 + .075)^1} + \frac{\$40,000}{(1 + .075)^2} + \frac{\$95,000}{(1 + .075)^3} + \frac{\$100,000}{(1 + .075)^4}$$
$$= \$37,209.30 + \$34,613.30 + \$76,471.25 + \$74,880.05$$
$$= \$223,173.90$$

By the table value:

$$PV = (\$40,000 \times .93023256) + (\$40,000 \times .86533261)$$
$$+ (\$95,000 \times .80496057) + (\$100,000 \times .74880053)$$
$$PV = \$223,173.91$$
$$NPV_{II} = \$223,173.90 - \$200,000 = \$23,173.90$$

Project I would incur a loss of \$5,403.85; project II would make a positive net value of \$23,173.90. Project I would be rejected; project II would be accepted.

1.2. INTERNAL RATE OF RETURN

Internal rate of return (IRR) is another method that helps to determine whether a proposed investment is worthwhile. The method is supposed to utilize the rate of return on invested capital that the proposed project is hoped to yield. The internal rate of return is sometimes called **profit rate** or **marginal efficiency of investment**. It is the rate that equates the cash outflows and inflows at their present value. In other words, it is the rate at which the net present value would equal zero since there would be no difference in the equality of the two cash flows.

$$NPV = \sum_{n=1}^{N} \frac{FV}{(1 + IRR)^n} - I_0 = 0$$

The primary criterion is that the internal rate of return has to be equal to or greater than the firm's required cost of capital. Computerized methods can now easily find the internal rate of return, but traditionally it has been found by mathematical trial and error. However, an equation was developed to get at least an initial estimate for the IRR, which can consequently be modified by trial and error to reach the exact value of the rate.

$$FV_1(1+r)^{-1} + FV_2(1+r)^{-2} + FV_3(1+r)^{-3} + \cdots + F_n(1+r)^{-n} - I_0 = 0$$

If we use a binomial expansion for one term, we obtain

$$FV_1(1-r) + FV_2(1-2r) + FV_3(1-3r) + \cdots + FV_n(1-nr) = I_0$$

We sum up the terms:

$$\sum_{k=1}^{n} FV_k(1-kr) = I_0$$

We rearrange the left term:

$$\sum_{k=1}^{n} FV_k - \sum_{k=1}^{n} k \cdot FV_k r = I_0$$

$$\sum_{k=1}^{n} FV_k - I_0 = \sum_{k=1}^{n} k \cdot FV_k r$$

$$\boxed{r = \frac{\sum_{k=1}^{n} FV_k - I_0}{\sum k \, FV_k}}$$

Example 1.2.1   A proposal for an investment project calls for $15,000 in initial capital. It promises that its returns will be as follows in the next five years, respectively: $3,600, $4,200, $5,500, $6,300, $7,500 (see table E1.2.1). $n = 5$; $FV_k = FV_1, FV_2, FV_3, FV_4, FV_5$; $I_0 = \$12,000$. What would be the internal rate of return?

$$\sum_{k=1}^{n} FV_k = \$3,600 + \$4,200 + \$5,500 + \$6,300 + \$7,500$$

$$= \$27,100$$

$$\sum_{k=1}^{n} k \cdot FV_k = \$91,200$$

$$r = \frac{\sum_{k=1}^{n} FV_k - I_0}{\sum_{k=1}^{n} k \cdot FV_k}$$

$$= \frac{\$27,100 - \$15,000}{\$91,200} = 13.27\%$$

TABLE E1.2.1

Year, k	FV_k	$k \cdot FV_k$
1	3,600	3,600
2	4,200	8,400
3	5,500	16,500
4	6,300	25,200
5	7,500	37,500
Σ	27,100	91,200

This is only an initial estimate of the internal rate of return, but we know from the formula that although it is a rough estimate, it is in the appropriate range. Now, we calculate the NPV at different rates to see if its value gets closer to the initial investment. In this way, we can pinpoint the exact internal rate of return (IRR) after several tries.

- At the rate of 18%, PV would be

$$PV = \frac{\$3,600}{(1+.18)^1} + \frac{\$4,200}{(1+.18)^2} + \frac{\$5,500}{(1+.18)^3} + \frac{\$6,300}{(1+.18)^4} + \frac{\$7,500}{(1+.18)^5}$$

$$= \$3,050.85 + \$3,016.37 + \$3,347.47 + \$3,249.47 + \$3,278.32$$

$$= \$15,942$$

- At 20%, PV would be $15,151.
- At 20.4%, PV would be exactly 15,000, which is exactly equal to the value of the initial investment ($15,000).

$$PV - I_0 = 0$$

$$\$15,000 - \$15,000 = 0$$

1.3. PROFITABILITY INDEX

Instead of taking the zero difference between the cash inflows and outflows, where the IRR has to make that difference nonexistent, another criterion is used to assess the worthiness of the new investment proposal. This third method is the **profitability index** (PI), defined as the ratio of the present value of the returns to the initial investment: in other words, the ratio of the inflows to the outflows in their current values,

$$PI = \frac{PV_{ci}}{I_0}$$

where PI is the profitability index, PV_{ci} is the present value of the cash inflows, and I_0 is the initial investment. The criterion for accepting a new investment proposal is that PI has to be at least equal to or greater than 1: $PI \geq 1$.

Example 1.3.1   In project I of Example 1.2.1, the present value of the cash inflows was calculated at $194,596, and the proposed capital to be invested was $200,000. If we follow the profitability index criterion, PI would be calculated as

$$PI = \frac{PV_{ci}}{I_0}$$

$$= \frac{\$194,596}{\$200,000}$$

$$= .97$$

Project I would be rejected based on the PI being less than 1. If we calculate PI for project II, we get

$$PI = \frac{\$223,174}{\$200,000}$$

$$= 1.12$$

and for the PI being greater than 1, project II would be accepted.

1.4. CAPITALIZATION AND CAPITALIZED COST

Capitalization of a fund (asset or liability) refers to the present value or cash equivalent of its unlimited number of periodic payments. For example, if a certain fund is invested now at a certain interest rate, we can assume that we would continue to collect periodic interest on that fund forever. Therefore, if we put this logic in reverse, we can realize that the current fund is, in fact, the present value for all of its periodic payments that are held in perpetuity. Capitalization is used to evaluate the cash equivalent of assets and liabilities that have periodic payments.

From a successful business management perspective, a firm should not only allocate funds to buy capital assets, but also allocate additional funds to maintain them throughout their useful lives, and allocate investment to replace them after they give their due service. The **capitalized cost** (K) of an asset is, therefore, the sum of :

- The asset's original cost, C
- The present value of its unlimited maintenance cost:

$$\frac{(C - S)}{(1 + r)^n - 1}$$

- The present value of its unlimited number of replacements, M/r :

$$K = C + \frac{C - S}{(1+r)^n - 1} + \frac{M}{r}$$

where K is the capitalized cost of an asset, C is the original cost of the asset, S is the scrap value of the asset after its useful life, and M is the annual maintenance cost of the asset. The capitalized cost calculations are often used in decision making to select the most economic alternatives of assets.

Example 1.4.1   A construction company is contemplating the purchase of heavy equipment. The decision maker narrowed down the alternatives to two of the best machines, which are described in Table E1.4.1. Which of the two machines should be purchased if the interest rate is $9\frac{1}{2}\%$?

We calculate the capitalized cost for both machines individually and choose the least costly as the better alternative.

$$K_1 = C_1 + \frac{C_1 - S_1}{(1+r)^n - 1} + \frac{M_1}{r}$$

$$= \$35,000 + \frac{\$35,000 - \$5,000}{(1+.095)^{10} - 1} + \frac{\$3,000}{.095}$$

$$= \$86,873.52$$

$$K_2 = C_2 + \frac{C_2 - S_2}{(1+r)^n - 1} + \frac{M_2}{r}$$

$$= \$39,000 + \frac{\$39,000 - \$4,000}{(1+.095)^{15} - 1} + \frac{\$2,500}{.095}$$

$$= \$77,379.25$$

Machine II should be purchased on the basis of having less capitalized cost.

Example 1.4.2   A town board was asked to estimate an endowment to build a children's playground. If the construction costs $50,000 and needs to be replaced

TABLE E1.4.1

	Machine I	Machine II
Initial cost ($)	35,000	39,000
Useful life (years)	10	15
Annual maintenance ($)	3,000	2,500
Scrap value ($)	5,000	4,000

every 10 years at an estimated cost of $40,000, and the maintenance cost is $15,000, how much would the endowment be if the interest rate is 12%?

The endowment total would be considered a capitalized cost and the replacement cost would be $C - S$.

$$K = \$50,000 + \frac{\$40,000}{(1 + .12)^{10} - 1} + \frac{\$15,000}{.12}$$

$$= \$193,994.72$$

The board will ask the donor to allocate $194,000.

Another application for capitalized cost is to determine the extent of improvement that can be made on asset performance or equipment productivity. Let's assume that we have a printing machine whose original cost was $65,000 with a scrap value estimated at $5,000 after 12 years. The machine productivity is 20,000 books a year and its maintenance cost is $3,000. The firm's engineer determines that installing an additional part can raise the machine's productivity to 30,000 books a year without affecting its maintenance or its useful age. How much can the firm spend economically to achieve the boost in productivity if the investment rate is 8%?

Here we can set an equation of ratios: the ratios of the capitalized costs of the machine to its productivity before and after the technological improvement. If K_b and K_a are the capitalized cost of the machine before and after the technological improvement, and P_b and P_a are the productivity of the machine before and after the technological improvement, then

$$\frac{K_b}{P_b} = \frac{K_a}{P_a}$$

We set up the capitalized costs where the subject of the question (how much we can spend) would be an addition (x) to the original cost in the calculation of the capitalized cost after the technological improvement. Then we would solve algebraically for x.

$$K_b = C_b + \frac{C_b - S_b}{(1 + r)^n - 1} + \frac{M_b}{r}$$

$$= \$65,000 + \frac{\$65,000 - \$5,000}{(1 + .08)^{12} - 1} + \frac{\$3,000}{.08}$$

$$= \$142,021$$

$$K_a = C_a + \frac{C_a - S_a}{(1 + r)^n - 1} + \frac{M_a}{r}$$

Note that $C_a = C_b + x$, where x is what should be spent on the technological improvement of the machine; $S_a = S_b$; and $M_a = M_b$ no change in the life of the machine and its residual value. Then

$$K_a = (C_b + x) + \frac{C_b + x - S_b}{(1+r)^n - 1} + \frac{M_b}{r}$$

$$= (\$65{,}000 + x) + \frac{\$65{,}000 + x - \$5{,}000}{(1 + .08)^{12} - 1} + \frac{\$3{,}000}{.08}$$

$$= \frac{\$227{,}271 + 3x}{1.5}$$

$$\frac{K_b}{P_b} = \frac{K_a}{P_a}$$

$$\frac{\$142{,}021}{20{,}000} = \frac{(\$227{,}271 + 3x)/1.5}{30{,}000}$$

$$x = \$30{,}758$$

The firm can spend \$30,758 to improve the machine and raise its productivity to 30,000 books a year.

1.5. OTHER CAPITAL BUDGETING METHODS

There are other ways to judge the worthiness of capital expenditures, ways which are not based on the time value of money. Most common of these ways are the following two:

The Average Rate of Return Method

Under the **average rate of return method**, the average rate of return (ARR) is calculated for alternative investment projects using data anticibates. The judging firm has to have its own minimum acceptable average rate of return for certain sorts of project to be used as a reference point. The decision of accepting or rejecting a certain proposed capital for allocation in the firm's next budget is made based on how the calculated average rate of returns meets that firm's established criterion. The average rate of return (ARR) is obtained by

$$\boxed{\text{ARR} = \frac{2 \cdot \text{APAT}}{C}}$$

where APAT is the average profit after taxes, which is a simple average calculated by dividing the total of the after-tax profits that are expected to be earned over a project's life (number of years of that proposed life). C is the proposed initial capital. The 2 in the formula came from originally dividing APAT by the average investment, which is, in turn, defined as the initial investment divided by 2.

The Payback Time Method

The **payback time method** considers the payback period, which refers to the number of years during which an initial investment can be recovered. The payback is calculated by dividing the initial investment (C) by the yearly cash inflows if the cash inflows are projected to be uniform. However, if the cash inflows are variable throughout the years of the project, the payback time would be whatever number of years is necessary to allow the variable cash inflows to accumulate until the initial investment is recovered. This, of course, would not make the payback calculation as cleancut as it is in the case of uniform cash inflows.

$$\boxed{\text{Payback} = \frac{C}{\text{YCI}}}$$

where C is the initial investment and YCI is the yearly cash inflows if it is uniform throughout the years.

Example 1.5.1   The Sunshine Company is considering the following two projects for capital allocations. Project X is asking for \$64,000 and project Y is asking for \$68,000 (see Table E1.5.1). Which of the two projects would win the company's approval? Use the average rate of return method, and the payback time method.

$$\text{ARR}_X = \frac{2 \cdot \text{APAT}}{C} = \frac{2(\$9,000)}{\$64,000} = 28\%$$

$$\text{ARR}_Y = \frac{2(\$8,900)}{\$68,000} = 26.2\%$$

$$\text{Payback}_X = \frac{C}{\text{YCI}} = \frac{\$64,000}{\$16,000} = 4$$

For payback$_Y$, the initial investment of \$68,000 would be recovered in the following manner: In the first 3-years, \$63,000 would be recovered

TABLE E1.5.1

	Project X (\$64,000)		Project Y (\$68,000)	
Year	Expected Profits (After (\$) Taxes)	Cash (\$) Inflows	Expected Profits (After (\$) Taxes)	Cash (\$) Inflows
1	9,500	16,000	21,500	40,800
2	10,200	16,000	9,000	12,200
3	10,200	16,000	5,500	10,000
4	8,000	16,000	4,500	10,000
5	7,100	16,000	4,000	8,000
	9,000	16,000	8,900	16,000

($40,800 + $12,200 + $10,000). The rest of the $5,000 to complete the initial investment of $68,000 would be recovered in the fourth year. Since the fourth year yield is $10,000, we can assume that the $5,000 would be recovered halfway through the year. This would make the payback time 3.5 years. Again, the decision would be up to the judging company and how it would set its criteria. Generally, the payback method is preferred by many business decision makers because of its use of cash inflows and their timing. The less time it takes to recover capital, the less time it takes to hold the money away from gaining returns and the less the risk of uncertainty. Still, both of these methods do not consider the time value of money directly, and that is their major flaw.

2 Depreciation and Depletion

All assets, even land to a certain degree, have a certain useful life during which they can provide services or revenues. Under normal circumstances, an asset's ability to provide useful and meaningful production tends to decrease throughout its useful life, down to a point when its value for production becomes insignificant. Some assets reach that point earlier, such as in the case of equipment and structures, which wear out prematurely, break down, or just become obsolete. The gradual loss of value in an asset's ability to produce is called **depreciation**. From a successful financial management perspective, the depreciation has to be compensated by gradual write-offs against the revenues that the asset generates during its useful life. In this way it becomes fairly possible to replace an asset at the time it retires by using those accumulated deductions that have been recovered through the asset's productive days. We can follow the value of an asset and its changes in a linear time line of its life from C to S (see Figure 2.1). When an asset is purchased new, its value is 100% and it is equal to the original cost (C). Throughout its useful life (n), its value decreases due to increasing depreciation. At the end of its life, the asset would have a residual or scrap value (S) equal either to zero or to a minimum percentage of its original value. During the asset's life, each period, say a year, would have its own portion of depreciation (R_k), where $0 \leq k \leq n$. Those portions of depreciation accumulate as the years go by. So the accumulated depreciation (D_k) would be none ($D_k = 0$) at the beginning and would be a full amount ($D_n = C - S$) at the end of the asset's life. The difference between the original cost and the accumulated depreciation at any point throughout (n) would be the book value (B_k) as it would be recorded at that point. So in this case, the book value at the very beginning would be just equal to the original cost ($B_0 = C$), and it would be down to the scrap value at the end of the useful life of the asset ($B_n = S$).

It is worthwhile mentioning here that the book value may not necessarily be equal to the market value of the asset. Internally, it is only necessary to record the portion of depreciation that has to be written off as an expense. This book value would be the same as the scrap value at the end of the asset's useful life. The big question now is: At what rate does the depreciation occur, and how do accountants figure out how much to write off each year? Well, there are several points of view and there are several methods of calculations in this regard. Some methods consider that depreciation occurs in a fixed rate for each year of the

Mathematical Finance, First Edition. M. J. Alhabeeb.

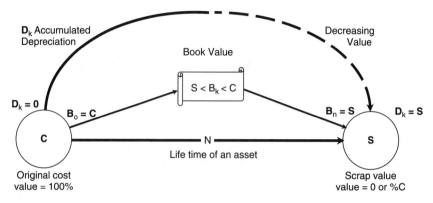

FIGURE 2.1

asset's life, and other methods use a variable rate, yet some other methods use the compound interest technique to figure out the rate of depreciation. The most common methods of calculating depreciation are described next.

2.1. THE STRAIGHT-LINE METHOD

The **straight-line method** is simple and straightforward. It assumes that depreciation occurs in equal amounts for each year of the asset's life, and therefore the accumulated depreciation is distributed evenly throughout the life span of the asset. So in this case, the accumulated depreciation (D_k) is a simple product of the number of years (k) and the rate of depreciation (R).

$$D_k = kR$$

where each R is

$$\boxed{R = \frac{C - S}{n}}$$

and therefore the book value B_k would be

$$B_k = C - D_k$$
$$B_k = C - kR$$

$$\boxed{B_k = C - \left(\frac{C - S}{n}\right)}$$

Example 2.1.1 Diving equipment was purchased for \$23,000. It has a useful life of 6 years, after which its value is estimated to be \$8,000. Use the straight-line method to figure out its full depreciation record.

$C = \$23,000$; $S = \$8,000$; $n = 6$ (see Figure E2.1.1 and Table E2.1.1).

$$R = \frac{C - S}{n}$$

$$= \frac{\$23,000 - \$8,000}{6}$$

$$= \$2,500$$

It is important to note that sometimes and for certain assets, such as machines and equipment, the depreciation may be expressed per unit of production or per hour of operation instead of per year. In such cases, n would be replaced by either the number of units of product (P) or the number of operating hours (H).

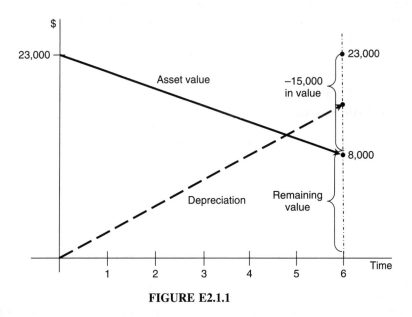

FIGURE E2.1.1

TABLE E2.1.1

Year, k (at the end)	Annual Depreciation, R ($)	Accumulated Depreciation, $D_k = kR$ ($)	Book Value, $B_k = C - D^k$ ($)
0	—	—	23,000
1	2,500	2,500	20,500
2	2,500	5,000	18,000
3	2,500	7,500	15,500
4	2,500	10,000	13,000
5	2,500	12,500	10,500
6	2,500	15,000	8,000

TABLE E2.1.2

Year, k (at the end)	Annual Production, AP (units)	Annual Depreciation, PR ($)	Accumulated Depreciation, D_p ($)	Book Value $C - D_p$ ($)
0	—	—	—	64,000
1	40,000	22,000	22,000	42,000
2	35,000	19,250	41,250	22,750
3	15,000	8,250	49,500	14,500
4	10,000	5,500	55,000	9,000
	100,000	55,000		

Example 2.1.2 A machine was purchased for $64,000 to produce a total of 100,000 units of a product. It is to retire after 4 years with a scrap value of $9,000. The machine productivity is distributed over the 4 years as follows (see Table E2.1.2):

Year 1: 40,000 units
Year 2: 35,000 units
Year 3: 15,000 units
Year 4: 10,000 units

Construct its depreciation schedule in consideration of its productivity using the straight-line method.

$$R = \frac{C - S}{P}$$

$$= \frac{\$64,000 - \$9,000}{100,000 \text{ units}}$$

$$= \$.55 \text{ depreciation per unit of production}$$

Example 2.1.3 A copying machine that costs $12,500 has a total useful life of 15,000 hours of work, after which it would still be worth $3,200 (see Table E2.1.3). If it is used 2,500 hours a year, what would the depreciation schedule look like?

$$\text{useful life in years} = \frac{15,000}{2,500} = 6 \text{ years}$$

$$R = \frac{C - S}{H}$$

$$R = \frac{\$12,500 - \$3,200}{15,000} = \$.62 \text{ depreciation per hour}$$

TABLE E2.1.3

Year, k (at the end)	Annual Hours of Service, AH	Annual Depreciation, AH $\cdot$ R ($\$$)	Accumulated Depreciation, D_n ($\$$)	Book Value, $C - D_n$ ($\$$)
0	—	—	—	12,500
1	2,500	1,550	1,550	10,950
2	2,500	1,550	3,100	9,400
3	2,500	1,550	4,650	7,850
4	2,500	1,550	6,200	6,300
5	2,500	1,550	7,750	4,750
6	2,500	1,550	9,300	3,200
	15,000	9,300		

2.2. THE FIXED-PROPORTION METHOD

The **fixed-proportion method** requires that the scrap value be positive. It cannot be used for assets which are used up completely so that their residual values become zero. Because it is one of the diminishing rate methods, depreciation is usually higher in the early years and lower in the later years of an asset's life. It is assumed that depreciation occurs as a fixed percentage (d), so that the depreciation in any year (R_k) is a constant proportion of the book value at the beginning of that year, which is practically equal to the book value at the end of the preceding year (B_{k-1}).

$$R_k = d \cdot B_{k-1}$$

The successive book values throughout an asset's useful life would be a geometric progression of a common ratio equal to $1 - d$ such that

$$B_k = C(1 - d)^k$$

and by the same logic, the scrap value would be

$$S = C(1 - d)^n$$

The accumulated depreciation (D_k) would normally be equal to the difference between the original cost (C) and the book value at any point:

$$D_k = C - B_k$$
$$D_k = C - C(1 - d)^k$$

$$\boxed{D_k = C[1 - (1 - d)^k]}$$

TABLE E2.2.1

Year	Depreciation, d (%)	Annual Depreciation ($)	Accumulated Depreciation ($)	Book Value ($)
0	0	0	0	12,600.00
1	35	4,410.00	4,410.00	8,190.00
2	35	2,866.50	7,276.50	5,323.50
3	35	1,863.23	9,139.73	3,460.27
4	35	1,211.09	10,350.82	2,249.18
5	35	787.21	11,138.03	1,461.97
6	35	511.69	11,649.72	950.28
7	35	332.60	11,982.32	617.68
8	35	216.19	12,198.51	401.49
9	35	140.52	12,339.03	260.97
10	35	91.34	12,430.37	169.63

Example 2.2.1 Construct the first 10 years of the depreciation schedule for a textile machine purchased for $12,600, if it depreciates 35% a year (see Table E2.2.1). Verify the following by using formulas:

(a) The annual depreciation in the 4th year
(b) The accumulated depreciation in the 7th year
(c) The book value in the 9th year
(d) When would the factory be able to sell this machine to the scrap yard for $72?

(a) The annual depreciation in the 4th year is $1,211.09:

$$R_k = d \cdot B_{k-1}$$
$$R_4 = d \cdot B_{4-1}$$
$$= d \cdot B_3$$
$$= .35(\$3,460.27)$$
$$= \$1,211.09$$

(b) The accumulated depreciation in the 7th year is $11,982.32:

$$D_k = C[1 - (1 - d)^k]$$
$$= \$12,600[1 - (1 - d)^k]$$
$$= \$12,600[1 - (1 - .35)^7]$$
$$= \$11,982.32$$

(c) The book value in the 9th year is $260.97:

$$B_k = C(1 - d)^k$$
$$B_9 = \$12,600(1 - .35)^9$$
$$= \$260.97$$

(d) If the scrap value is $72, we can find n:

$$S = C(1 - d)^n$$
$$72 = \$12,600(1 - .35)^n$$
$$= \$12,600(.65)^n$$
$$\frac{72}{\$12,600} = (.65)^n$$
$$\log \frac{72}{\$12,600} = n \log(.65)$$
$$\frac{\log(72/\$12,600)}{\log .65} = n$$
$$12 = n$$

So at the end of the 12th year of the machine's life it would be sold as scrap for $72.00.

Example 2.2.2   What is the fixed percentage of depreciation (d) for a piece of equipment purchased at $25,000 which has a scrap value of $3,500 after 10 years?

Since we have C, S, and n, we can use

$$S = C(1 - d)^n$$
$$\$3,500 = \$25,000(1 - d)^{10}$$
$$\frac{\$3,500}{\$25,000} = (1 - d)^{10}$$
$$.14 = (1 - d)^{10}$$
$$(.14)^{1/10} = 1 - d$$
$$d = 1 - .82$$
$$= .18 \text{ or } 18\%$$

2.3. THE SUM-OF-DIGITS METHOD

The calculation technique in the **sum-of-digits method** is similar to the technique of the rule of 78, which was explained earlier. It uses a fraction in the denominator which is the sum of certain digits. Just like the rule of 78, which allows charging higher interests in early periods and lower interest in later periods, this method would use a diminishing rate of depreciation where a higher portion of depreciation is written off earlier than later. The rate of depreciation for each year is a fraction of the denominator (dd) which is the sum of the digits representing the years of an asset's useful life (n). It is determined by

$$dd = \frac{n(n+1)}{2}$$

while the numerators of the fractions for all years are the number of years in reverse order, such that the years from 1 to n are lined up as from n to 1:

1	2	3	$- - - - \rightarrow$	$n-1$	$n-2$	n
n	$n-1$	$n-2$	$\leftarrow - - - -$	3	2	1

So the depreciation rates for the first year (R_1) and the second year (R_2), down to the last year of the asset's life (n), are calculated by

$$R_1 = \frac{n}{dd} \cdot D_n \qquad \text{where} \quad D_n = C - S$$

$$R_2 = \frac{n-1}{dd} \cdot D_n$$

$$R_3 = \frac{n-2}{dd} \cdot D_n$$

$$\vdots$$

$$R_{n-2} = \frac{3}{dd} \cdot D_n$$

$$R_{n-2} = \frac{2}{dd} \cdot D_n$$

$$R_n = \frac{1}{dd} \cdot D_n$$

We can generally consider any year (k) in the useful life of an asset and calculate its depreciation share by:

$$R_k = \left[\frac{n-k+1}{dd}\right](C - S)$$

and for the full depreciation for all years (D_n), we sum up the R_k's as

$$D_n = \sum_k R_k = \frac{dd}{dd}(C - S)$$

$$D_n = \sum_k R_k = C - S$$

which is the difference between the original cost (C) and the scrap value (S).

Example 2.3.1 An equipment system costs $23,000 and has an estimated useful life of 8 years (see Figure E2.3.1 and Table E2.3.1). Its scrap value is estimated at $4,500. Construct its depreciation schedule using the sum-of-digits method.

$$D_n = C - S$$
$$= \$23,000 - \$4,500$$
$$= \$18,500$$

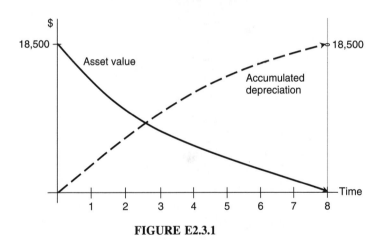

FIGURE E2.3.1

TABLE E2.3.1

Year	Depreciation Fraction	Annual Depreciation ($)	Accumulated Depreciation ($)	Book Value ($)
0	0	0	0	18,500.00
1	2/9	4,111.11	4,111.11	14,388.89
2	7/36	3,597.22	7,708.33	10,791.67
3	1/6	3,083.33	10,791.66	7,708.34
4	5/36	2,569.44	13,161.10	5,138.90
5	1/9	2,055.55	15,416.65	3,083.35
6	1/12	1,541.66	16,958.31	1,541.69
7	1/18	1,027.77	17,986.08	513.92
8	1/36	513.92	18,500.00	0

$$dd = \frac{n(n+1)}{2}$$

$$= \frac{8(8+1)}{2}$$

$$= 36$$

$$R_1 = \frac{n}{dd} \cdot D_n$$

$$= \frac{8}{36}(\$18{,}500)$$

$$= \$411.11 \tag{1}$$

$$R_2 = \frac{n-1}{dd} \cdot D_n$$

$$= \frac{8-1}{36}(\$18{,}500)$$

$$= \$3{,}597.22 \tag{2}$$

$$R_3 = \frac{n-2}{dd} \cdot D_n$$

$$= \frac{8-2}{36}(\$18{,}500)$$

$$= \$3{,}083.22 \tag{3}$$

$$R_4 = \frac{n-3}{dd} \cdot D_n$$

$$= \frac{8-3}{36}(\$18{,}500)$$

$$= \$2{,}569.44 \tag{4}$$

$$R_5 = \frac{n-4}{dd} \cdot D_n$$

$$= \frac{8-4}{36}(\$18{,}500)$$

$$= \$2{,}055.55 \tag{5}$$

$$R_6 = \frac{n-5}{dd} \cdot D_n$$

$$= \frac{8-5}{36}(\$18{,}500)$$

$$= \$1{,}541.66 \tag{6}$$

$$R_7 = \frac{n-6}{dd} \cdot D_n$$

$$= \frac{8-6}{36}(\$18,500)$$

$$= \$1,027.77 \tag{7}$$

$$R_8 = \frac{n-7}{dd} \cdot D_n$$

$$= \frac{8-7}{36}(\$18,500)$$

$$= \$513.88 \tag{8}$$

2.4. THE AMORTIZATION METHOD

The **amortization method** considers the time value of money by amortizing the depreciation charges so that each of the annual charges includes not only a certain share of an asset's cost but also interest on the book value for each depreciating year. The full depreciation during the useful life of an asset would be treated as if it is a present value of an ordinary annuity whose annual payments represents the annual depreciation charges. In this case, the present value of the full depreciation (D_n) is the difference between the original cost and the discounted scrap value, $C - S(1+r)^{-n}$, where n is the useful life of the asset. Let's recall the formulas for the payments of an ordinary annuity when the current value of a fund is given:

$$A = \frac{CV \cdot r}{1 - (1+r)^{-n}}$$

and the equivalent table formula was

$$A = CV \cdot \frac{1}{a_{\overline{n}|r}}$$

Now we can replace the annuity periodic payment (A) by the periodic depreciation allowance (D_k). Also, we can replace the current value of an annuity fund (CV) by the full depreciation (D_n). Given that $D_n = C - S(1+r)^{-n}$, we can obtain

$$D_k = \frac{[C - S(1+r)^{-n}]r}{1 - (1+r)^{-n}}$$

and for the table formula, we can obtain

$$D_k = C - S(1+r)^{-n} \cdot \frac{1}{a_{\overline{n}|r}}$$

Example 2.4.1   A machine that costs \$27,000 has a useful life of 10 years and a final scrap value of \$2,000 (see Table E2.4.1). Calculate the annual depreciation charges using the amortization method, and construct the full depreciation schedule if the interest rate for investment is $8\frac{1}{2}\%$.

$C = \$27,000$; $S = \$2,000$; $n = 10$ years; $r = .085$.

$$D_k = \frac{[(C - S)(1 + r)^{-n}]r}{1 - (1 + r)^{-n}}$$

$$= \frac{[\$27,000 - \$2,000(1 + .085)^{-10}].085}{1 - (1 + .085)^{-10}}$$

$$= \$3,980.20$$

By the table method:

$$D_k = (C - S)(1 + r)^{-n} \cdot \frac{1}{a_{\overline{n}|r}}$$

$$= \$27,000 - \$2,000(1 + .085)^{-10} \cdot \frac{1}{a_{\overline{10}|.085}}$$

$$= \$26,115.43(.15240771)$$

$$= \$3,980.20$$

It is important to remember that the annual depreciation figures in column (2) include the interest on the actual depreciation charges as they appear in column (4) as the depreciation principal or net depreciation allowances.

TABLE E2.4.1

(1)	(2)	(3)	(4)	(5)	(6)
	Annual	Interest on	Depreciation	Accumulated	Book
Year,	Depreciation,	Depreciation	Principal	Depreciation	Value
k	D_k	$[(6) \times r]$	$[(2) - (3)]$	[from (4)]	$[C - (5)]$
0	—	—	—	—	27,000.00
1	3,980.20	2,295.00	1,685.20	1,685.20	25,314.80
2	3,980.20	2,151.76	1,828.44	3,513.64	23,486.36
3	3,980.20	1,996.34	1,983.86	5,497.50	21,502.50
4	3,980.20	1,827.71	2,152.49	7,649.99	19,350.01
5	3,980.20	1,644.75	2,335.45	9,985.44	17,014.56
6	3,980.20	1,446.24	2,533.96	12,519.40	14,480.60
7	3,980.20	1,230.85	2,749.35	15,268.75	11,731.25
8	3,980.20	997.16	2,983.40	18,251.79	8,748.21
9	3,980.20	743.60	3,236.60	21,488.39	5,511.61
10	3,980.20	468.49	3,511.71	25,000.00	2,000.00
	39,802.00	14,802.00	25,000.00		

2.5. THE SINKING FUND METHOD

Under the **sinking fund method,** depreciation is treated as a sinking fund to accumulate enough funds so that to replace an asset at the end of its useful life. In this case, depreciation charges act as deposits to the sinking fund with their earned interest. The depreciation charges would be calculated in the same way as calculating the payments of an ordinary annuity when the future value is known. In this case, the future value would be the full amount of depreciation $(C - S)$.

Let's recall the formula for the payment of the ordinary annuity when the future value is known:

$$A = \frac{FV \cdot r}{(1 + r)^n - 1}$$

and if the table value is used, the formula would be

$$A = \frac{FV}{S_{\overline{n}|r}}$$

Now, let's replace the annuity payment (A) by the annual depreciation (R_k), and the future value of annuity (FV) by the depreciation throughout the asset's life $(C - S)$:

$$R_k = \frac{(C - S)r}{(1 + r)^n - 1}$$

and if the table value is used, the formula would be

$$R_k = \frac{C - S}{S_{\overline{n}|r}}$$

Now the accumulated depreciation (D_k) at any point during the useful life of the asset or at the end of k years would be equal to the accumulated value of the sinking fund for the same period of time. Therefore, D_k would be

$$D_k = R_k \left[\frac{(1 + r)^k - 1}{r} \right]$$

and if the table value is used, the formula would be

$$D_k = R_k \cdot S_{\overline{k}|r}$$

TABLE E2.5.1[a]

(1)	(2)	(3)	(4)	(5)	(6)
	Sinking Funds		Annual	Accumulated	Book
Year,	Deposits,	Interest	Depreciation	Depreciation	Value
k	R_k	$[(5) \times 0.7]$	$[(2) + (3)]$	[from (4)]	$(C - S)$
0	—	—	—	—	20,000.00
1	1,559.48	—	1,559.48	1,559.48	18,440.52
2	1,559.48	109.16	1,668.64	3,228.12	16,771.88
3	1,559.48	225.97	1,785.45	5,013.57	14,986.43
4	1,559.48	350.95	1,910.43	6,923.99	13,076.00
5	1,559.48	484.68	2,044.16	8,968.15	11,031.85
6	1,559.48	627.77	2,187.25	11,155.40	8,844.60
7	1,559.48	780.88	2,340.36	13,495.76	6,504.24
8	1,559.48	944.70	2,504.18	16,000.00	4,000.00
	12,475.84	3,524.16	16,000.00		

[a] Some final entries are rounded off.

Substituting for R_k, we get:

$$D_k = \frac{C - S}{S_{\overline{n}|r}} \cdot S_{\overline{k}|r}$$

Notice that if R_k is not already calculated, and if the second formula above is used, there would be two table values, $S_{\overline{n}|r}$ and $S_{\overline{k}|r}$.

Example 2.5.1 Use the sinking fund method to calculate the depreciation charges and construct the entire depreciation table for a structure costing $20,000 whose value decreases to $4,000 after 8 years given that the investment rate is 7% (see Table E2.5.1).

$$R_k = \frac{(C - S)r}{(1 + r)^n - 1}$$

$$= \frac{(\$20,000 - \$4,000).07}{(1 + .07)^8 - 1}$$

$$= \$1,559.48$$

or

$$R_k = \frac{C - S}{S_{\overline{n}|r}}$$

$$= \frac{\$20,000 - \$4,000}{S_{\overline{8}|0.7}} = \frac{\$16,000}{10.25980257}$$

$$= \$1,559.48$$

2.6. COMPOSITE RATE AND COMPOSITE LIFE

In reality, firms deal with many assets of many types. A useful way to deal with the depreciation of assets is to calculate the collective depreciation where assets are grouped by similar types and close categories. The composite rate method can be helpful in computing the depreciation charges of a group of certain assets. The **composite rate** is obtained by dividing the total annual depreciation charges of a group of assets by the combined original costs of those assets. Given that the individual depreciation charges are calculated by the straight-line method:

$$R_{\text{comp}} = \frac{\sum_{k=1}^{m} R_k}{\sum_{k=1}^{m} C_k}$$

where $k = 1, 2, 3, \ldots, m$, where m is the number of assets in the group.

Example 2.6.1   Table E2.6.1 shows the depreciation information for five pieces of equipment in the same category. Find the composite rate of depreciation to the group.

$$R_{\text{comp}} = \frac{\sum_{k=1}^{m} R_k}{\sum_{k=1}^{m} C_k} \qquad m = 5$$

$$= \frac{\$20,700}{\$176,000}$$

$$= 11.79\%$$

The composite rate method would make it much easier to apply the rate next year on the same group of assets if there are no significant changes in the conditions and circumstances that may alter the value of the assets. Let's suppose, for example, that the original value of this group of assets becomes \$176,500 due to a minor additional cost incurred during the year. Still, a rate of 11.79% can be valid to apply, and the depreciation charges for the group would be

$$\$176,500 \times .1179 = \$20,809$$

which is very close to the current depreciation charges of \$20,750.

Another composite measure of depreciation is the **composite life**, which refers to the average useful life of a group of assets. However, it is not calculated as a simple average of years. Its calculation depends on the method used to obtain the annual depreciation charges. If the depreciation charges are equal due to use of the straight-line method, the composite life (L_{comp}) would be obtained by

TABLE E2.6.1

(1) Equipment No.	(2) Original Cost, C_k ($)	(3) Useful Life (years)	(4) Scrap Value ($)	(5) Annual Depreciation, R_k ($)	(6) Total Depreciation, W_k ($)
1	20,000	5	2,000	3,600	18,000
2	23,000	6	2,000	3,500	21,000
3	37,000	8	3,000	4,250	34,000
4	41,000	8	5,000	4,500	36,000
5	55,000	10	6,000	4,900	49,000
	176,000	37	18,000	20,750	158,000

dividing the total depreciation charges or the wearing values ($\sum W_i$) by the total annual depreciation charges ($\sum R_i$).

$$L_{\text{comp}} = \frac{\sum_{k=1}^{m} W_k}{\sum_{k=1}^{m} R_k}$$

But if the annual depreciation charges are variable due to using the sinking fund method, the composite life would be equal to the time needed for the total annual deposits of the sinking fund to mature to what is equal to the total depreciation charges of the entire group of those assets. The matter then becomes finding (n) of an annuity formula, where the combined sinking fund deposits for the group of assets ($\sum R_i$) be the payment (A), and the total depreciation charges ($\sum W_i$) would be the future value (FV): So the term formula

$$n = \frac{\ln\left[\dfrac{\text{FV} \cdot r}{A} + 1\right]}{\ln(1 + r)}$$

would be

$$n = \frac{\ln\left[\dfrac{(\sum W_k) r}{\sum R_k} + 1\right]}{\ln(1 + r)}$$

Example 2.6.2   Calculate the composite life in Example 2.6.1. The total of the last column (6), the total depreciation for all five pieces of equipment is the combined wearing value $\sum W_i$, and the total of column (5) is the combined annual depreciation for all equipment.

$$L_{\text{comp}} = \frac{\sum_{k=1}^{5} W_k}{\sum_{k=1}^{5} R_k}$$

$$= \frac{\$158,000}{\$20,750}$$

$$= 7.6 \text{ years}$$

Example 2.6.3 Use the sinking fund method on the equipment in Example 2.6.2 to calculate the composite life if the investment rate is 6%.

First we have to calculate the five depreciation deposits (R_i). Let's use the table method:

$$R_k = \frac{C - S}{S_{\overline{n}|r}}$$

$$R_1 = \frac{\$20,000 - \$2,000}{S_{\overline{5}|.06}} = \frac{\$18,000}{5.63709296} = \$3,193.14$$

$$R_2 = \frac{\$23,000 - \$2,000}{S_{\overline{6}|.06}} = \frac{\$21,000}{9.97531854} = \$3,010.62$$

$$R_3 = \frac{\$37,000 - \$3,000}{S_{\overline{8}|.06}} = \frac{\$34,000}{9.89746791} = \$3,435.22$$

$$R_4 = \frac{\$41,000 - \$5,000}{S_{\overline{8}|.06}} = \frac{\$36,000}{9.89746791} = \$3,637.29$$

$$R_5 = \frac{\$55,000 - \$6,000}{S_{\overline{10}|.06}} = \frac{\$49,000}{13.18079494} = \$3,717.53$$

$$\sum_{k=1}^{5} R_k = \$16,993.80$$

Since $\sum_{k=1}^{5} W_k$ from Table E2.6.1 was $158,000, we can apply the n formula:

$$L_{\text{comp}} = n = \frac{\ln\{[(\sum W_k) \cdot r / \sum R_k] + 1\}}{\ln(1 + r)}$$

$$n = \frac{\ln\{[(\$158,000)(.06)/\$16,993.80] + 1\}}{\ln(1 + .06)}$$

$$= 7.6 \text{ years}$$

2.7. DEPLETION

Natural resources as financial assets depreciate through the gradual and systematic use of their reserve capacities. Typical examples are oil and gas, minerals, and timber. As we have seen, depreciation comprises the wearing out of productive assets due to the making of products during the assets' useful life; removal and using up a natural resource is a similar concept, called **depletion**.

Also, as the depreciation charges are made to replace productive assets after their useful life is over, the net annual income from any natural resource subject to depletion must be discounted by an allowance of annual depletion. Calculation of the depletion is similar to calculation of the depreciation, which is expressed as production units. So the depletion rate per unit is calculated and multiplied by the number of units of production during a year to obtain the annual depletion.

$$DP = \frac{C - S}{P}$$

where DP is the depletion rate per unit of product and P is the total production.

Example 2.7.1 A gravel pit was purchased for $40,000, and its value after excavation was estimated at $4,000 (see Table E2.7.1). It had the capacity to yield at least 90,000 truckloads of gravel in 5 years distributed as 15,000, 21,000, 20,000, 18,000, and 16,000 truckloads, after which it was exhausted. Construct the depletion schedule.

$$DP = \frac{C - S}{P}$$

$$= \frac{\$40,000 - \$4,000}{90,000}$$

$$= .40 \quad \text{depletion per truckload}$$

Depletion can also be recovered using the sinking fund technique. If someone investing in a depleting resource wants to recover the resource value after it has been depleted, he can set aside a portion of his annual income from the resource and deposit it in a sinking fund to accumulate, at the time of depletion, to an amount that can recover the original value of the depleted resource. In this case,

TABLE E2.7.1

(1) Year	(2) Annual Production, AP	(3) Annual Depletion [(2) × DP]	(4) Accumulated Depletion [from (3)]	(5) Book Value [C − (4)]
1	—	—	—	40,000
2	15,000	6,000	6,000	34,000
3	21,000	8,400	14,400	25,600
4	20,000	8,000	22,400	17,600
5	18,000	7,200	29,600	10,500
	16,000	6,400	36,000	4,000

his net annual income (NI) would be equal to his income before depletion (I) minus what he deposited into the sinking fund (R_k).

$$NI = I - R_k$$

We can also rearrange this to

$$I = NI + R_k$$

$$I = NI + \frac{C - S}{S_{\overline{n}|i}}$$

Since NI can be obtained by multiplying the original cost (C) by the yield rate (r), we can plug that in:

$$\boxed{I = C(r) + \frac{C - S}{S_{\overline{n}|i}}}$$

Notice that r is the yield rate and i is the investment rate for the sinking fund.

Were the resource to be depleted completely, to the point where there would be no scrap value, the formula becomes

$$I = C(r) + \frac{C}{S_{\overline{n}|i}}$$

$$\boxed{I = C\left(r + \frac{1}{S_{\overline{n}|i}}\right)}$$

Example 2.7.2   A depleting resource has an initial cost of $450,000 and a residual value of $22,000. It can produce up to 324,000 tons of raw material, 40,000 tons of which can be produced in the first year. Assume that the production of raw material continues at the same rate as that of the first year and that a sinking fund to recover the resource yields $7\frac{1}{2}\%$ interest.

(a) Find the depletion charge for the first year.
(b) Calculate the annual income required to give a yield rate of 14%.

(a) $$DP = \frac{\$450,000 - \$22,000}{324,000 \text{ tons}} = \$1.32 \qquad \text{depletion rate per ton}$$

$$D_{1st \ year} = 40,000 \text{ tons} \times \$1.32 = \$52,800$$

$$\text{useful life} = \frac{324,000 \text{ tons}}{40,000 \text{ tons}} = 8 \text{ years}$$

(b)
$$I = C(r) + \frac{C - S}{S_{\overline{n}|i}}$$

$$= \$450,000(.14) + \frac{\$450,000 - \$22,000}{S_{\overline{8}|.075}}$$

$$= 63,000 + \frac{\$428,000}{10.44637101}$$

$$= \$103,971$$

Unit IV Summary

Investors have to have a solid basis on which to make investment decisions, and investing firms usually get more offers of new projects than they have money to commit to every project that sounds promising. The matter then becomes what criteria investors can rely on to help them decide how to budget their capital and assure a reasonable degree of profitability. In this unit we examined three major techniques used to assess the potential worthiness of an investment. First was net present value, which compares the discounted prospective cash inflows with the investment committed. The second analytical technique was the internal rate of return, which equates the initial investment with the present value of the returns expected. The third technique was the profitability index, which was basically a ratio of the present value of the cash inflows to the cash outflows.

Next we discussed capitalization and capitalized cost for their relevance to the need to maintain capital investments throughout their useful life and replace them as soon as possible after that useful life is over. This idea introduced the concepts of depreciation and depletion and various methods of assessing and calculating depreciation charges. The discussion went into detail over the straight-line method and its varieties of expressive depreciation by units of production or hours of operations. We also reviewed the fixed proportion method, the sum-of-digits method, the amortized method, and the sinking fund method. Then we discussed composite rate and composite life to deal with the depreciation of many assets at the same time. A related subject was the special aspect of asset depreciation in the depletion that is specific to resources that can be used up completely through gradual removal of their elements. The concluding subjects were capital budgeting methods that do not use the time value of money: the average rate of return method and the payback time method.

Mathematical Finance, First Edition. M. J. Alhabeeb.
© 2012 John Wiley & Sons, Inc. Published 2012 by John Wiley & Sons, Inc.

List of Formulas

Net present value

$$NPV = \sum_{n=1}^{N} FV(1+r)^{-n} - I_0$$

Internal rate of return

$$NPV = \sum_{n=1}^{N} \frac{FV}{(1+IRR)^n} - I_0 = 0$$

$$r = \frac{\sum_{k=1}^{n} FV_k - I_0}{\sum_{k=1}^{n} k \cdot FV_k} \qquad \text{for rough estimate of IRR}$$

Profitability index

$$PI = \frac{PV_{ci}}{I_0}$$

Capitalized cost

$$K = C + \frac{C-S}{(n+r)^n - 1} + \frac{M}{r}$$

$$\frac{K_b}{P_b} = \frac{K_a}{P_a}$$

Depreciation

Straight-line method:

$$D_k = kR$$

$$R = \frac{C-S}{n}$$

Mathematical Finance, First Edition. M. J. Alhabeeb.
© 2012 John Wiley & Sons, Inc. Published 2012 by John Wiley & Sons, Inc.

$$B_k = C - D_k$$

$$B_k = C - kR$$

$$B_k = C - \frac{C - S}{n}$$

Fixed-proportion method:

$$R_k = d \cdot B_{k-1}$$

$$B_k = C(1 - d)^k$$

$$S = C(1 - d)^n$$

$$D_k = C - B_k$$

$$D_k = C[1 - (1 - d)^k]$$

Sum-of-digits method:

$$dd = \frac{n(n + 1)}{2}$$

$$R_1 = \frac{n}{dd} \cdot D_n$$

$$R_2 = \frac{n - 1}{dd} \cdot D_n$$

$$R_3 = \frac{n - 2}{dd} \cdot D_n$$

$$\vdots$$

$$R_{n-2} = \frac{3}{dd} \cdot D_n$$

$$R_{n-1} = \frac{2}{dd} \cdot D_n$$

$$R_n = \frac{1}{dd} \cdot D_n$$

$$R_k = \frac{n - k + 1}{dd} \cdot D_n$$

$$D_n = \sum R_k = C - S$$

Amortization method:

$$D_k = \frac{[(C - S)(1 + r)^{-n}]r}{1 - (1 + r)^{-n}}$$

$$D_k = (C - S)(1 + r)^{-n} \cdot \frac{1}{a_{\overline{n}|r}}$$

Sinking fund method:

$$R_k = \frac{(C - S)r}{(1 + r)^n - 1}$$

$$R_k = \frac{C - S}{S_{\overline{n}|r}}$$

$$D_k = R_k \left[\frac{(1 + r)^k - 1}{r} \right]$$

$$D_k = R_k \cdot S_{\overline{k}|r}$$

$$D_k = \frac{C - S}{S_{\overline{n}|r}} \cdot S_{\overline{k}|r}$$

Composite rate

$$R_{\text{comp}} = \frac{\sum_{k=1}^{m} R_k}{\sum_{k=1}^{m} C_k}$$

Composite life

$$L_{\text{comp}} = \frac{\sum_{k=1}^{m} W_k}{\sum_{k=1}^{m} R_k}$$

$$n = \frac{\ln \left[\dfrac{(\sum W_k)r}{\sum R_k} + 1 \right]}{\ln(1 + r)}$$

Depletion:

$$DP = \frac{C - S}{P}$$

$$I = C(r) + \frac{C - S}{S_{\overline{n}|i}}$$

$$I = C \left(r + \frac{1}{S_{\overline{n}|i}} \right)$$

Other capital budgeting methods

Average rate of return:

$$ARR = \frac{2 \cdot APAT}{C}$$

Payback time:

$$Payback = \frac{C}{YCI}$$

Exercises for Unit IV

1. A project proposal stated that it would provide at least $20,000 in annual returns for the next 3 years but requires an initial investment of $50,000. Will its approval be northwhile if the cost of capital is 8%?

2. Find the net present value for a project that would yield $25,000, $35,000, $30,000, and $40,000 in the first 4 years of operation if the interest rate is $9\frac{1}{2}\%$ and the initial investment is $80,000.

3. A capital budgeting committee is to choose which of the following two projects is worth investing $150,000 in when the cost of capital is $6\frac{1}{2}\%$.

Year:	1	2	3	4	5
Revenue of project I ($)	30,000	55,000	62,000	69,000	73,000
Revenue of project II ($)	80,000	71,000	65,000	53,000	32,000

4. Find the internal rate of return for a project requiring an investment of $250,000, where the project promises to provide $85,000 in the first year, $70,000 in the second year, and $100,000 in the third.

5. An investor plans to earn $30,000 next year and $15,000 the year after from two investments, $20,000 now and $20,000 next year. What will be the internal rate of return?

6. Calculate the profitability index (PI) for a project whose present cash flow value is $500,000 with an initial investment of $430,000.

7. Will the headquarters of a franchise reject or accept a project for opening a new branch if the present value of its cash flow is $320,000 and the capital requested is $405,000?

8. A startup landscaping company has to determine the most prudent investment in new trucks and mowers. They narrow their choice to two sets. Which set will be better given that money is worth $7\frac{1}{4}\%$?

Mathematical Finance, First Edition. M. J. Alhabeeb.
© 2012 John Wiley & Sons, Inc. Published 2012 by John Wiley & Sons, Inc.

Capital Required	Set I	Set II
Initial cost ($)	68,000	65,000
Life span (years)	12	11
Maintenance ($)	5,000	4,500
Residual value ($)	7,000	6,000

9. A machine has a value of $65,000 and is estimated to operate up to 6 years, ending with a scrap value of $7,000. Construct a depreciation table using the straight-line method.

10. A mechanical installation in an office will cost $23,000. Its useful life is be 4 years, after which it has to be removed and thrown away, but the removal cost will be $1,200. Prepare a depreciation schedule using a straight-line method.

11. A machine that costs $65,000 will depreciate to $5,000 in 10 years. Calculate its book value at the end of the sixth year and the depreciation expenses for the seventh year using the fixed-proportion method.

12. How long will it take for equipment valued at $35,000 to depreciate to less than half of its original value if it normally depreciates to $5,000 in 14 years? Use the fixed-proportion method.

13. An asset has a value of $44,000 and a scrap value of $6,500 after 9 years of useful life. Set up a depreciation table using the sum-of-digits method.

14. A piece of equipment costs $72,000. Its useful life is estimated as 15 years, after which it would be declared as scrap with a value of $6,500. Construct a depreciation table and calculate the book value at the end of the 10th year using the sum-of-digits method.

15. Using the amortization method, construct a depreciation table for a machine whose value is $22,000 with a useful life of 7 years and a scrap value of $3,450. The interest rate is 9%.

16. An asset has a value of $1,500 and a useful life of 5 years. Its trade-in value is $300. Use the amortized method to construct a depreciation schedule assuming that the interest rate is 5%.

17. Using the sinking fund method and assuming that the rate of interest is 8%, prepare a depreciation schedule for an asset worth $42,000, dropping to $4,300 in 5 years.

18. Find the composite rate of depreciation for a local firm's equipment, which cost $500,000 with an annual depreciation estimated at $85,000.

19. Calculate the composite rate and composite life through use of the straight-line method of depreciation for the following group of assets:

Asset	Original Cost ($)	Scrap Value ($)	Useful Life (years)
1	2,600	200	4
2	3,800	450	12
3	6,400	380	11
4	5,750	590	8
5	8,100	1,200	10

20. Use the sinking fund method of depreciation to calculate the composite life for the group of assets in Exercise 19 assuming an interest rate of $6\frac{1}{2}\%$.

21. A local coal mine is purchased for $380,000 on the basis of its estimated reserve of 200,000 tons of coal. In the first year of operation, the mine produced one-fourth of its reserve. Calculate the total depletion and first-year depletion deduction given that the land can be salvaged for $10,000.

22. Construct a depletion schedule for an oil field that is purchased for $780,000 and promises to produce 25,000, 40,000, 55,000, 45,000, and 35,000 barrels of oil in 5 years, after which the salvage value of the field is estimated at $12,000.

23. Two investment proposals are submitted for approval of $25,000 each in initial investment. Their estimated profits after taxes are:

Year	Proposal 1 ($)	Proposal 2 ($)
1	2,500	6,000
2	3,300	4,900
3	4,000	4,200
4	5,200	3,150
5	6,100	2,000

Make a capital budgeting decision using the average rate of return method.

24. Suppose that another couple of proposals is submitted for capital allocation. They have the following cash inflows:

Year	Proposal 1 ($)	Proposal 2 ($)
1	10,000	5,500
2	10,500	5,500
3	4,200	5,500
4	1,900	5,500
5	1,580	5,500

Make a capital budgeting decision based on the payback period criteria if both projects require a $25,000 initial investment.

UNIT V

Mathematics of the Break-Even Point and Leverage

1 Break-even Analysis

Break-even analysis is an important technical tool for business performance and profit planning that utilizes restructuring the fundamental relationships between costs and revenues. It is also called **cost-volume profit analysis**. It offers the business planner or manager a plausible approximation of the point when profits start to be collected. This point occurs on the realization of the stage in which the total cost of production has been recovered by revenues from product sales. **Break-even analysis** is therefore, a process to determine the amount of products that must be produced and sold before any profit can be earned. It can also be the determination of the amount of revenue that can be collected before earning any profit. The reliance on finding the revenue instead of the production size can be more practical and convenient if the firm produces or sells multiple products. For example, General Motors can easily determine how many Chevy Malibus should be sold before that plant begins to earn profits, but this method cannot easily be used for Wal-Mart because Wal-Mart stores sell tens of thousands of products. It is, therefore, more appropriate to determine how much sales revenue should be collected before a certain Wal-Mart store begins to earn profits.

Technically, the break-even analysis is to find the point that refers to both the break-even quantity of product, and the break-even revenue of sales. Called the **break-even point**, this is the point at which the profit would be zero and the total cost would equal the total revenue. Geometrically, it is the point of intersection between the total cost and total revenue curves. Once we locate this point, we would know that all production before that point incurs some loss and that any product produced and sold after that point would yield some profits.

1.1. DERIVING BEQ AND BER

If the total cost (C) includes both fixed cost (FC) and variable cost (VC), then

$$C = FC + VC$$

Since the variable cost changes with the size of production, let's consider (v) as the variable cost per unit of production, and if the production is Q, then

$$VC = vQ \quad \text{and} \quad C = FC + vQ \tag{1}$$

Mathematical Finance, First Edition. M. J. Alhabeeb.
© 2012 John Wiley & Sons, Inc. Published 2012 by John Wiley & Sons, Inc.

Also, the revenue (R) will depend on how many units of product are sold. If we consider p as the price per unit of product, then

$$R = pQ \tag{2}$$

Profit (Pr), sometimes called **operating profit** or **EBIT** (earnings before interests and taxes), is the difference between total revenue and total cost:

$$Pr = R - C \tag{3}$$

$$= pQ - (FC + vQ)$$

$$= pQ - FC - vQ$$

$$= Q(p - v) - FC$$

Since profit is zero at the break-even point,

$$0 = Q(p - v) - FC$$

$$Q(p - v) = FC$$

$$Q = \frac{FC}{p - v}$$

This Q is the quantity of products at the break-even point, and it is called the **break-even quantity** (BEQ):

$$\boxed{BEQ = \frac{FC}{p - v}}$$

where FC is the fixed operating cost, p is the price per unit of product, and v is the variable operating cost per unit of a product.

Similarly, we can find the revenue at the break-even point.

We start with equation (3):

$$Pr = R - C$$

$$= R - (FC + vQ)$$

$$= R - FC - vQ$$

We substitute for the Q value obtained in equation (2):

$$R = pQ$$

$$\frac{R}{p} = Q$$

$$Pr = R - FC - v \cdot \frac{R}{p}$$

At the break-even point, Pr = 0:

$$0 = R - \text{FC} - v \cdot \frac{R}{p}$$

$$\text{FC} = R\left(1 - \frac{v}{p}\right)$$

$$R = \frac{\text{FC}}{1 - (v/p)}$$

This revenue is the revenue at the break-even point, called the **break-even revenue** (BER):

$$\boxed{\text{BER} = \frac{\text{FC}}{1 - (v/p)}}$$

By knowing the break-even quantity or revenue, a firm can:

1. Determine and control its operations to sustain the proper level to cover all operating costs.
2. Assess and control its ability and timing to earn profits at different levels of production and volumes of sale.

Before we apply the break-even technique, we should be able to know the operational definitions of the variables involved, especially to distinguish between the fixed and variable costs, which would be the first task to perform before solving any BEQ or BER problems.

1.2. BEQ AND BER VARIABLES

Fixed Cost

Fixed costs are the costs that are not associated with the size of production or sale of products. They are a function of time, not sales or production, and therefore are incurred whether or not the firm produces anything or sells any product. Typical examples of fixed costs are rent, lease payments, insurance premiums, regular utilities, executive and clerical salaries, and debt services. All these are expenses that have to be paid regardless of how much the firm produces or sells. Geometrically, they are represented by a horizontal line.

Variable Cost

Variable costs are any costs associated with the size of production or sale. They fluctuate up and down in a positive relationship with the volume of production. Typical examples of variable costs are production material, production labor, and production utilities, such as the cost of electricity and gas consumed by the productive machines and equipment, as well as merchandise insurance, transportation and storage. The more a firm produces, the more the variable costs incur. In this sense, variable cost would be a product of the size of production (Q) and the variable cost per unit of a product (v).

Contribution Margin

The **contribution margin** is the amount of profit earned on each unit sold above and beyond the break-even quantity, and similarly, it would be the amount of loss the firm would incur on each unit produced below the break-even point. Technically, it is equal to the unit price of the product discounted for the unit variable cost. Mathematically, CM would be

$$\text{CM} = p - v$$

and that is, in fact, the denominator of the BEQ formula.

Example 1.2.1   Aroma is a coffee shop in the downtown area. It carries a fixed operating cost of $2,500 and a variable operating cost of 49 cents per cup. Its famous coffee sells for $1.79 a cup. How many cups of coffee does this business have to sell, and how much revenue does it have to collect before starting to get any profit?

$$\begin{aligned}
\text{BEQ} &= \frac{\text{FC}}{p - v} \\
&= \frac{2,500}{1.79 - .49} \\
&= 1,923 \text{ cups of coffee} \\
\text{BER} &= \frac{\text{FC}}{1 - v/p} \\
&= \frac{2,500}{1 - (.49/1.79)} \\
&= \$3,442
\end{aligned}$$

Also,

$$\begin{aligned}
\text{BER} &= \text{BEQ} \cdot p \\
&= 1,923 \times 1.79 \\
&= \$3,442
\end{aligned}$$

Example 1.2.2   Table E1.2.2 shows the cost data as they appear in the records of Modern Books, a company specialized in custom book binding and sells its service for $30 per book. Calculate the break-even quantity and the break-even revenue.

TABLE E1.2.2

Item	Frequency	Cost ($)
Rent	Monthly	2,800
Property taxes	Semiannually	1,665
Insurance	Quarterly	1,112
Administrative salaries	Monthly	6,580
Employee benefits	Annually	5,312
Wages	Per book	3.00
Paper	Per book	2.15
Cardboard	Per book	1.35
Glue, tape, thread	Per book	.55
Leather	Per book	1.95
Ink and paint	Per book	.95
Shipping and handling service	Per book	2.00

The first task is to separate fixed and variable costs based on our understanding of the concepts:

- *Fixed costs*: rent, property taxes, insurance administrative salaries, and employee benefits
- *Variable costs*: wages, paper, cardboard, glue, tape, and thread, leather, ink and paint, and shipping and handling

The second task is to unify the frequencies of the cost items. It would be a standard to convert every fixed-cost item into an annual and have every variable cost item expressed per unit.

Fixed costs:

$$
\begin{aligned}
\text{Rent} &= \$2,800 \times 12 = \$33,600 \\
\text{Property taxes} &= \$1,665 \times 2 \ = 3,300 \\
\text{Insurance} &= \$1,112 \times 4 \ = 4,448 \\
\text{Administrative salaries} &= \$6,580 \times 12 = 78,960 \\
\text{Employee benefits} &= \$5,312 \times 1 \ = 5,312 \\
\hline
\text{Total} &= \qquad\quad = \$125,620
\end{aligned}
$$

Variable costs:

$$
\begin{aligned}
\text{Wages} &= \$3.00 \\
\text{Paper} &= 2.15 \\
\text{Cardboard} &= 1.35 \\
\text{Glue, tape, thread} &= .55 \\
\text{Leather} &= 1.95 \\
\text{Ink and paint} &= 4.95 \\
\text{Shipping and handling} &= 2.00 \\
\hline
\text{Total} &= \$11.95
\end{aligned}
$$

$$BEQ = \frac{FC}{p - v}$$
$$= \frac{125,620}{30 - 11.95}$$
$$= 6,959 \text{ books}$$

$$BER = \frac{FC}{1 - v/p}$$
$$= \frac{125,620}{1 - (11.95/30)}$$
$$= \$208,786$$

Also,

$$BER = BEQ \cdot p$$
$$= 6,959.56(30.00)$$
$$= 208.786$$

1.3. CASH BREAK-EVEN TECHNIQUE

Sometimes there are some noncash charges that a company has to deal with as a significant part of its operating fixed cost. Often, these charges are the depreciation charges that have to be deducted from the operating fixed cost to prevent overestimation of the break-even point. In such a case, the formula to calculate the break-even point would be adjusted to the **cash break-even quantity** (CBEQ), which is equal to

$$\boxed{CBEQ = \frac{FC - NC}{p - v}}$$

where NC represents any noncash charges constituting a sizable portion of the fixed cost.

Example 1.3.1   The records of the Riverbent Company indicate a $5,700 fixed cost and depreciation charges of $1,767, which is a little more than one-third of the fixed cost (see Figure E1.3.1). Suppose that the variable cost per unit is $2.30 and the product sells for $7.00. What is the company's cash break-even quantity, and how does it compare to the regular break-even quantity?

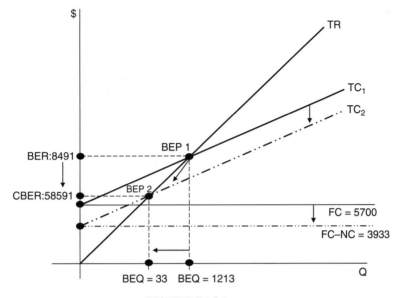

FIGURE E1.3.1

$$CBEQ = \frac{FC - NC}{p - v}$$

$$= \frac{5,700 - 1,767}{7.00 - 2.30}$$

$$= \$837$$

$$CBER = CBEQ(p) = 837(7.00) = \$5,859$$

If we calculate the regular BEQ, we would not discount the depreciation from the fixed cost. In this case, the BEQ will be

$$BEQ = \frac{FC}{p - v}$$

$$= \frac{5,700}{7.00 - 2.30}$$

$$= 1,213 \text{ unit}$$

$$BER = BEQ(p) = 1,213(7.00) = \$8,491$$

which is a case of overstating the break-even point. Excluding the noncash charges reduces the fixed cost and total cost, resulting in lowering the break-even point from BEP1 to BEP2 in the graph above. This would lead to a lower cash break-even quantity (CBEQ) and a lower cash break-even revenue (CBER).

1.4. THE BREAK-EVEN POINT AND THE TARGET PROFIT

If a business owner has a specific objective to achieve, such as a certain profit specified in advance, that specific profit figure would be called a **target profit**. It can be preset by making it part of the fixed cost, so that based on the size of the break-even quantity and revenue it can be determined. The break-even quantity and revenue formulas would then have to be adjusted by adding the target profit (TP) to the fixed cost in the numerator, and the new BEQ_{tp} and BER_{tp} would be

$$\boxed{BEQ_{tp} = \frac{FC + TP}{p - v}}$$

and

$$\boxed{BER_{tp} = \frac{FC + TP}{1 - (v/p)}}$$

Example 1.4.1   Let's assume that the Riverbent Company in Example 1.3.1 decides to collect at least $6,000 profit as a first step (see Figure E1.4.1). What will be the break-even quantity and the break-even revenue?

$$TP = \$6,000$$

$$\begin{aligned}
BEQ_{tp} &= \frac{FC + TP}{p - v} \\
&= \frac{5,700 + 6,000}{7.00 - 2.30} \\
&= 2,490 \text{ units}
\end{aligned}$$

$$\begin{aligned}
BER_{tp} &= \frac{FC + TP}{1 - (v/p)} \\
&= \frac{5,700 + 6,000}{1 - (2.30/7.00)} \\
&= \$17,425
\end{aligned}$$

or

$$\begin{aligned}
BER_{tp} &= 2,489.36(7.00) \\
&= \$17,425
\end{aligned}$$

This case is opposite to that of cash break-even. Here a target profit is added to the fixed cost, pushing the total cost up from TC_1 to TC_2 and, as a result,

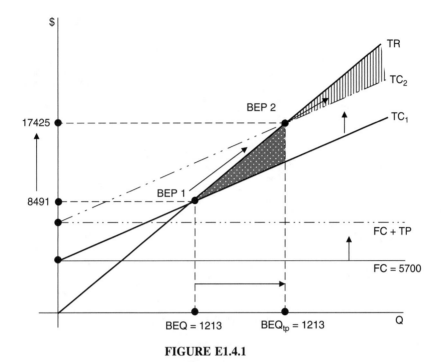

FIGURE E1.4.1

moving up both the break-even quantity and the break-even revenue. The second break-even point has a different meaning this time. It is no longer the point at which the profit is zero as was the first point, but it is the point beyond which profits would move higher than the target achieved.

1.5. ALGEBRAIC APPROACH TO THE BREAK-EVEN POINT

Corenza's home business produces dolls at a variable cost of $20 per doll and a fixed cost of $300. If it sells the doll for $50 a piece, we can write the cost and revenue equations in the following way (see Figure 1.1):

$$C = \text{FC} + vQ$$
$$= 300 + 20Q$$
$$R = pQ$$
$$= 50Q$$

At the break-even point, cost would equal revenue:

$$C = R$$
$$300 + 20Q = 50Q$$

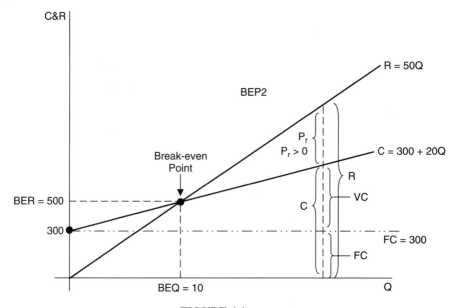

FIGURE 1.1

$$300 = 50Q - 20Q$$
$$= 30Q$$
$$Q = \frac{300}{30}$$
$$= 10 \qquad \text{this is the BEQ}$$
$$R = 50Q$$
$$= 50(10)$$
$$= 500 \qquad \text{this is the BER}$$

Break-Even Time

Since the break-even point has been expressed by production size and revenue amount, it can also be expressed by time, especially if a firm knows its production capacity, so we must be sure that it is stable and consistent and can be measured by time units. By the spirit of the time value of money, it seems interesting to identify the break-even point in terms of time. Let's suppose that a firm knows its production rate, and let's consider such a rate as q. If the time required to produce a certain size of production is t, and if Q refers to the break-even quantity, then we can write

$$Q = qt \qquad (4)$$

Suppose that the Corenza family business can produce only 2 dolls a day. Their break-even time would be 5 days:

$$Q = qt$$
$$10 = 2t$$
$$t = \frac{10}{2}$$
$$= 5 \text{ days}$$

In reality, products take some time to be sold, which means that there is a lag of time between having a product produced and available for sale and actually collecting a revenue from its sale. So, let's denote that lag of time by t_L, and let's go back to the original cost and revenue equations and substitute for the Q value in equation (4):

$$C = FC + vQ$$
$$C = FC + vqt \qquad (5)$$
$$R = pQ$$
$$R = pqt$$

and we can factor in the time lag in the revenue equation:

$$R = pq(t - t_L) \qquad (6)$$

Now let's equate C and R, equations (5) and (6), under the break-even condition and solve for t as the break-even time:

$$C = R$$
$$FC + vqt = pq(t - t_L)$$
$$FC + vqt = pqt - pqt_L$$
$$FC + pqt_L = pqt - vqt$$
$$FC + pqt_L = qt(p - v)$$
$$t = \frac{FC + pqt_L}{q(p - v)}$$

This t is the break-even time (BET), which is expressed in terms of the fixed cost (FC), unit price of the product (p), production rate (q), time lag (t_L), and variable cost per unit (v).

$$\boxed{BET = \frac{FC + pqt_L}{q(p - v)}}$$

Example 1.5.1   Suppose that the Corenza family decides to produce beach-craft souvenirs at the rate of 4 pieces a day at a $28 variable cost per piece. Suppose that its fixed cost stays the same, at $300. Each souvenir will sell for $60, but it will take 3 days for the revenue to start coming in. When will this business break even: at what quantity and at what revenue?

$$\text{BET} = \frac{FC + pqt_L}{q(p - v)}$$

$$= \frac{300 + 60(4)(3)}{4(60 - 28)}$$

$$= 7.97 \text{ or } 8 \text{ days to break even (this is } t).$$

$$R = pq(t - t_L)$$

$$= 60(4)(8-3)$$

$$\text{BER} = \$1,200$$

$$\text{BEQ} = \frac{\text{BER}}{P}$$

$$= \frac{1,200}{60}$$

$$= 20 \text{ dolls}$$

Example 1.5.2   Find BET, BER, and BEQ for Goodtract, a tire manufacturing firm that has the following data (see Figure E1.5.2):

$$\text{Fixed cost} = \$80,000$$

$$\text{Variable cost per tire} = \$20$$

$$\text{Daily production rate} = 100 \text{ tires}$$

$$\text{Selling price per tire} = \$80$$

$$\text{Time lag for revenue} = 20 \text{ days}$$

First, we construct the cost and revenue equations when a time element is involved.

$$C = FC + vqt$$

$$= 80,000 + 20(100)t$$

$$= 80,000 + 2,000t$$

$$R = pq(t - t_L)$$

$$= 80(100)(t - 20)$$

$$= 8,000t - 160,000$$

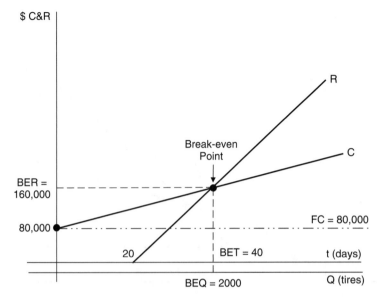

FIGURE E1.5.2

At the break-even point, $C = R$:

$$80,000 + 2,000t = 8,000t - 160,000$$

$$240,000 = 6,000t$$

$$t = \frac{240,000}{6,000}$$

$$\text{BET} = t = 40 \text{ days to break even}$$

$$\text{BER} = R = 8,000t - 160,000$$

$$= 8,000(40) = 160,000$$

$$= \$160,000$$

$$\text{BEQ} = \frac{\text{BER}}{P}$$

$$= \frac{160,000}{80}$$

$$= 2,000 \text{ tires}$$

1.6. THE BREAK-EVEN POINT WHEN BORROWING

Very often a just starting business owner does not have enough capital to start production. Resorting to personal or business loans would probably be the proper

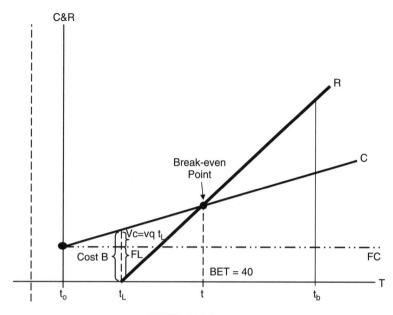

FIGURE 1.2

course of action. If a business uses debt to finance production, the break-even point would be more meaningful when the business wants to liquidate the loan. That would be at the time when profit equals the debt interest payment (see Figure 1.2). Beyond that point, any additional profits collected would continue to be higher than the interest. Let's assume that a firm needs enough funds to cover both its fixed and variable costs of production. The fund that will finance the production will be borrowed, and its amount (B) will be equal to

$$B = FC + vqt_L \tag{7}$$

It would satisfy both the need to cover the fixed cost (FC) and the need to cover the variable cost (vqt_L), which would be determined by the unit variable cost (v) times the production rate (q) times the proper period of time. The time in this case would be the lag time (t_L) defined by the period between starting the production and starting to receive revenues. Considering a simple interest method, the total interest would be determined by

$$I = P \cdot r \cdot t$$

Therefore, considering B in equation 7 as the principal, the total interest on such a loan would be:

$$I = (FC + vqt_L)rt_b \tag{8}$$

where r is the interest rate and t_b is the time for borrowing or the time to maturity.

Now, we can set up the equality between interest (I) and profit (Pr) and solve for t, which is the break-even point at which equality between profit and the interest payment on the loan will occur.

$$I = \text{Pr}$$

$$I = R - C$$

$$(\text{FC} + vqt_L)r(t + t_0) = pq(t - t_L) - (vqt + \text{FC})$$

The interest side is determined by equation (8), where the time of maturity is determined by adding the time between receiving the loan and starting production (t_0) to the time (t) which would be at some point after production when that equality between profit and interest occurs. The revenue and cost part of the equation are determined according to equations (5) and (6):

$$(\text{FC} + vqt_L)(rt + rt_0) = pq(t - t_L) - (vqt + \text{FC})$$

$$\text{FC} \cdot rt + \text{FC}rt_0 + vqt_Lrt + vqt_Lrt_0 - pqt + pqt_L + vqt + \text{FC} = 0$$

$$t(\text{FC} \cdot r + vqt_Lr - pq + vq) + rt_0(\text{FC} + vqt_L) + pqt_L + \text{FC} = 0$$

$$t[r(\text{FC} + vqt_L) - q(p - v)] = -rt_0(\text{FC} + vqt_L) - pqt_L - \text{FC}$$

$$\boxed{t = \frac{\text{FC} + pqt_L + rt_0[\text{FC} + vqt_L]}{q(p - v) - r[\text{FC} + vqt_L]}}$$

where t is the break-even point between the profit and the cost of interest on borrowing.

Example 1.6.1   Consider a firm that produces cell phones. Its fixed cost is $60,000 and the variable cost per phone is $15. The firm can produce 200 phones a day and can sell them for $50 each but does not collect revenue until 40 days after production begins. Suppose that the firm obtains a loan at 10% interest in order to cover all costs. It receives the money 30 days before production and is to pay it off in 2 years. Calculate the cost of interest and determine the time of break-even (t).

FC $= 60,000$; $v = 15$; $q = 200$; $p = 50$; $t_L = 40$ days; $r = 10\%$; $t_b = 2$ years; $t_0 = 30$ days.

$$I = (\text{FC} + qvt_L)rt_b$$

$$= [60,000 + 15(200)(40)](.10)(2)$$

$$= 36,000$$

$$t = \frac{\text{FC} + pqt_L + rt_0(\text{FC} + vqt_L)}{q(p - v) - r(\text{FC} + vqt_L)}$$

$$= \frac{60{,}000 + 50(200)(40) + (.10/365)(30)[60{,}000 + 15(200)(40)]}{200(50 - 15) - (.10/365)[60{,}000 + 15(200)(40)]}$$

$$= 66.40 \text{ days or } 67 \text{ days}$$

The profit on the 67 days after the start of production would be

$$\text{Pr} = R - C$$
$$= pq(t - t_L) - (vqt + \text{FC})$$
$$= 50(200)(67 - 40) - [15(200)(67) + 60{,}000]$$
$$= 270{,}000 - 261{,}000$$
$$= 9{,}000$$

The interest on the 67 days after the start of production would be

$$I = (\text{FR} + vqt_L)rt$$
$$= [60{,}000 + 15(200)(40)] \left(\frac{.10}{365} \right) (67)$$
$$= 3{,}304.10$$

1.7. DUAL BREAK-EVEN POINTS

As we have seen so far, the revenue and cost functions have been either linear or approximated as linear, and the graph of the two functions consists, of straight lines intersecting one time at one point, forming a single break-even point (see Figure 1.1). Usually to the left of the break-even point, the total cost line is above the total revenue to indicate the losses area, and to the right of the point, the total cost line runs below the total revenue line, indicating the profit area.

In other cases, either both or one of the functions would be nonlinear, a case that might result in having the two curves intersect with each other in more than one point, often in two points, resulting in the creation of two break-even points. The area formed between the two curves up and down, and between the two break-even points left and right, is the area of gain. Profit usually begins at the lower right after the first break-even point, which is called the **lower break-even point** and increases until it reaches the maximum at some point of production. After that, profit begins to decrease until it gets to zero again at the second break-even point, called the **upper break-even point**. Beyond that point, losses are incurred again. Let's consider the following revenue and cost functions:

$$R = -2Q^2 + 86Q$$
$$C = 16Q + 140$$

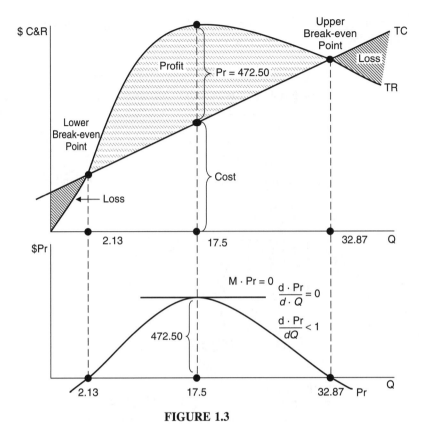

FIGURE 1.3

and let's:

1. Locate the two break-even points.
2. Determine at what level of production maximum profit is reached.
3. Calculate the maximum profit.

$$Pr = R - C$$
$$= -2Q^2 + 86Q - 16Q - 140$$
$$= -2Q^2 + 70Q - 140$$

At the break-even point(s), the profit would be zero, so we set the equation to zero:

$$Pr = -2Q^2 + 70Q - 140 = 0$$

and solve for Q as the break-even point. The function is quadratic and Q will have two values. The function is of the following form:

$$Y = ax^2 + bx + c$$

We can solve for Q by following the quadratic formula:

$$x = \frac{-b \pm \sqrt{b^2 - 4ac}}{2a}$$

In the case of our function above, those values are $a = 2$, $b = 70$, and $c = -140$.

$$Q = \frac{-b \pm \sqrt{b^2 - 4ac}}{2a}$$

$$Q = \frac{-70 \pm \sqrt{(70)^2 - 4(-2)(-140)}}{2(-2)}$$

$$= \frac{-70 \pm \sqrt{3,780}}{-4}$$

$$Q_1 = \frac{-70 + 61.48}{-4} = 2.13 \qquad \text{lower break-even point}$$

and

$$Q_2 = \frac{-70 - 61.48}{-4} = 32.87 \qquad \text{upper break-even point}$$

The maximum profit occurs at a production size of Q_v, which can be obtained as the vertex of the quadratic function. The vertex point can be found by

$$Q_v = \frac{-b}{2a}$$

$$= \frac{-70}{2(-2)}$$

$$= 17.5$$

Substituting for this amount of production in the profit function gives us the maximum profit (Pr^m):

$$Pr^m = -2Q_v^2 + 70Q_v - 140$$

$$= -2(17.5)^2 + 70(17.5) - 140$$

$$= 472.50$$

To prove that this value of the function is the maximum, two checks have to be verified:

1. The first derivative of the function has to have a value of zero when Q_v is plugged in:

$$\frac{d\text{Pr}}{dQ} = 0$$

The first derivative of the profit function refers to the marginal profit, and the zero indicates that the tangent at the maximum point is a horizontal line.

$$\text{Pr}^m = -2Q_v^2 + 70Qv - 140$$

$$\frac{d\text{Pr}^m}{dQ_v} = -4Q_v + 70$$

$$= -4(17.5) + 70$$

$$= 0$$

2. Since both the maximum point at the top of the down-opened parapola and the minimum point at the bottom of the up-opened parapola have horizontal line tangents, we need a second check to refer to the maximum condition. That is the second derivative of the profit function, which has to have a negative value when Q_v is plugged in:

$$\frac{d^2\text{Pr}}{dQ^2} < 0$$

The negativity refers to the fact that the function is going down. If it were a minimum, it would be going up.

$$\frac{d\text{Pr}}{dQ} = -4Q_v + 70$$

$$\frac{d^2\text{Pr}}{dQ^2} = -4$$

This condition can be expressed simply by the fact that the coefficient of x^2, which is a in the quadratic function, has to be negative.

1.8. OTHER APPLICATIONS OF THE BREAK-EVEN POINT

The concept of the break-even point can be applied in many situations and for different purposes. Excluding the equilibrium point between supply and demand, which is probably the most significant classic application, the break-even point may come second in terms of its applications in finance and economics. Let's look at a couple of those applications.

The Break-Even Point and the Stock Selling Decision

Stocks are bought and sold at different prices and, typically, commissions are paid through the transactions. The break-even point concept can be useful in finding the break-even price beyond which an investor can sell her stocks and make a gain. In this sense, a break-even analysis can once again be considered a valuable tool in decision making.

Suppose that a certain number (K) of a stock is purchased at a purchase price per share (x), and a commission rate on purchase (ip) is paid. We can then write the cost of stock purchase equation (C) as

$$C = xK + ip(xK)$$
$$C = xK(1 + ip) \tag{1}$$

In the same manner, we can write the revenue of stock sale equation (R) if the same number of those stocks (K) is sold at a selling price per share of y and a similar commission on sale (is) is paid:

$$R = yK - is\,(yK)$$
$$R = yK(1 - is) \tag{2}$$

Now we can find the selling break-even price per share (y_b) by equating the cost of buying with the revenue of selling [equations (1) and (2)]:

$$C = R$$
$$xK(1 + ip) = yK(1 - is)$$

Cancel K on both sides and solve for y:

$$\boxed{y_b = \frac{x(1 + ip)}{1 - is}}$$

Example 1.8.1 Wayne purchases 50 shares of Pepsi-Cola at $95 per share and pays a 3.5% commission. What is the break-even selling price if the commission rate stays the same for selling? Verify the answer.

$x = 95; K = 50; ip = 3.5\%; is = 35\%.$

$$\begin{aligned}
y_b &= \frac{x(1 + ip)}{1 - is} \\
&= \frac{95(1 + .035)}{1 - .035} \\
&= \$101.89
\end{aligned}$$

This is the break-even price for Wayne. He does not make any gain in selling his stock unless the selling price is higher than $101.89.

To verify: If he sells at $101.89, his revenue will be

$$R = yK$$
$$= 101.89(50)$$
$$= 5,094.50$$

He must pay a 3.5% sales commission and his net revenue (R_n) will be

$$R_n = R - \text{is} \cdot yK$$
$$= 5,094.50 - .035(101.89)(50)$$
$$= 4,916.2$$

The stock has already cost him

$$C = xK$$
$$= 95(50)$$
$$= 4,750$$

He paid 3.5% commission on the purchase, which is added to his cost, making the total cost

$$C_t = C + \text{ip} \cdot xK$$
$$= 4,750 + .035(95)(50)$$
$$= 4,916.2$$

So it is verified that a selling price of $101.89 would make Wayne even, neither gaining nor losing, since his total cost of purchase would turn out to be equal to his net revenue of sale. He would, then, know better than not to sell unless the price per share goes up beyond the $101.89.

The Break-Even Point and the Early Retirement Decision

Some people may want to retire early. If they happen to entertain such a thought, they are probably aware that their early retirement pension will come at a certain discount, as compared to the regular pension that they would receive if they chose to wait until the regular retirement age. One should always be concerned about how much pension money would be lost by choosing early retirement. That, of course, will depend not only on how much the discount would be but also on how early the retirement would be. A logical way of thinking that through is to contemplate how long it would take for the early retirement pension to catch up

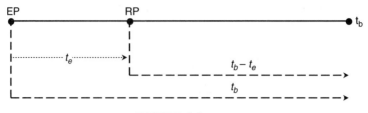

FIGURE 1.4

with the regular pension. In other words, what would be the break-even number of years to make the discounted early retirement pension equal to the regular in-time retirement pension. The answer lies in a consideration of what would be gained by waiting for the regular retirement age and what would be lost if early retirement were chosen. That rate, in fact, defines the break-even time, as we will see below.

Let's assume that a company is offering an early retirement pension (EP) which is less than the regular pension (RP) of an in-time retirement. Let's also assume that t_b is the break-even time, a time in the future when the discounted EP would catch up with the regular pension. Suppose that t_e is how early the retirement would be. That is the difference in time between regular and early retirement. Now we can construct the total revenue of early retirement (TR_e) and the total revenue of regular in-time retirement (TR_{in}) after a look at the time line in Figure 1.4.

$$TR_e = EP \cdot t_b$$

$$TR_{in} = RP(t_b - t_e)$$

At t_b, the two pensions would be equal:

$$TR_e = TR_{in}$$

$$EP \cdot t_b = RP(t_b - t_e)$$

$$EP \cdot t_b = RP \cdot t_b - RP \cdot t_e$$

$$EP \cdot t_b - RP \cdot t_b + RP \cdot t_e = 0$$

$$RP \cdot t_e = RP \cdot t_b - EP \cdot t_b$$

$$RP \cdot t_e = t_b(RP - EP)$$

$$\boxed{t_b = \frac{RP \cdot t_e}{RP - EP}}$$

Example 1.8.2 Rosemary is considering retiring next year when she turns 63. Her company is offering \$72,000 a year for her early retirement. But if she waits for the age of 65, her pension would be $90,000. Find the break-even time.

RP $= 90,000$; EP $= 72,000$; $t_e = 2$ years.

$$t_b = \frac{RP \cdot t_e}{RP - EP}$$

$$= \frac{90,000(2)}{90,000 - 72,000}$$

$$= 10$$

This means that the early retirement pension of $72,000 would become equal to the in-time retirement when Rosemary is 73. So it would be much better to work 2 more years if she can.

In reality, most pension plans allow pension payments to be adjusted annually for cost of living increases. In such cases, the break-even time formula above would be adjusted to allow the formula to obtain the break-even time, similar to the formula used to obtain the maturity time of an annuity. The new formula for the break-even time for early retirement when the pension is adjusted annually by a certain rate (r) would be

$$t_b^{adj} = \frac{\ln\left[\dfrac{(RP-EP)}{RP(1+r)^{-t_e}-EP}\right]}{\ln(1+r)}$$

Example 1.8.3   Steve has applied for early retirement at age 63, when he would get a $32,000 pension. Due to his deteriorating health, his wife urges him not to work 2 more years to get the full pension of $40,000. His company adjusts pensions by 2.75% cost-of-living increases. How old would he be when the two pensions equal each other?

$$t_b^{adj} = \frac{\ln\left[\dfrac{(RP-EP)}{RP(1+r)^{-t_e}-EP}\right]}{\ln(1+r)}$$

$$= \frac{\ln[(40,000 - 32,000)/40,000(1+.0275)^{-2} - 37,000]}{\ln(1+.0275)}$$

$$= 11.3$$

Steve would be a little older than 74 $(63 + 11.3)$ years old.

TABLE 1.1   Sensitivity of BEQ and BER Toward FC, v, p, and CM

For an increase ↑ (+) in:	BEQ and BER would:
Fixed cost	(+)↑
Variable cost per unit	(+)↑
Price per unit	(−)↓
Contribution margin	(−)↓

1.9. BEQ AND BER SENSITIVITY TO THEIR VARIABLES

BEQ and BER sensitivity refers to the change in the value of the break-even quantity and break-even revenue in response to the change in their component variables. Table 1.1 shows (by a plus or minus sign and arrow) how the break-even quantity and revenue respond to an increase in each of the related variables, holding the other variables unchanged. It should all be understood as a simple mathematical fact of the characteristics of the ratio if BEQ and BER are considered ratios in a sense that each is an amount divided by anther amount, and the outcome would depend on the typical relationships between the numerator and the denominator.

1.10. USES AND LIMITATIONS OF BREAK-EVEN ANALYSIS

As a technical analysis, the break-even method shed a lot of light on several business problems, which brought about its wide use and increasing popularity. At the heart of business decision making, there have been many primary uses of the break-even technique, such as:

1. Evaluating the potential capacity of a firm to cover all of its operating costs to be able to make the desired profits.
2. Assessing the way in which the profit relates to the sale and measuring its responsiveness to fluctuation in the sale levels.
3. Providing a measure of knowledge of the business potential risk, especially when it comes to the variability of the investment returns on assets and the degree of the business operating leverage.
4. Providing a measure of understanding of the possibilities of launching a new product or expanding the business further. This is especially valuable for decision making in small businesses that would naturally be ambitious to move beyond achieving profits to widening the range of product lines and expanding the entire business vertically and horizontally.

Despite all of these benefits and uses, the break-even technique has some significant limitations, such as:

1. In most technical cases, the analysis assumes linearity in both the cost and revenue functions. In reality and due to the relationships of product price and variable cost to the level of sale, and due to many other effects, the cost and revenue functions end up being essentially nonlinear.

2. The discrete and abstract nature of the division between fixed and variable costs can make complete sense in theory, but it is not easy in practice. Many types of expenses stand between the fixed and variable expenses. Some have the nature of semivariables, and some are basically cross-listed.

3. Break-even analysis would seem to be perfect for a single-product firm, but it gets more complicated in a multiproduct business, even if the break-even point is converted to the break-even revenue. The complication stems from the difficulty or even the impossibility of assigning costs to different product lines and various products produced by the same production facility at the same time.

4. Break-even analysis is a short-term analysis in terms of the horizon of its variables. This would entail overlooking many of the costs and benefits that occur in the long run, such as those for advertising, research, and development.

2 Leverage

Break-even analysis provides a tool to calculate the break-even point, after which a firm would begin to collect profits as it produces and sells more products. The crucial question, then, becomes to what extent production and sale would have to increase to achieve a certain level of profits. This is a question of profit sensitivity to the variability of product sales—and that is what **leverage** is all about. There are three types of leverage: operating leverage, financial leverage, and total leverage.

2.1. OPERATING LEVERAGE

Operating leverage refers to the responsiveness of a change in profits due to a change in sales. Specifically, it is the potential use of fixed operating costs to magnify the effects of a change in sales on operating income or earnings before interest and taxes (EBIT). Fixed operating costs are those items at the top of the balance sheet, such as leases, executive salaries, property taxes, and the like.

The **degree of operating leverage** (DOL) measures the responsiveness of profit as a percentage change in operating income (OY) relative to a percentage change in sales (S):

$$DOL = \frac{\%\Delta OY}{\%\Delta S}$$

$$DOL = \frac{[(OY_2 - OY_1)/OY_1] \times 100}{[(S_2 - S_1)/S_1] \times 100}$$

$$\boxed{DOL = \frac{(OY_2 - OY_1)/OY_1}{(S_2 - S_1)/S_1}}$$

Example 2.1.1 If the operating income of a small business increases from \$884 to \$1,680 as it expands sales of its product from \$4,200 to \$5,880, what is the

Mathematical Finance, First Edition. M. J. Alhabeeb.
© 2012 John Wiley & Sons, Inc. Published 2012 by John Wiley & Sons, Inc.

degree of operating leverage?

	S	OY	
S_1	4,200	884	OY$_1$
S_2	5,880	1,680	OY$_2$

$$\text{DOL} = \frac{(OY_2 - OY_1)/OY_1}{(S_2 - S_1)/S_1}$$

$$= \frac{(1,680 - 884)/884}{(5,880 - 4,200)/4,200}$$

$$= \frac{.90}{.40}$$

$$= 2.25$$

This means that for every 1% change in sales, there will be a 2.25% change in operating income (profit or EBIT). In other words, as sales increased by 40%, profits rose by 90%, as shown in the denominator and numerator of the DOL, respectively. It is also shown in Figure E2.1.1. A move from Q_1 to Q_2 on the units of sale (or R_1 to R_2 on the dollars of revenue) results in a move from Prof$_1$ to Prof$_2$. This effect would also hold in the other direction, meaning that a reduction in sales (units of product or dollars of revenue) by 40% would result in a drop in profits by 90%. We can calculate a move in sales in the opposite direction, from Q_1 to Q_0, or in revenue from R_1 to R_0, which is about 40% less (420 to 252 units or \$4,200 to \$2,520) and see the profit drop 90%, from Prof$_1$ = 884 to Prof$_0$ = 88.4.

$$\text{DOL} = \frac{(884 - 88.4)/884}{(4,200 - 2,520)/4,200}$$

$$= \frac{.90}{.4}$$

$$= 2.25$$

So the DOL can actually be considered a multiplier of the effect that a change in sales will have on profit. In other words, DOL is a concept of elasticity. It is the elasticity of profit with respect to sales. In such a sense we can express a partial change in profit (π) relative to a partial change in output (Q):

$$\text{DOL} = \frac{\partial \pi / \pi}{\partial Q / Q}$$

This degree of leverage can be obtained at any level of output. If the fixed cost is constant, the change in profit ($\partial \pi$) will be

$$\partial \pi = \partial Q(p - v) \tag{1}$$

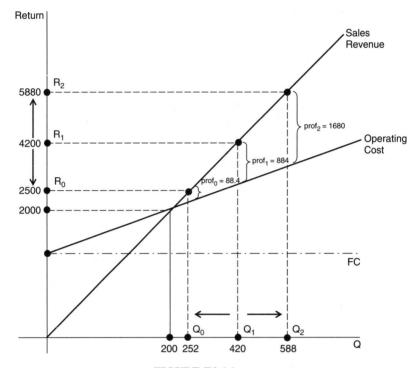

FIGURE E2.1.1

The profit (π) is

$$\pi = R - C$$
$$= PQ - (\text{FC} + vQ)$$
$$= PQ - \text{FC} - vQ$$
$$= Q(P - v) - \text{FC} \qquad (2)$$

Substituting equations (1) and (2) in the DOL equation above yields

$$\text{DOL} = \frac{\partial Q(P - v)}{Q(P - v) - \text{FC}} \cdot \frac{Q}{\partial Q}$$

Canceling ∂Q, we get

$$\boxed{\frac{Q(P - v)}{Q(P - v) - \text{FC}}}$$

which is the degree of operating leverage at any level of output.

Example 2.1.2   Find and interpret the degree of operating leverage (DOL) for a firm that has the following data: FC = $2,500; variable cost per unit = $5; unit price = $10; units of product sold = 1,000

$$\text{DOL} = \frac{Q(P - v)}{Q(P - v) - FC}$$

$$= \frac{1,000(10 - 5)}{1,000(10 - 5) - 2,500}$$

$$= 2$$

A degree of operating leverage of 2 means that for every 1% change in sales, operating income will change by 2%.

2.2. OPERATING LEVERAGE, FIXED COST, AND BUSINESS RISK

A major observation can be made on the DOL formula above. This observation is related to the fixed operating cost and its effect on DOL. Mathematically, since the term $Q(p - v)$ is present in both the numerator and denominator, the role of fixed cost (FC) becomes crucial. Any increase in FC will make the denominator lower and the DOL higher, and any decrease in FC will make the denominator higher and the DOL lower. We can then conclude that the higher a firm's fixed operating cost relative to its variable cost, the greater the degree of leverage and ultimately, the higher the profit. Let's check this by observing the change in DOL if the fixed cost in Example 2.1.2 increases from 2,500 to 4,000:

$$\text{DOL} = \frac{1,000(10 - 5)}{1,000(10 - 5) - 4,000}$$

$$= 5$$

A 5-degree leverage means that the profit will increase by 5% instead of 2% when sales increase by 1%. Table 2.1 and Figure 2.1 show how the DOL and the profit (Pr) change as the fixed cost increases notably across three firms selling the same product for the same price. However, as we have seen before, operating leverage also works in the opposite direction. This means that if we take the figures in Example 2.1.2, a decrease in sales of 1% would be more damaging by lowering profits by 5%. So the change in fixed cost would really make the situation more sensitive in both ways, which can translate into presenting a source of business risk. Such a potential risk can come from two factors:

1. In the case of seeking more profits through an increase in the fixed cost, the firm would take the risk of not being able to cover all of the high cost and the additional risk of not being able to maintain the increase in sales.

TABLE 2.1 Three Firms Selling a Product at the Same Price with Three Different Cost Functions

Firm	Q	R	Cost FC + vQ	Pr	A·Pr	$DOL = \dfrac{Q(p-v)}{Q(P-v-FC)}$
Firm 1						
B-even →	1,000	3,000	3,000	0	0	FC = $1,000; vc = $2.00;
	1,500	3,500	4,000	500	500	$P = \$3.00$
	2,000	6,000	5,000	1,000	500	
	2,500	7,500	6,000	1,500	500	$DOL = \dfrac{3,500(3-2)}{3,500(3-2)-1,000}$
	3,000	9,000	7,000	2,000	500	$= 1.4$
⟶	3,500	10,500	8,000	2,500	500	
	4,000	12,000	9,000	3,000	500	
Firm 2	1,000	3,000	4,000	−1,000	—	FC = $2,250; vc = $1.75;
B-even →	1,800	5,400	5,400	0	−1,000	$P = \$3.00$
	2,500	7,500	6,625	875	625	
	3,000	9,000	7,500	1,500	625	$DOL = \dfrac{3,500(3-1.75)}{3,500(3-1.75)-2,250}$
⟶	3,500	10,500	8,375	2,125	625	$= 2.06$
	4,000	12,000	9,250	2,750	625	
	4,500	13,500	10,125	3,375	625	
Firm 3	1,000	3,000	5,000	−2,000	—	FC = $3,750; vc = $1.25;
B-even →	2,143	6,429	6,429	0	−2,000	$P = \$3.00$
	2,500	7,500	6,875	625	625	
	3,000	9,000	7,500	1,500	875	$DOL = \dfrac{3,500(3-1.25)}{3,500(3-1.25)-3,700}$
⟶	3,500	10,500	8,125	2,375	875	$= 2.58$
	4,000	12,000	8,750	3,250	875	
	4,500	13,500	9,375	4,125	875	

2. In the case of a slight drop in the sales, the firm would take a risk of finding its profit dropping significantly. It may drop to a point that threatens a reasonable recovery.

So the ever-increasing temptation to benefit from new technology, to modernize production, and to replace labor-intensive technology with capital-intensive technology would probably increase efficiency but require incurring a lot more fixed cost, which, if done, would have to be done with much more caution and consideration by financial managers, who have to weigh the benefits of increasing profits carefully and rationally, considering all the associated risks.

2.3. FINANCIAL LEVERAGE

Financial leverage is the responsiveness of the change in a firm's earnings per share (EPS) to the change in its operating income (profit) or earnings before interest and taxes (EBIT). Similar to operating leverage, financial leverage expresses

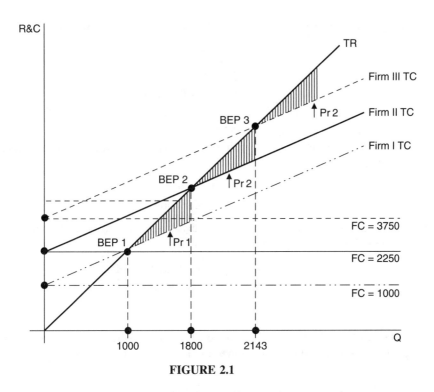

FIGURE 2.1

the potential use of fixed financial charges to magnify the effects of changes in operating income (OY) on earnings per share (EPS). The fixed financial charges here are specifically: interest on debt and dividends for preferred stock. These are the liabilities at the lower part of the balance sheet. In other words, financial leverage is mainly about financing the business, totally or partially, by debt, and it is no surprise that the financial leverage approach is sometimes popularly called OPM (**other people's money**) and is also called **trading on equity** or **debt financing**. Using debt in business investment typically arises from rational reasons, such as:

1. The returns on debt investment are higher than the returns on equity investment.
2. Debt capital is most likely to be readily available.
3. Using debt capital usually does not affect the voting situation in a firm.

In a nutshell, if a firm has a high fixed cost of financing, it would be considered heavy on financial leverage and would probably enjoy high returns on investment but would, in turn, face potential financial risks.

Similar to operating leverage, financial leverage has its own degree, the **degree of financial leverage** (DFL). It measures the responsiveness of the change in earnings per share (EPS) relative to the change in operating income (OY):

$$DFL = \frac{\%\Delta EPS}{\%\Delta OY}$$

$$DFL = \frac{(EPS_2 - EPS_1)/EPS_1}{(OY_2 - OY_1)/OY_1}$$

Note that EPS refers to the net earnings that would be distributed to common stockholders:

$$EPS = \frac{net\ income}{no.\ shares\ of\ common\ stock}$$

and if there are any preferred stocks, their dividends have to be deducted from net income before obtaining the EPS:

$$EPS = \frac{net\ income - preferred\ stock\ dividends}{no.\ shares\ of\ common\ stock}$$

Example 2.3.1   Calculate the degree of financial leverage in the Sureluck Company, which has the data shown in Table E2.3.1 for two consecutive years.

First, we need to calculate operating income (OY) and earnings per share (EPS):

$$operating\ income = gross\ income - operating\ expenses$$
$$for\ year\ 1 = \$30,250 - \$10,100 = \$20,150$$
$$for\ year\ 2 = \$32,900 - \$10,940 = \$21,960$$

For EPS, we need to calculate net income by deducting interest and taxes from the operating income:

$$net\ income = operating\ income - (interest\ and\ taxes)$$
$$for\ year\ 1 = \$20,150 - (\$766 + \$3,400 = \$15,984$$
$$for\ year\ 2 = \$21,960 - (\$870 + \$3,490) = \$17,600$$

TABLE E2.3.1

Item	Year 1	Year 2
Gross income ($)	30,250	32,900
Operating expenses ($)	10,100	10,940
Interest ($)	766	870
Income taxes ($)	3,400	3,490
Number of shares	30,000	30,000

$$EPS_1 = \frac{15,984}{30,000} = .53$$

$$EPS_2 = \frac{17,600}{30,000} = .59$$

	OY	EPS	
OY_1	20,150	.53	EPS_1
OY_2	21,960	.59	EPS_2

$$DFL = \frac{(EPS_2 - EPS_1)/EPS_1}{[(OY_2 - OY_1)/OY_1]}$$

$$= \frac{(.59 - .53)/.53}{(21,960 - 20,150)/20,150}$$

$$= 1.26$$

This 1.26 DL means that for every 1% change in Sureluck's operating income there would be a 1.42 change in its earnings per share. This would be true for both the increase or decrease in the EPS that would follow, respectively, an increase or decrease in operating income.

Example 2.3.2 Let's assume that the Sureluck Company in Example 2.3.1 distributed about 43% of its net income as dividends for preferred stocks. How would that affect the DFL?

The change would be in the calculation of the EPS. The preferred stock dividends have to be deducted from net income before dividing the net income among the common shareholders.

$$\text{preferred stock dividends} = 43\% \text{ of net income}$$

$$\text{for year } 1 = .43 \times 15,984$$

$$= 6,873$$

$$\text{for year } 2 = .43 \times 17,600$$

$$= 7568$$

$$EPS_1 = \frac{15,984 - 6,873}{30,000} = .3037$$

$$EPS_2 = \frac{17,600 - 7,568}{30,000} = .3344$$

$$DFL = \frac{(.3344 - .3037)/.3037}{(21,960 - 20,150)/20,150}$$

$$= 1.125$$

There is another formula for DFL at a base level of operating income. It is more direct when preferred stock dividends are paid. With this formula we would not have to calculate EPS.

$$DFL = \frac{OY}{OY - \left[I + \left(\dfrac{D_{ps}}{1 - T}\right)\right]}$$

where OY is the operating income, I is the interest, D_{ps} is the dividends for preferred stocks, and T is the income tax rate.

If we calculate the DFL of Example 2.3.2, we need to know the income tax rate. But we can, for this purpose, assume that it is based on the relation of taxes to income. So let's just assume that the tax rate in this example was 16.8% for year 1 and 15.9% for year 2. We can now calculate the degree of financial leverage individually for each year based on the operating income:

$$DFL = \frac{OY}{OY - \left[I + \left(\dfrac{D_{ps}}{(1 - T)}\right)\right]}$$

$$DFL_1 = \frac{20,150}{20,150 - \left[3,400 + \left(\dfrac{6,873}{1 - .168}\right)\right]}$$

$$= 2.37$$

$$DFL_2 = \frac{21,960}{21,960 - \left[3,490 + \left(\dfrac{7,568}{1 - .159}\right)\right]}$$

$$= 2.32$$

Example 2.3.3 In this example we follow the impact of financial leverage on earnings per share throughout three possible alternatives of financial plans. Let's assume that a firm needs $100,000 to expand its business, and let's assume that the board of directors has the following alternative plans to finance this expanding project.

   I: 100% equity financing. The entire $100,000 would be obtained by selling 1,000 shares at $100 each.

  II: $\frac{2}{3}$ equity financing and $\frac{1}{3}$ debt financing. That is, 66% of the $100,000 ($66,000) would be obtained internally, and 34% ($34,000) would be obtained by borrowing at 9.5% interest.

 III: $\frac{1}{3}$ equity financing and $\frac{2}{3}$ debt financing. That is, 34% ($34,000) would be obtained internally and 66% ($66,000) would be a business loan at 9.5% interest.

TABLE E2.3.3 Financial Leverage and Earnings per Share in Three Financing Plans

Financing Plan	Total Capital ($)	Equity Part, EQ ($)	Debt Part, D ($)	Operating Income, OY ($)	Interest, I 9.5% ($)	OY − I ($)	Taxes, T 40%	Net Income, NY (OY − I − T) ($)	Return on Equity, NY/EQ (%)	EPS, NY/no. shares	DFL
Plan I	100,000	100,000	0	20,000	0	20,000	8,000	12,000	12	2.40	
OY 25% ↑	100,000	100,000	0	25,000	0	25,000	10,000	15,000	15	3	1.25
OY 25% ↓	100,000	100,000	0	15,000	0	15,000	6,000	9,000	9	1.8	1
Plan II	100,000	66,000	34,000	20,000	3,230	16,770	6,708	10,062	15.2	2.01	
OY 25% ↑	100,000	66,000	34,000	25,000	3,230	21,770	8,708	13,062	19.8	2.61	1.49
OY 25% ↓	100,000	66,000	34,000	15,000	3,230	11,770	4,708	7,062	10.7	1.41	1.19
Plan III	100,000	34,000	66,000	20,000	6,270	13,730	5,492	8,238	24.2	1.65	
OY 25% ↑	100,000	34,000	66,000	25,000	6,270	18,730	7,492	11,238	33	2.25	1.82
OY 25% ↓	100,000	34,000	66,000	15,000	6,270	8,730	3,492	5,238	15.4	1.05	1.45

In Table E2.3.3 we see the data for these three financial plans. We are assuming that the operating income is 20% of the capital employed and that taxes are 40% of the earnings after interest. We are allowing the operating income to increase and decrease by 25% in each plan. The number of shares is assumed to be 5,000.

Data show that as the firm uses debt to finance the expansion project, returns on equity rises in increasing rates from 15.2% in plan II to 24.2% in plan III: that is, by 60%. Also, as operating income increases by 25%, return on equity increases by 30% (from 15.2% to 19.8%) in plan II and by 36% (from 24.2% to 33%) in plan III. It is also evident that an increase in leverage causes EPS to increase. That is, the increase in EPS relative to a 1% increase in operating income is shown by the DFL increasing throughout the plans from 1.25 to 1.49 to 1.82. Conversely, the decrease in leverage causes the EPS to decline. Also, the effects on EPS of both the income and the decrease in income are more dramatic as we move from plan II to plan III, which uses higher debt.

2.4. TOTAL OR COMBINED LEVERAGE

We have seen that changes in sales revenues have caused greater changes in a firm's operating income, which, in turn, has been reflected in changes in the earnings per share. It is logical to conclude that using both operating leverage and financial leverage would strongly affect the firm's earnings per share. Therefore, the combined effects of both types of leverage is what we call the **total leverage**. It is defined as the potential use of both operating and financial fixed costs to magnify the effect of changes in sales on a firm's earnings per share. Figure 2.2 shows the connection between sales and earnings per share through changes in operating income or EBIT utilizing the two types of leverage, operating and financial.

We can see that there is an ultimate link between the changes in sale revenue and the variations in earnings per share that can be visualized as the line of combined or total leverage. The degree of combined leverage (DCL) can be obtained as a product of the degree of operating leverage (DOL) and the degree

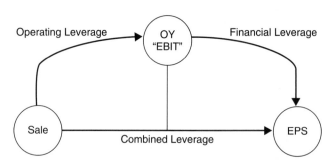

FIGURE 2.2

of financial leverage (DFL):

$$DCL = DOL \cdot DFL$$

$$DCL = \frac{\%\Delta OY}{\%\Delta S} \cdot \frac{\%\Delta EPS}{\%\Delta OY}$$

Canceling out $\%\Delta OY$, we get

$$DCL = \frac{\%\Delta EPS}{\%\Delta S}$$

$$DCL = \frac{[(EPS_2 - EPS_1)/EPS_1] \times 100}{[(S_2 - S_1)/S_1] \times 100}$$

$$\boxed{DCL = \frac{EPS_2 - EPS_1}{EPS_1} \cdot \frac{S_1}{S_2 - S_1}}$$

Example 2.4.1 In the last two years, a small firm managed to increase its sales revenue from \$88,000 to \$96,000. Its earnings per share have also been increasing, from 1.05 to 1.56. What would be the degree of combined leverage?

$$\boxed{DCL = \frac{EPS_2 - EPS_1}{EPS_1} \cdot \frac{S_1}{S_2 - S_1}}$$

$$= \frac{1.56 - 1.05}{1.05} \left(\frac{88,000}{96,000 - 88,000} \right)$$

$$= 5.34$$

This means that for every 1% change in the firm's sales, earnings per share have changed by 5.34%.

We can also use the following formula to obtain the degree of combined or total leverage:

$$\boxed{DCL = \frac{Q(p - v)}{Q(p - v) - FC - I - [D_{ps}/(1 - T)]}}$$

where Q is a given size of production, P is the price of a unit of production, v is the variable cost per unit of production, FC is the fixed cost, I is the interest, D_{ps} is the dividends for preferred stocks, and T is the tax rate.

Example 2.4.2 A firm faces a \$6,000 fixed cost and a \$1.50 variable cost per unit. It pays \$8,210 as interest on its debt, pays income taxes at the rate of 36%, and pays \$10,000 in dividends for preferred stock. What would be the degree of its combined leverage if it sells 15,000 units at \$6.50 each?

$$DCL = \frac{Q(p-v)}{Q(p-v) - FC - I - [D_{ps}/(1-T)]}$$

$$= \frac{15,000(6.50 - 1.50)}{15,000(6.50 - 1.50) - 6,000 - 8,210 - [10,000/(1 - .36)]}$$

$$= 1.66$$

Unit V Summary

The break-even point is a technique to guide businesses to the level of sales that is required to cover all operating costs and assess a firm's position on profitability. Technically, the break-even point is where sales revenues equal the total cost of production. It is the point where profit is zero. From here it becomes obvious that all production and sales before that point yield only loss, and profits begin to be collected when sales pass that point. The point can be expressed by size of production, amount of revenue, and time. It is also expressed in cash in cases where noncash charges such as depreciation are deducted from the fixed cost. The break-even revenue and production can also be obtained when a certain amount of profit is predetermined. Some important applications of the break-even point were explained in this unit, such as the break-even calculation when a firm incurs some borrowing; when the revenue and cost functions, or at least one of them is nonlinear, so that they produce more than one break-even point; the break-even and stock-selling decisions; and break-even in the early retirement decision.

Break-even quantity and revenue sensitivity to their component variables, and the limitations of the break-even analysis, were also addressed. Leverage was naturally a subject closely related to break-even analysis. Three types of leverage were explained as they are related to the use of fixed cost and business risk. Operating leverage, defined as the potential use of fixed operating costs that would magnify the effects of changes in sales on operating income, leads to a positive relationship between fixed cost and degree of leverage. The higher the fixed costs, the higher the degree of operating leverage. This degree of operating leverage is, in turn, related directly to the level of business risk.

The second type of leverage is financial leverage. It is defined as the potential use of fixed financial costs, such as interest on business debt and dividends for preferred stock, to magnify the effect of changes in operating income on business earnings per share. This also leads to a direct positive relationship between operating income and earnings per share. The higher the fixed financial costs, the greater the degree of financial leverage, which is also related directly to the level of business risk.

The third type of leverage in total leverage, a combination of operating leverage and financial leverage. It is defined as the potential use of both types of

Mathematical Finance, First Edition. M. J. Alhabeeb.
© 2012 John Wiley & Sons, Inc. Published 2012 by John Wiley & Sons, Inc.

fixed costs, operating and financial, to magnify the effects of changes in sales on business earnings per share: The higher the total fixed cost, the greater the degree of total leverage and the greater the level of business risk. That increasing level of business risk comes as the price of a big benefit, which is the increase in earnings on equity and earnings per share.

List of Formulas

Break-even quantity:

$$\text{BEQ} = \frac{\text{FC}}{p - v}$$

Cash break-even quantity:

$$\text{CBEQ} = \frac{\text{FC} - \text{NC}}{p - v}$$

Break-even quantity with target profit:

$$\text{BEQ}_\text{tp} = \frac{\text{FC} - \text{TP}}{p - v}$$

Break-even revenue:

$$\text{BER} = \frac{\text{FC}}{1 - (v/p)}$$

$$\text{BER} = \text{BEQ} \cdot p$$

Contribution margin:

$$\text{CM} = p - v$$

Break-even revenue with target profit:

$$\text{BER}_\text{tp} = \frac{\text{FC} + \text{TP}}{1 - (v/p)}$$

Break-even time:

$$\text{BET} = \frac{\text{FC} + pqt_L}{q(p - v)}$$

Mathematical Finance, First Edition. M. J. Alhabeeb.
© 2012 John Wiley & Sons, Inc. Published 2012 by John Wiley & Sons, Inc.

Break-even time between profit and cost of interest in borrowing:

$$t = \frac{FC + pqt_L + rt_0(FC + vqt_L)}{q(p - v) - r(FC + vqt_L]}$$

Break-even for stock buying and selling:

$$y_b = \frac{x(1 + ip)}{1 - is}$$

Break-even for early retirement:

$$t_b = \frac{RP \cdot t_e}{RP - EP}$$

Break-even for adjusted pension:

$$t_b^{adj} = \frac{\ln\left[(RP - EP)/RP(1 + r)^{-t_e} - EP\right]}{\ln(1 + r)}$$

Degree of operating leverage:

$$DOL = \frac{OY_2 - OY_1}{OY_1} \cdot \frac{S_1}{S_2 - S_1}$$

$$DOL = \frac{Q(p - v)}{Q(p - v) - FC}$$

Degree of financial leverage:

$$DFL = \frac{EPS_2 - EPS_1}{EPS_1} \cdot \frac{OY_1}{OY_2 - OY_1}$$

$$DFL = \frac{OY}{OY - \left[I + \left(\dfrac{D_{ps}}{1 - T}\right)\right]}$$

Degree of combined (total) leverage:

$$DCL = \frac{EPS_2 - EPS_1}{EPS_1} \cdot \frac{S_1}{S_2 - S_1}$$

$$DCL = \frac{Q(p - v)}{Q(p - v) - FC - I - [D_{ps}/(1 - T)]}$$

Exercises for Unit V

1. Find the break-even quantity for a firm whose fixed operating cost is $5,700 and variable operating cost is $1.95 per unit, given that its product sells for $7.00 per unit.

2. Calculate the break-even revenue for the firm above if the fixed cost stays the same at $5,700 but the variable cost increases to $2.65 per unit and the product is sold at $9.00 now.

3. Find both the break-even quantity and the break-even revenue for a small business if its unit product is sold for $29 in the market.

Item	Frequency	Cost ($)
Rent	Annually	50,400
Property taxes	Quarterly	1,200
Insurance	Semiannually	2,340
Salaries	Weekly	3,800
Employee benefits	Annually	12,000
Wages	Per unit	6.75
Material	Per unit	7.50
Transportation	Per unit	2.25
Shipping	Per unit	1.12

4. Suppose that the firm in Exercise 3 has depreciation changes totaling 15% of its fixed cost. What would be the cash break-even quantity (CBEQ)?

5. Find the break-even quantity and break-even revenue for the firm in Exercise 4 if management has set up $55,000 as a target profit that must be obtained.

6. Steve is opening a small wood shop to build custom mailboxes. He has the following costs:

   Fixed cost: $3,000 a month

   Wages for two assistants who work 40 hours a week,

   40 weeks a year earning $12 an hour

Mathematical Finance, First Edition. M. J. Alhabeeb.
© 2012 John Wiley & Sons, Inc. Published 2012 by John Wiley & Sons, Inc.

Material: $10 per box

Handling: $1.00 per box

If he sells his boxes at $30 each, calculate the shop's:

(a) Contribution margin

(b) Break-even quantity

(c) Break-even revenue

(d) Break-even quantity and revenue if he wants to make $5,000 in profit
Also draw the break-even graph.

7. Let's assume that Steve in Exercise 6 also decided to make birdhouses. He can make 7 birdhouses a day at a $3 variable cost per unit. Given that his fixed cost stays at $3,000 and that he can sell each birdhouse at $10 but would need to wait for 10 days to collect his revenue, calculate:

(a) Break-even time

(b) Break-even revenue

(c) Break-even quantity

8. Calculate the point of break-even for a toy company that produces 1,000 toys a day at a fixed cost of $50,000 and a variable cost of $8 per toy and for a market price of $15. The company's sales revenue would not be collected until 30 days after case production, but the company already obtained a loan 50 days before production at $8\frac{1}{2}$ % interest and a maturity of 3 years.

9. Find the break-even selling price for an investor who purchases 150 shares of a stock at $35 each and pays a 4% commission. Assume that the selling commission is at the same rate of purchase commission. Verify your answer.

10. Determine the break-even point for a person who contemplating retiring 4 years early and receiving $62,000 as opposed to $71,000 if he retires on time.

11. Suppose that the firm of the person in Exercise 10 adjusts pensions by 3.25% for the cost of living. How long would he wait to have the early pension and the full pension equal each other?

12. Find the degree of operating leverage (DOL) for the following firm through its change in sale (S) and operating income (OY) between years 1 and 2.

Year	OY	S
1	15,000	65,000
2	185,000	90,000

13. Determine the degree of operating leverage for a firm that has the percentage change in sales equal to 35% and the percentage change in operating income is 46%.

14. Find the degree of financial leverage for a firm given the following data:

Item	Year A	Year B
No. shares	50,000	50,000
Gross income ($)	72,000	79,000
Operating income ($)	22,500	23,900
Interest ($)	5,620	6,850
Income taxes ($)	20,160	22,120

15. Calculate the degree of combined leverage (DCL) if you find that DOL $= 4.8$, $\%\Delta OY = 57$, and $\%\Delta EPS = 43$.

UNIT VI

Mathematics of Investment

1 Stocks

In addition to their ordinary internal approach to maintaining a solid equity by retaining earnings, business firms, especially large corporations, take an additional external approach to finance their capital needs. To fund their startup operations, maintain their continuous spending, and fund their expansion projects, corporations may raise the needed capital by issuing and selling stocks and bonds. Two major types of securities serve as negotiable instruments of ownership. Investors who are interested in long-term investment buy and sell stocks and bonds in the market and hope to earn money through receiving dividends as their shares of a firm's profits, as well as through making capital gains when their investment values appreciate. To ensure the safety of their investment and take the best path to earning good returns, investors would have to be studious and smart in following the frequent fluctuations of prices and trends in their stocks and bonds. In this chapter we focus on mathematical operations related to stocks, and continue in the following chapters to address bonds and other types of securities.

For corporations, stocks are a major external source of equity capital that have claims on a firm's income and assets secondary to claims by their regular business creditors. For an investor, stocks are shares of ownership in the assets and earnings of a firm, and therefore a source of individual income. There are two major types of stocks; common stocks and preferred stocks. **Common stocks** are the most basic form of business ownership. They represent the great ability that a business has to increase its funding, and at the same time they place a minimum constraint on the firm, and for that, common stockholders are compensated with higher dividends and voting rights, compared with the holders of preferred stock. Voting rights in a corporation means that a stockholder would share in making major policy decisions, election of officers, and evaluation of management and progress. The power of voting would be commensurate with the number of shares that a stockholder holds. Common stocks can be owned privately or publicly and may come in various classes and categories.

Preferred stocks are of the fixed-income type. Stockholders of preferred stocks receive, as a dividend, a fixed percentage or amount of a firm's earnings. The firm is obligated to pay those dividends even before it pays its common stockholders. However, this priority privilege for preferred stocks is countered by the lack of voting rights. Other perceived disadvantages of preferred stocks are their relatively higher costs and the sensitivity of their market price to the

Mathematical Finance, First Edition. M. J. Alhabeeb.
© 2012 John Wiley & Sons, Inc. Published 2012 by John Wiley & Sons, Inc.

fluctuation of interest rate. However, one big advantage of preferred stocks, on the firm's side, is that they tend to increase the firm's leverage, as we will see in detail later.

1.1. BUYING AND SELLING STOCKS

The primary objective for the common stockholders is to earn money, not only by receiving dividends but also by trading and making capital gains through buying and selling the right stock at the right time. The basic premise here is that they would like to buy at a time of **undervalued stocks**. That is when the true value of the stock is expected to be higher than its market value. They would also sell at the time of **overvalued stocks**, meaning that the true value of a stock is considered to be less than what it is sold for in the market.

For that reason, an investor has to be aware not only of the market changes and price fluctuations but also of the commissions and brokerage fees and the way they are calculated. In the stock exchange market, brokerage fees are set based on the market value of stocks and the number of shares purchased or sold. The brokerage rate is usually a combination of fixed and variable costs. The fixed part would increase with the amount of purchase or sale, and the variable part, which is usually a percentage, decreases as the amount of the purchase or sale, goes up, as shown in Table 1.1.

Example 1.1.1   Dale purchased 350 shares of Country Pizza's common stock at $25.75 per share but later sold 200 shares at $29.25 each. What would his capital gain and his investment rate of return be?

1. Total cost of investment:

$350 \times \$25.75 = \$9,012.50$	initial investment
$\$9,012.50 \times .006 = \$54.08$	brokerage % on purchase
$\$70 + \$54.08 = \$124.08$	total brokerage fee
$\$9,012.50 + \$124.08 = \$9,136.58$	cost of total investment

TABLE 1.1

Amount of Purchase or/Sale ($)	Brokerage Rate	
	Initial Charge ($) +	%
→ 2,500	25	.015
2,501–6,000	50	.007
6,001–22,000	70	.006
22,001–50,000	90	.004
50,001–500,000	150	.002
500,000 →	250	.001

2. Net sale:

$$200 \times \$29.25 = \$5,850 \qquad \text{total sale}$$

$$\$5,850 \times .007 = \$40.95 \qquad \text{brokerage \% on sale}$$

$$\$50 + \$40.95 = \$90.95 \qquad \text{total brokerage on sale}$$

$$\$5,850 - \$90.95 = \$5,759.05 \qquad \text{net sale}$$

3. Cost of 200 shares sold:

$$\$9,136.50 \times \frac{200}{350} = \$5,220.85$$

4. Capital gain:

$$\$5,759.05 - \$5,220.86 = \$538.19 \qquad \text{capital gain}$$

5. Rate of return:

$$\frac{\$538.19}{\$5,220.85} = 10.3\%$$

Note that we did not get the cost of the 200 shares sold initially because that would place the brokerage fee at the second line of the chart. But in reality, he already paid the fees on the entire 350 shares purchased based on the third line of the chart. That was why we just calculated the share of these 200 sold out of what he paid for the total number of shares he purchased (350).

Example 1.1.2 Suppose that a new investor decided to dedicate $5,000 to purchase a certain stock and pay its fees. If he was told that the commission would be $50 plus .007, find (1) how many shares he would get if that stock sells for $28.58, and (2) what would be his yield if that stock pays $2.33 dividend per share.

First, we have to determine his net investment; that is, we exclude the brokerage fee from his $5,000.

$$NI + .007NI + 50 = \$5,000$$

$$NI(1 + .007) = \$4,950$$

$$NI = \frac{\$4,950}{1.007} = \$4,915.60 \qquad \text{net investment}$$

Since the stock price per share is $28.50, we would find the number of shares:

$$\frac{\$4,915.60}{\$28.58} = 172 \text{ shares}$$

His total dividends would be

$$172 \times \$2.33 = \$400.76$$

The yield on his investment is

$$\frac{\$400.76}{\$5,000} = 8\%$$

1.2. COMMON STOCK VALUATION

The estimation of stock value is needed whether or not the stock is actively traded in the stock exchange. It is important to find the price or value of securities such as the stocks, not only for buying and selling purposes, but also for other financial purposes such as estate tax valuation and new-share insurance. The main premise in consideration of the current value of a share of common stock is that it should be estimated by the present value of its future cash flows, which are the dividends that would come in the future. If an investor purchases a certain stock for a price of P_0, she would expect to receive dividends D on this investment as long as she keeps holding the shares. Stocks do not have any maturity date, so the future of receiving dividends is open. Another investor's expectation would be the price appreciation (PA) of her stock prices in the market that would bring a capital gain through selling at a higher price. The formal rate of return that investors expect to receive on their investment is called the **market capitalization rate** (MCR), or **expected rate** (Er), or just the **rate of return** (r). It is determined by

$$\text{MCR} = \text{Er} = r = \frac{D + PA}{P_0}$$

where D is the expected dividend per share and PA is the stock price appreciation that would be obtained by getting the difference between the current purchase price P_0 and the expected price (P_1) of that stock a year after the purchase.

$$PA = P_1 - P_0$$

We can, therefore, adjust the formula to

$$\boxed{r = \frac{D + P_1 - P_0}{P_0}}$$

That is, an investor's expected annual return. It is weighing two typical types of returns, the expected cash dividend per share and the capital gain, where both are weighed relative to the original purchase price of stock (P_0).

Example 1.2.1   Gill purchases 50 shares of stock in a local firm at $75.00 per share. He expects to get a dividend of $4.00 per share at the end of the year. He also expects to sell his shares at $80.00 each. What is his expected rate of return (r)?

$$r = \frac{D + P_1 - P_0}{P_0}$$

$$= \frac{\$4 + \$80 - \$75}{\$75}$$

$$= 12\%$$

We can also get this rate if we consider all shares.

$50 \times \$75 = \$3,750$	purchase price for investment
$50 \times \$80 = \$4,000$	sale price for investment
$50 \times \$4 = \$200$	total dividend expected

$$r = \frac{\$200 + \$4,000 - \$3,750}{\$3,750} = 12\%$$

Mathematically, we can obtain any variable in a formula in terms of the other variables available. Let's assume that we are given a forecast for next year's price of a stock (P_1), its dividend (D_1), and the market capitalization rate (r). Shouldn't that mean that we can calculate the current price of that stock (P_0)? Technically, yes. The current price (P_0) is

$$P_0 = \frac{D_1 + P_1}{1 + r}$$

Basically, this means that we discounted the future returns of the shares into their current value. We can also formulate the price next year (P_1) in terms of the expected dividend for the year after (D_2) and the price of that year (P_2):

$$P_1 = \frac{D_2 + P_2}{1 + r}$$

and if we plug this value of P_1 into the P_0 formula above, we get

$$P_0 = \frac{1}{1 + r}(D_1 + P_1)$$

$$= \frac{1}{1 + r}\left(D_1 + \frac{D_2 + P_2}{1 + r}\right)$$

$$= \frac{1}{1 + r}\left[D_1 + \frac{1}{1 + r}(D_2 + P_2)\right]$$

$$= \frac{D_1}{1 + r} + \frac{D_2 + P_2}{(1 + r)^2}$$

and if we continue to get our estimation for the next year after (year 3), we get

$$P_0 = \frac{D_1}{1+r} + \frac{D_2 + P_2}{(1+r)^2} + \frac{D_3 + P_3}{(1+r)^3}$$

and similarly for any number of years in the future, such as k, and if we substitute for P_2 in terms of P_3, and for P_3 in terms of P_4, and so on, all subsequent P's would disappear and we end up with

$$P_0 = \frac{D_1}{1+r} + \frac{D_2}{(1+r)^2} + \cdots + \frac{D_k + P_k}{(1+r)^k}$$

and the final summation would be

$$P_0 = \sum_{t=1}^{k} \frac{D_t}{(1+r)^t} + \frac{P_k}{(1+r)^k}$$

This means that the current price of a stock is actually equal to the sum of all discounted dividends for a number of future years, defined by the period k. It is noteworthy to say here that as k continues to increase and approaches infinity, the value of the last price would approach zero, and for this reason, the last price can be eliminated and the formula becomes

$$P_0 = \sum_{t=1}^{\infty} \frac{D_t}{(1+r)^t}$$

This means that the present value of a common stock is just like the present value of any other asset. It is defined by the summation of the discounted stream of future dividends. However, this formula assumes a case of zero growth in dividends. In other words, the dividends stay constant throughout future years. It implies that all D's are the same, and therefore the formula can be written as

$$P_0 = D_1 \sum_{t=1}^{\infty} \frac{1}{(1+r)^t}$$

and since $\sum_{t=1}^{\infty} \left[1/(1+r)^t \right]$ is just the table present value interest factor (PVIF) for any r and any t, we can rewrite the formula as

$$P_0 = D_1(\text{PVIF}_{r,t})$$

$$\boxed{P_0 = \frac{D_1}{r}}$$

Example 1.2.2   If Bright Paint stock pays a $14.95 dividend per share and it is expected to be constant indefinitely, what would the value of this stock be if the required return is $11\frac{1}{2}\%$?

$$P_0 = \frac{D_1}{r}$$

$$= \frac{\$14.95}{.115}$$

$$= \$130.00$$

But if the dividend grows in a constant rate such as g, the current value of the common stock would be adjusted by $D + Dg = D(1 + g)$:

$$P_0 = \frac{D_1(1+g)}{1+r} + \frac{D_2(1+g)^2}{(1+r)^2} + \cdots + \frac{D_k(1+g)^k}{(1+r)^k}$$

and we end up with

$$\boxed{P_0 = \frac{D_1}{r - g}} \qquad r > g$$

where r, the required return or the market capitalization rate, is assumed to be larger than g, the expected constant growth rate for dividends. This formula is called the **Gordon formula** after M. J. Gordon, who along with E. Shapiro, published an article entitled "Capital Equipment Analysis: The Required Rate of Profit," which was published in *Management Science* in 1956. This formula was developed originally by J. B. Williams in his book *The Theory of Investment Value*, published by Harvard University Press in 1938, but stayed unpopular until Gordon and Shapiro rediscovered it 18 years later.

Since $D_1 = D_0 + D_0 g$, $D_1 = D_0(1 + g)$, we can write the foregoing formula as

$$\boxed{P_0 = \frac{D_0(1 + g)}{r - g}}$$

Example 1.2.3   Suppose that an investor wishes to get a 12% yield from a stock whose estimated growth is 8%. Suppose also that the firm selling the stock has distributed 60% of its earnings as dividends, that its earnings per share is $3.20, and that it sells the stock for $40.00 per share. What would the value of the stock be, and what would be the expected return?

First, we get the dividend as 60% of the earnings per share.

$$D_0 = \$3.20(.60) = \$1.92$$

$$P_0 = \frac{D_0(1 + g)}{r - g}$$

$$= \frac{\$1.92(1 + .08)}{.12 - .08}$$

$$= \$51.84 \qquad \text{what the current stock price should be}$$

$$r = \frac{D}{P_0} + g$$

$$= \frac{\$1.92}{\$40} + .08$$

$$= 12.8\%$$

Note that we used the actual price $40 for P_0, not the estimated current value ($51.84). Moreover, it is worthwhile to note that this r (the market capitalization rate that would often serve as the cost of common stock), would be obtained directly from the Gordon formula:

$$\boxed{r = \frac{D_1}{P_0} + g}$$

Example 1.2.4 The common stock for Big Book Company sells for $55 and its dividend in 2010 was $4.30, which has been growing through the last five years as shown in Table E1.2.4. What would the cost of this common stock be?

From the past history of dividends, we can calculate g, the rate of growth from 2005 to 2009:

$$g = \sqrt[n]{\frac{FV}{CV}} - 1$$

$$= \sqrt[4]{\frac{\$4.00}{\$2.95}} - 1$$

$$= .08$$

TABLE E1.2.4

Year	Dividend ($)
2009	4.00
2008	3.75
2007	3.33
2006	3.15
2005	2.95

$$r = \frac{D_1}{P_0} + g$$

$$= \frac{\$4.30}{\$55} + .08$$

$$= 15.8\%$$

There is another way to find g value. Given that D_1/P_0 is actually the dividend yield, g can be found by multiplying the return on equity (ROE) by the plowback ratio (Plow):

$$g = (ROE)(Plow)$$

where ROE is the ratio of the earnings per share (EPS) to the equity per share (Eq).

$$ROE = \frac{EPS}{Eq}$$

and Plow is the complementary amount of the payout ratio (Pout):

$$Plow = 1 - Pout$$

and the payout ratio (Pout) is the ratio of the expected dividend (D_1) to the earnings per share (EPS).

$$Pout = \frac{D_1}{EPS_1}$$

Example 1.2.5   A common stock is selling for \$35.00 per share, and its expected dividend at the end of next year is \$1.15 per share. The earning per share in this firm is \$2.30, and the book equity per share is \$14.50. What would the market capitalization rate or the cost of this stock be?

$$ROE = \frac{EPS}{Eq}$$

$$= \frac{\$2.30}{\$14.50} = 16\%$$

$$Pout = \frac{D_1}{EPS_1}$$

$$= \frac{\$1.15}{\$2.30} = .50$$

$$Plow = 1 - Pout$$

$$= 1 - .50 = .50$$

$$g = \text{ROE} \cdot \text{Plow}$$

$$= .16(.50) = .08$$

$$r = \frac{D_1}{P_0} + g$$

$$= \frac{\$1.15}{\$35} + .08$$

$$= 11.3\%$$

1.3. COST OF NEW ISSUES OF COMMON STOCK

The cost of a new issue of common stock is obtained by the same market capitalization rate formula but also by consideration of:

1. An underpriced stock to be sold at a price below its current market price (P_0). This may seem necessary to the firm in order to sell a new issue of the stock. The underpricing amount (Un) would be the difference between current price (P_0) and new price (P_n).

$$\text{Un} = P_0 - P_n$$

2. The flotation cost (FC), which is the cost of issuing and selling the new stock.

So if both items above are subtracted from the stock price, we would get what is called the **net proceed** (N_n), which would replace P_0 in the new formula of the cost of a new issue of stock (r_n).

$$\boxed{r_n = \frac{D}{N_n} + g}$$

Example 1.3.1   Suppose that there is a new issue of the stock in Example 1.2.5. The firm decides to underprice it by \$1.55 per share and calculated the flotation cost at \$.75 per share. Determine the cost of the new issue of this common stock. $P_0 = 55$; Un $= 1.55$; Fc $= \$.75$; $D_1 = \$4.30$; $g = .08$.

$$N_n = P_0 - (\text{Un} + \text{Fc})$$

$$= 55 - (\$1.55 + \$.75)$$

$$= \$52.70$$

$$r_n = \frac{D}{N_n} + g$$

$$= \frac{\$4.30}{\$52.70} + .08$$

$$= 16.2\%$$

Note that the cost of the new issue is greater than the cost of the current stock because of dividing the dividend by the new net proceeds, which is less than the current price.

1.4. STOCK VALUE WITH TWO-STAGE DIVIDEND GROWTH

The Gordon formula assumes that the expected growth rate for dividend (g) is lower than the required return or market capitalization rate (r). But suppose that at some point, earnings and dividends go up to exhibit a very high growth, sufficient for g to be greater than r. In such a case, Gordon's formula would produce a negative value for the stock, and that is why we should estimate the value of stock using the two-stage growth formula, which addresses the growth in two stages and estimates the stock value by combining the present value of dividends from year 1 to n with the present value of dividends from $n + 1$ to infinity.

$$P_0 = D_0 \cdot \frac{1 - [(1 + g_1)/(1 + r)]^n}{r - g_1}(1 + g_1) + D_{n+1}\left(\frac{1}{r - g_2}\right)\left(\frac{1}{1 + r}\right)^n$$

Example 1.4.1 Find the current price per share of the common stock at SURE Corporation. Annual dividends are expected to grow by 25% for the next 5 years and slow back to 5% after that. Given that the dividend is \$37.00 per share and the required rate of return is 15%.

$g_1 = .25; g_2 = .05; D_0 = \$37.00; r = .15; n = 5.$

$$P_0 = D_0 \cdot \frac{1 - [(1 + g_1)/(1 + r)]^n}{r - g_1}(1 + g_1) + D_{n+1}\left(\frac{1}{r - g_2}\right)\left(\frac{1}{1 + r}\right)^n$$

$$= 37\left[\frac{1 - [(1 + .25)/(1 + .15)]^5}{.15 - .25}\right](1 + .25)$$

$$+ 37(1 + .25)^5(1 + .05)\left(\frac{1}{.15 - .05}\right)\left(\frac{1}{1 + .15}\right)^5$$

$$= \$239.23 + \$589.46 = \$828.69$$

1.5. COST OF STOCK THROUGH THE CAPM

The **capital asset pricing model** (CAPM) addresses the relationship between the required return (or in this context, the cost of common stock) and the nondiversifiable portion of the overall risk that a firm may face. The nondiversifiable risk is the external part of the asset's risk, which is attributable to out-of-firm circumstances and conditions and which affect all firms in the industry, therefore cannot be reduced or eliminated by the usual remedy of portfolio diversification. It is expressed in the model by beta (β) as an index to measure the variation of an asset's return in response to the changes in the market return. According to this model, the required return or cost of common stock (r_c) is obtained by

$$r_c = R_f + \beta(R_m - R_f)$$

The cost is actually a combination between the risk-free rate of return (R_f), which is measured by the return on a U.S. Treasury bill, and the risk premium $\beta(R_m - R_f)$ which is the difference between the market return (R_m) and the risk-free return (R_f) modified by beta (β).

Example 1.5.1   Use the CAPM to estimate the cost of Joyland's common stock if its beta equals 1.70 and the market return is 12%. Given the risk-free rate of 8%:

$$R_f = 8\%, R_m = 12\%, \text{ and } \beta = 1.70.$$
$$r_c = R_f + \beta(R_m - R_f)$$
$$= .08 + 1.70(.12 - .08)$$
$$= 14.8\%$$

1.6. OTHER METHODS FOR COMMON STOCK VALUATION

The P/E Multiples Method

In the **P/E multiples method**, the value of a common stock in a firm (V_c) would be measured according to use of the price/earning ratio (P/E_i) of the industry, which is often published by some standard financial reports. The value (V_c) would be obtained by multiplying the firm's expected earnings per share (EPS$_e$) by the industry's P/E multiples:

$$V_c = \text{EPS}_e(P/E_i)$$

This method is more suited for firms that are not publicly traded.

Example 1.6.1   A train service corporation expects its earnings per share to be $3.10 at a time that the price/earnings ratio in the land transportation industry is rated at 9. What would be a good estimation of the value of a share in the train service corporation?

$$V_c = \text{EPS}_e(P/E_i)$$
$$= \$3.10(9)$$
$$= \$27.90$$

The Book Value per Share Method

The **book value per share method**, is based on a hypothetical situation by which one would think that a good estimation of the value of a share in a firm would be the stockholder's per capita share of all the firm's net proceeds, if all assets are liquidated and all liabilities, including the preferred stocks, are paid. Assets and liabilities have their accounting value already recorded in the firm's books, which is why this method is called the **book value method**.

Example 1.6.2   The ABC Company's books show that the total assets are $2.7 million and total liabilities are $1.2 million. The company's preferred stock dividends would reach $650,000 and it has 85,000 common stock shares. What is the value of a share?

$$V_c = \frac{\$2,700,000 - (\$1,200,000 + \$650,000)}{85,000}$$

$$= \$10.00$$

The Liquidation Value per Share Method

The **liquidation value per share method** is exactly the same as the book value method except that this one is actual, not hypothetical. It is about the value of a share in a firm that is actually being liquidated.

Example 1.6.3   A firm's team responsible for handling the firm's assets liquidation reported that $2.3 million would remain after they pay all liabilities, including the preferred stock. The firm plans to distribute this net proceeds among the 110,000 shares of common stocks. The value of a share would, therefore, be

$$V_c = \frac{\$2,300,000}{110,000}$$

$$= \$20.91$$

1.7. VALUATION OF PREFERRED STOCK

Since preferred stock does not have any maturity date and it pays a fixed dividend for as long as the stock is outstanding, the payment of dividends can be considered a perpetuity, and for that we can write the formula of its value (P_p) as

$$P_p = D_p \left(\frac{1}{r_p} \right)$$

$$\boxed{P_p = \frac{D_p}{r_p}}$$

Example 1.7.1   Wide Range Company pays a fixed annual dividend of $7.50 to their preferred stockholders. What is the value of this preferred stock if the required rate of return is estimated at 14.5%?

$$
\begin{aligned}
P_p &= \frac{D_p}{r_p} \\[2mm]
&= \frac{\$7.50}{.145} \\[2mm]
&= \$51.72
\end{aligned}
$$

1.8. COST OF PREFERRED STOCK

The **cost of preferred stock** is represented by the required rate of return (r_p), which can be obtained by rearranging the preceding formula:

$$
r_p = \frac{D_p}{P_p}
$$

But instead of using the value of the preferred stock (P_p), we have to use the net proceeds (N_p), which would be P_p after subtracting any flotation cost (FC).

$$
N_p = P_p - \text{FC}
$$

and the right formula becomes

$$
\boxed{r_p = \frac{D_p}{N_p}}
$$

Example 1.8.1   If XYZ Corporation pays out 9% of its stock par value of $96 as a dividend for preferred stock given that the cost to issue and sell the stock is $4.00, what is the cost of the stock?

$$
\begin{aligned}
D_p &= .09(\$96) = \$8.64 \\[2mm]
N_p &= \$96 - \$4 = \$92 \\[2mm]
r_p &= \frac{D_p}{N_p} \\[2mm]
&= \frac{\$8.64}{\$92} \\[2mm]
&= 9.4\%
\end{aligned}
$$

2 Bonds

Bonds are one of the major securities that are traded in the capital market. A **bond** is a long-term investment instrument that is usually issued and sold by big businesses and governments to a diverse group of lenders for the purpose of raising large amounts of capital. Bonds are written promises by the borrower (governments or corporations) to the lender (the investors, individuals or institutions) to pay:

1. The full amount borrowed on or before a certain date in the future called a **maturity date** or **redemption date**. This amount is also called the **redemption value**. It is often the amount that is printed on the face of the bond, called the **face value** or **par value** of the bond. It usually comes in a denomination of $1,000 or its multiples.

2. A fixed set of cash payoffs calculated as the semiannual interest on the face value, using a certain interest rate called the **bond rate**, **coupon rate**, or **contract rate**, and it is usually stated in annual form. These interest payments are paid throughout the maturity period and until the amount borrowed is redeemed.

The maturity period is often 10 years and more. It is usually divided into semiannual payments and the interest is paid on certain days called **payment days** or **coupon days**.

2.1. BOND VALUATION

Since the investor receives interest payments throughout the maturity period and receives the principal back on the redemption date, the value of the bond at the time of purchase is assessed as the present value of the future payments of interest plus the present value of the redemption amount. Suppose that a bond with a par value of $1,000.00 matures in 10 years, earning a 8% bond rate. The annual interest would be

$$\$1,000 \times .08 = \$80$$

Mathematical Finance, First Edition. M. J. Alhabeeb.
© 2012 John Wiley & Sons, Inc. Published 2012 by John Wiley & Sons, Inc.

But since the interest payments are made on semiannual terms, each payment would be $40 for the next 20 terms. The last term would be paying the 20th interest payment of $40 plus the original amount borrowed, $1,000: that is, $1,040.

Year 1		Year 2		$\longrightarrow$	Year 9		Year 10	
term 1	term 2	term 3	term 4	$\longrightarrow$	term 17	term 18	term 19	term 20
40	40	40	40	$\longrightarrow$	40	40	40	40 1,000

The value of this bond at the time of purchase (B_0) would be equal to the present value of all the future cash flows in the table above. So if we discount all interest payments and the redemption amount back to the current value, using a semiannual rate of .04 (.08/2) and a maturity of 20 (10 × 2), we should get $1,000 as a current value:

$$B_0 = 40 \left[\frac{1}{(1 + .04)^1} \right] + 40 \left[\frac{1}{(1 + .04)^2} \right] + \cdots + 40 \left[\frac{1}{(1 + .04)^{20}} \right]$$
$$+ 1,000 \left[\frac{1}{(1 + .04)^{20}} \right]$$
$$= 38.462 + 36.98 + \cdots + 18.256 + 474.64$$
$$= \$1,000$$

Figure 2.1 shows how all cash flows of the bond, including the redemption value plus all the interest payments are brought back in time from the future to the present to form the discounted value in the current time. Suppose that the bond in the example is to be purchased in 2011 and to get redeemed in 2021. But if the investor wants to assess the value of this investment by comparing its return to returns of other alternatives, the comparison would require getting the present value of the same set of cash flows using an interest rate that differs from the bond rate stated. This interest rate is usually the prevailing rate of return on comparable investment. It is considered the desired or required rate of return (i). It is often called the **current rate** or the **yield to maturity** (YTM).

So let's assume that at the time of considering this bond as an investment purchase, the market rate of return on similar securities is 10%, which would be what the investor is giving up if buying the bond at 8%. In this case we can obtain the value of the bond of 8% coupon rate by discounting the previous cash flows at the yield rate of 10%, which gives us a semiannual rate of .05.

$$B_0 = 40 \left[\frac{1}{(1 + .05)^1} \right] + 40 \left[\frac{1}{(1 + .05)^2} \right] + \cdots$$
$$+ 40 \left[\frac{1}{(1 + .05)^{20}} \right] + 1,000 \left[\frac{1}{(1 + .05)^{20}} \right]$$
$$= 38.09 + 36.28 + \cdots + 15.07 + 376.90$$
$$= \$875.38$$

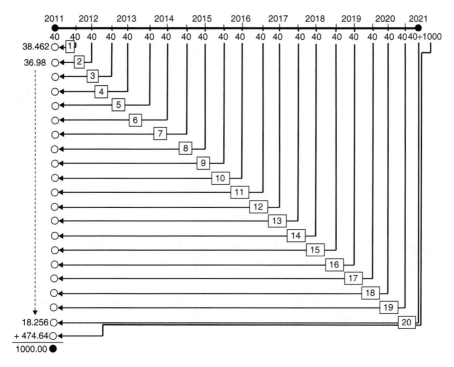

FIGURE 2.1

Notice that the value of the bond turned out to be less than before when using a higher rate of return. The reason is that we are reversing the direction from the future to the present when we discount. The more money grows forward, the more it shrinks backward.

We can use the following general formula to obtain the value of a bond with a required yield rate (i):

$$B_0 = I \left[\sum_{t=1}^{n} \frac{1}{(1+i)^t} \right] + M \frac{1}{(1+i)^n}$$

where B_0 is the value of the bond or the purchase price at the required or desired yield rate, I is the interest payment per semiannual term, i is the required or desired interest yield rate, t is any semiannual term, n is the number of semiannual terms in the maturity period, and M is the redemption value of the bond.

Knowing that the term $\sum_{t=1}^{n} 1/(1+i)^t$ is the accumulated present value interest factor (PV/FA), that the term $1/(1+i)^n$ is the present value interest factor (PVIF), and that both of them are the same table values that we dealt with before, we can rewrite the formula above in the following way, which would allow us to use the table values:

$$B_0 = I(\text{PV}/\text{FA}_{i,n}) + M(\text{PVIF}_{i,n})$$

and it can also be written, using $a_{\overline{n}|i}$ and v^n, as we have shown earlier. The formula can then also be written this way:

$$B_0 = I(a_{\overline{n}|i}) + M(v^n)$$

The more the semiannual terms, the more tedious the manual calculations become and therefore utilizing the table values comes in handy and makes it much easier.

Example 2.1.1   A $1,000 bond is maturing in 15 years. Its semiannual coupon is at $9\frac{1}{2}\%$. Find the purchase price that would yield 11%.

The annual coupon rate is $9\frac{1}{2}\%$, so the semiannual rate is $4\frac{3}{4}\%$ or .0475. $I = \$1,000(.0475) = \$47.50$. The redemption value (M) is $1,000, the annual yield is 11%, and the semiannual yield is $5\frac{1}{2}\%$ or .055.

$$\begin{aligned}
B_0 &= I(a_{\overline{n}|i}) + M(v^n) \\
&= \$47.50(a_{\overline{30}|.055}) + \$1,000(v^n) \\
&= \$47.50(14.534) + \$1,000(.2006) \\
&= \$890.96
\end{aligned}$$

Example 2.1.2   If the face value of a bond is $2,000, its coupon rate is $12\frac{1}{4}\%$, and if it is redeemable at par at the end of 10 years, what would its current value be if the investor wants the yield to be 10%?

The bond rate is $12\frac{1}{4}\%$ annually, so the semiannual rate is .06125. The interest payment would be

$$\$2,000 \times .06125 = \$122.50$$

The redemption value (M) is $2,000, the annual required rate of return is 10% and the semiannual rate is 5%.

$$\begin{aligned}
B_0 &= I(a_{\overline{n}|i}) + M(v^n) \\
&= \$122.50(a_{\overline{20}|.05}) + \$2,000(v^n) \\
&= \$122.50(12.462) + \$2,000(.3769) \\
&= \$2,280.40
\end{aligned}$$

2.2. PREMIUM AND DISCOUNT PRICES

In Examples 2.1.1 and 2.1.2 we saw that when the required yield was larger than the bond rate, the value or purchase price was lower than the face value of the bond, and when the yield rate was lower than the bond rate, the value or purchase price was higher than the face value.

$$\text{if}\quad i > r: B_0 \downarrow$$

$$\text{if}\quad i < r: B_0 \uparrow$$

If the $1,000 bond at $9\frac{1}{2}\%$ bond rate is sold for $890.96 to yield 11% return, it is said to be sold at discount, and the discount amount is the difference between the face value and the purchase price:

$$\$1,000 - \$890.96 = \$109.04 \qquad \text{discount}$$

Similarly, if the $2,000 bond at a $12\frac{1}{4}\%$ bond rate is sold for $2,280,40 to give a 10% yield, the bond is said to be sold at premium, and the premium amount would be the difference between the purchase price and the face value:

$$\$2,280.40 - \$2,000 = \$280.40 \qquad \text{premium}$$

So when the purchase price is different from the face value of the bond, the difference, discount or premium, is carried by the buyer (investor). We can look at the discount of $109.04 this way:

$$\frac{\$1,000 \times .095}{2} = \$47.50 \qquad \text{semiannual interest based on } 9\tfrac{1}{2}\%$$

and

$$\frac{\$1,000 \times .11}{2} = \$55.00 \qquad \text{semiannual interest based on } 11\%$$

$$\$55.00 - \$47.50 = \$7.50 \qquad \text{difference in interest per term}$$

If we discount this difference for the entire maturity (30 terms) at a yield rate of 11% (semiannual 5.5%), we get

$$\$7.50(a_{\overline{30}|.055})$$

$$47.50(14.534) = \$109.04 \qquad \text{discount}$$

which says that the discount is the present value of the differences in interest throughout the entire maturity period and that it must be subtracted from the par

value (redemption amount) to form the current value or purchase price of the
bond:

$$M - D_s = B_0$$

$$\$1{,}000 - \$109.04 = \$890.96$$

In the same manner, we can look at the premium of $280.40 in the following
way:

$$\frac{\$2{,}000 \times .1225}{2} = \$122.50 \qquad \text{semiannual interest based on } 12\tfrac{1}{4}\%$$

$$\frac{\$2{,}000 \times .10}{2} = \$100.00 \qquad \text{semiannual interest based on } 10\%$$

$$\$122.50 - \$100 = \$22.50 \qquad \text{difference in interest per term}$$

Discounting this difference at a yield rate of 10% or 5% semiannually for 20
terms gives us

$$\$22.50(a_{\overline{20}|.05})$$

$$\$22.50(12.462) = \$280.40 \qquad \text{premium}$$

This also says that the premium is, in fact, the present value of the difference in
interest throughout the maturity period of 20 terms and therefore must be added
to the full (redemption) value to form the purchase price of the bond:

$$M + P_m = B_p$$

$$\$2{,}000 + \$280.40 = \$2{,}280.40$$

Generally, we can write the formula of discount (D_s) and premium (P_m) as the
present value of the differences between face value and the purchase price of a
bond:

$$D_s = (M_i - M_r)a_{\overline{n}|i}$$

$$\boxed{D_s = M(i - r)a_{\overline{n}|i}} \qquad \text{when} \quad i > r$$

$$P_m = (M_r - M_i)a_{\overline{n}|i}$$

$$\boxed{P_m = M(r - i)a_{\overline{n}|i}} \qquad \text{when} \quad i < r$$

In conclusion, a bond may sell at a value that is less than its par value when
the required rate of return is greater than the rate stated on the bond. It would
be a case of selling at a discount. It may also sell at a value higher than its face
value when the required rate of return is lower than the bond rate. That would
be selling at a premium.

Example 2.2.1   A bond with a face value of $5,000 is redeemable in 8 years at a bond rate of 7.5%. If it is to be purchased to yield 6%, would it be purchased at a premium or discount, and what would the purchase price be?

First let's get the semiannual rates:

$$\text{semiannual bond rate} = .075/2 = .0375$$

$$\text{semiannual yield} = .06/2 = .03$$

Since the required yield is less than the bond rate, the bond would sell at a premium.

$$P_m = M(r - i)a_{\overline{n}|i}$$

$$= \$5,000(.0375 - .03)a_{\overline{8}|.03}$$

$$= \$5,000(.0075)(7.019)$$

$$= \$263.24 \quad \text{premium}$$

$$B_p = M + P_m$$

$$= \$5,000 + \$263.24 = \$5,263.24$$

Example 2.2.2   A $2,000 bond pays $10\frac{1}{2}$% bond interest twice a year for 20 years, at the end of which it will be redeemable at par value. If the investor wants to get a 12% yield, would the bond sell at a premium or a discount, and for how much?

$$\text{semiannual bond rate} = .105/2 = .0525$$

$$\text{semiannual yield} = .12/2 = .06$$

Since the yield is higher than the coupon rate, the bond would sell at a discount.

$$D_s = M(i - r)a_{\overline{22}|.06}$$

$$= \$2,000(.06 - .0525)(11.4699)$$

$$= \$172.05 \quad \text{discount}$$

$$B_d = M - D_s$$

$$= \$2,000 - \$172.05 = \$1,827.95$$

2.3. PREMIUM AMORTIZATION

As mentioned earlier, a premium is recovered gradually throughout the interest payments, and therefore the investor would not receive the premium price at redemption. Only the original redemption value would be received in the final term. For that reason the recorded premium price on the investor's book should be reduced as the fraction of premium is received semiannually. In other words,

TABLE 2.1 Premium Amortization Schedule

(1) Semiannual Term	(2) Interest Payment [(par) (r)]	(3) Yield .0225 [.0225(5)]	(4) Amortization of Premiums [(2) − (3)]	(5) Book Value of the Bond [(5) − (4)]
				3,163.06
1	90	71.17	18.83	3,144.23
2	90	70.74	19.25	3,124.97
3	90	70.31	19.69	3,105.28
4	90	69.87	20.13	3,085.15
5	90	69.42	20.58	3,064.57
6	90	68.95	21.05	3,043.52
7	90	68.48	21.52	3,022.00
8	90	67.99	22.00	3,000.00
	720	556.94	163.06	

the book value of the premium price has to be amortized and reduced to be equal to the original redemption value when the redemption date comes.

Let's consider a bond of \$3,000 at 6% redeemable in 4 years and purchased to provide a $4\frac{1}{2}\%$ yield. The semiannual bond rate is .03 and the semiannual yield is .0225.

$$P_m = M(r - i)a_{\overline{n}|i}$$
$$= \$3,000(.03 - .0225)a_{\overline{8}|.0225}$$
$$= \$3,000(.0075)(7.2472)$$
$$= \$163.06$$
$$B_p = \$3,000 + \$163.06 = \$3,163.06$$

The amortization schedule in Table 2.1 shows how the premium price of \$3,085.16 is gradually reduced to the redemption value of \$3,000 by the end of the 8th semiannual term of the maturity time of 4 years.

$$I = \$3,000(.03) = \$90$$

To construct the amortization table we carried out the following steps:

1. The first entry was the premium price \$3,163.06 as the first book value.
2. We multiplied the book value by the yield rate to get the first entry in column (3):

$$\$3,163.08 \times .0025 = \$71.17$$

3. We subtracted $71.17 from the interest of the first term (90) to get the first entry on column (4):

$$\$90 - \$71.17 = \$18.83$$

4. We got the second entry of the book value by subtracting the amortization amount of $18.83 from the previous book value:

$$\$3,163.06 - \$18.83 = \$3,144.23$$

5. We multiplied this new book value of $3,144.23 by the yield rate as in step 2 and repeat the sequence until you get the following results:
   (a) The last book value must equal the face value of the bond: in this case, $3,000.
   (b) The total of column (4) must equal the premium: in this case, $163.06.
   (c) The accumulated yield [total of column (3)] must equal the difference between the total interest and the premium: in this case:

$$\$720 - \$163.06 = \$556.94.$$

2.4. DISCOUNT ACCUMULATION

Unlike the case of the premium price, the discount price is less than the face value, but the investor would still receive the face value at redemption. Therefore, the total deficit of the discount would be recovered gradually and little by little through the interest payments. This would require the book value to increase gradually from the discount price until it becomes equal to the face value of the bond on the redemption day, and that is what the discount accumulation is all about.

Let's consider a bond whose face value is $8,000 redeemable in 3 years at 5%, but it is purchased to yield 8%.

$$\text{semiannual bond rate} = .05/2 = .025$$
$$\text{semiannual yield} = .08/2 = .04$$

Since the yield is greater than the bond rate, the bond is purchased at discount.

$$D_s = M(I - r)a_{\overline{n}|i}$$
$$= \$8,000(.04 - .025)a_{\overline{6}|.04}$$
$$= \$8,000(.015)(5.2421)$$
$$D_s = \$629.05 \qquad \text{discount}$$
$$B_d = \$8,000 - \$629.05 = \$7,370.95 \qquad \text{purchase price}$$

TABLE 2.2   Discount Accumulation Schedule

(1) Semiannual Term	(2) Interest Payment [(par) (r)]	(3) Yield .04 [.04(5)]	(4) Accumulation of Discount [(3) − (2)]	(5) Book Value of the Bond [(4) + (5)]
				7,370.95
1	200	294.84	94.84	7,465.79
2	200	298.63	98.63	7,564.42
3	200	302.58	102.58	7,666.99
4	200	306.68	106.68	7,773.67
5	290	310.95	110.95	7,884.62
6	200	315.38	115.38	8,000.00
	1,200	1,829.05	629.05	

$$I = \$8,000(.025)$$

$$= \$200 \qquad \text{semiannual interest payment}$$

To construct Table 2.2 we carried out the following steps:

1. The first entry was the discount price $7,370.95 as the first recording of the book value.

2. We multiplied the book value by the yield rate to get the first entry in the column (3).

$$\$7,730.95 \times .04 = \$294.84$$

3. We subtracted the interest of the first term ($200) from $294.84 to get the first entry in column (4) ($94.84).

$$\$294.84 - \$200 = \$94.84$$

4. We added this $94.84 to the first book value to get the second book value.

$$\$7,370.95 + \$94.84 = \$7,465.79$$

5. We multiplied this new book value by the yield rate as in step 2 and repeated steps 2 to 5 until all the deficits were accumulated to complete the book value in the face value of $3,000.00 as the last recorded book value.

6. We checked that:
   (a) The total of column (4) is equal to the amount of the discount ($629.05).
   (b) The total of column (3) is the sum of the discount and the total interest.

$$\$1,829.05 = \$629.05 + \$1,200$$

2.5. BOND PURCHASE PRICE BETWEEN INTEREST DAYS

The bond price (B_0) that we established earlier assumed that an investor would buy on one of the interest payment dates of the maturity time. But, in reality, that may or may not happen. In case the purchase occurs on any day in between these established coupon days, we need to know two more bond price terms: purchase price between dates (B_{bd}), sometimes called the **flat price**, and the **quoted price** (B_q). We also need to distinguish between the interest at the bond rate and the interest at the yield rate, both of which are involved in the calculation.

$$B_{bd} = B_0 + Y_{bp}$$

So the bond purchase price between dates (B_{bd}) is a sum of the purchase price (B_0) calculated on the day of payment that immediately precedes the day of purchase, and the portion of interest (Y_{bp}) on the purchase price (B_0) and for the time between the purchase date and the day of payment before it (bp). This interest is calculated based on the yield rate (i):

$$Y_{bp} = B_0(i)b_p$$

The quoted price (B_q), sometimes called the **net price**, is

$$B_q = B_{bd} - I_{ac}$$

where I_{ac} is the interest on the face value (M) and for the time between the purchase date and the payment date before it (bp), but calculated based on the bond or coupon rate (r). It is called the **accrued interest** and represents the interest portion that belongs to the seller.

$$I_{ac} = M(r)b_p$$

Example 2.5.1 A bond of $1,000 at 8% maturing in 10 years on February 2020 (see Figure E2.5.1). Its interest has been paid every August 15 and February 15 starting August 15, 2010. Find the purchase price and the quoted price if the bond is purchased on December 15 to make an 11% yield.

The semiannual bond rate $= .08/2 = .04$; the yield rate $= .11/2 = .055$; $n = 20$; $M = \$1,000$; $I = \$1,000(.04) = 40$.

$$B_0 = I(a_{\overline{n}|i}) + M(v^n)$$

$$= 40(a_{\overline{n}|i}) + \$1,000(v^n)$$

$$= 40(11.9504) + \$1,000(.3427)$$

$$= \$820.72 \qquad \text{purchase price on the interest date}$$

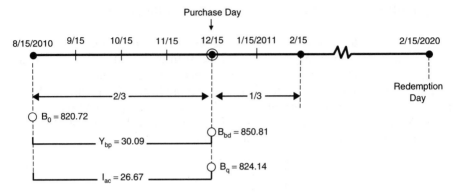

FIGURE E2.5.1

$$Y_{bp} = B_0(i)b_p$$
$$= \$820.72(.055)(\tfrac{1}{3})$$
$$= \$30.09$$
$$B_{bd} = \$820.72 + \$30.09$$
$$= \$850.81$$
$$I_{ac} = M(r)b_p$$
$$= \$1,000(.04)(\tfrac{1}{3})$$
$$= \$26.67 \qquad \text{accrued interest}$$
$$B_q = B_{bp} - I_{ac}$$
$$= \$850.81 - \$26.67$$
$$= \$824.14$$

Example 2.5.2   A bond of \$5,000 at 5% matures in 6 years and 4 months (see Figure E2.5.2). What would the flat price be to yield 7%? Also find the seller's share of the accrued interest and the bond net price.

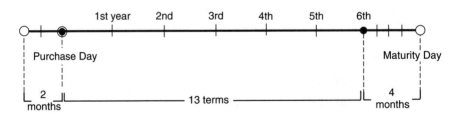

FIGURE E2.5.2

The flat price is the purchase price between dates (B_{bd}), the seller's share is the accrued interest (I_{ac}), and the net price is the price quoted (B_q). Since the redemption date is 4 months after the sixth year, we can conclude that the interest date immediately preceding the purchase date would have been 2 months before the purchase date. We can conclude that the maturity term is 13, which is 6 years × 2 terms, and the 4 months are written as the 13th term.

$M = \$5,000$; $r = 5\%$; semiannual rate $= .05/2 = .025$; $i = .07$; semiannual rate $= .07/2 = .035$; $n = 13$ terms; bp $= 2$months $= \frac{1}{3}$ of a term.

$$I = M(r)$$
$$= \$5,000(.025) = \$125$$
$$B_0 = I(a_{\overline{13}|.035}) + M(v^n)$$
$$= \$125(10.3027) + \$5,000(.6394)$$
$$= \$4,484.84 \qquad \text{purchase price on the payment date}$$
$$Y_{bp} = B_0(i)\, b_p$$
$$= \$4,484.84(.035)\left(\tfrac{1}{3}\right)$$
$$= \$52.32$$
$$B_{bd} = B_0 + Y_{bp}$$
$$= \$4,484.84 + \$52.32$$
$$= \$4,537.16 \qquad \text{purchase price between dates(flat price)}$$
$$I_{ac} = M(r)\, b_p$$
$$= \$5,000(.025)\left(\tfrac{1}{3}\right)$$
$$= \$41.67 \qquad \text{accrued interest or seller's share}$$
$$B_q = B_{bd} - I_{ac}$$
$$= \$4,537.15 - \$41.67$$
$$= \$4,495.49 \qquad \text{net or quoted price}$$

Note that the price between dates (B_{bd}) can also be obtained by what is called the **practical method**, which is based on a simple interest formula:

$$\boxed{B_{bd} = B_0[1 + i(\text{bp})]}$$

where bp is the portion of time before purchase, as we have seen. In our example, this should give us the same result:

$$B_{bd} = \$4,484.84\left[1 + (.035)\left(\tfrac{1}{3}\right)\right]$$
$$= \$4,537.16$$

2.6. ESTIMATING THE YIELD RATE

Bonds for sale in the investment market are usually listed at their quoted or net price. As we have seen before, investors can calculate the value of the prospective bond at the yield rate desired, mainly to compare to the price quoted. If the value they calculate is greater or at least equal to the quoted price, they would know that their purchase would bring a return equal to or greater than the desired yield rate. If the yield is not known, it can be approximated for the purpose of that comparison. Three methods can be used to approximate the yield rate.

The Average Method

The **average method**, also called the **bond salesman's method**, uses the average investment as the base of comparison for the annual interest income. So the yield rate (YR) is obtained by dividing the annual interest income (AII) by the average annual investment (AAI).

$$\boxed{YR = \frac{AII}{AAI}}$$

Annual interest income is obtained by applying the annual bond rate (r) on the face value (M) and adjusting it for the premium or discount in their average sense. The average premium (P_m/n) would be subtracted and the average discount (D_s/n) would be added.

$$AII = M(r) - \frac{P_m}{n}$$

or

$$AII = M(r) + \frac{D_s}{n}$$

The **average annual investment** (AAI) is the average between the face value (M) and the price quoted, B_q:

$$AAI = \frac{M + B_q}{2}$$

Example 2.6.1   If a \$4,000 at 5.5% coupon rate is to be purchased at a quoted price of \$4,350 to be redeemed in 14 years, what is the yield rate?

$$P_m = B_q - M$$
$$= \$4,350 - \$4,000 = \$350$$

$$\text{AII} = M(r) - \frac{P_m}{n}$$

$$= \$4,000(.055) - \frac{\$350}{14}$$

$$= \$195$$

$$\text{AAI} = \frac{M + B_q}{2}$$

$$= \frac{\$4,000 + \$4,350}{2}$$

$$= \$4,175$$

$$\text{yield rate(YR)} = \frac{\text{AII}}{\text{AAI}}$$

$$= \frac{\$195}{\$4,175}$$

$$= .0467 = 4.67\%$$

It is noteworthy to mention that the average annual investment can be obtained by getting the actual amounts of investments which are connected to either the premium price or the discount price in an arithmetic progression whose common difference is the average premium (P_m/n) or the average discount (D_s/n). So, to illustrate, we can get the investment amounts in Example 2.6.1 starting with the premium price of \$4,350 and reducing it by the average premium $\$350/14 = 25$, down to the face value of \$4,000. Therefore, the arithmetic progression of investment would be \$4,350, \$4,325, \$4,300, \$4,275, \$4,250, \$4,225, \$4,200, \$4,175, \$4,150, \$4,125, \$4,100, \$4,075, \$4,050, \$4,025, \$4,000. The actual average of these amounts is their total divided by their number:

$$\frac{\$62,625}{15} = \$4,175$$

The Interpolation Method

According to the mathematical **interpolation method**, the yield rate can be approximated by association and comparison with related variables. We start with two close yield rates, use them to calculate their associated purchase price, and then set up an interpolation table and solve for the unknown yield rate.

Example 2.6.2   Let's try to use Example 2.6.1 to approximate the yield rate by interpolation. We have a quoted price of \$4,350 that is a premium price. It indicates immediately that the yield rate would be lower than the bond rate of 5.5% or 2.75% semiannually. If we look at the PVIFA table, we would realize that the closest values of rates on the table are 2.5% and 2.25%. Let's use those rates to get their associated purchase premium prices:

$$P_m = M(r - i)a_{\overline{n}|i}$$
$$B_p = M + P_m$$

For 2.5% with $n = 28$ and $M = \$4000$:

$$P_m = \$4,000(.0275 - .0225)a_{\overline{28}|.0225}$$
$$= \$4,000(.005)(20.6078)$$
$$= \$412.16$$
$$B_p = \$4,000 + \$412.16$$
$$= \$4,412.16$$

For 2.25% with $n = 28$ and $M = \$4000$:

$$P_m = \$4,000(.0275 - .025)a_{\overline{28}|.025}$$
$$= \$4,000(.0025)(19.9648)$$
$$= \$199.65$$
$$B_p = \$4,000 + \$199.65$$
$$= \$4,199.65$$

Now we have three prices ($4,412.16, $4,350.00, $4,199.65) and two rates (2.25% and 2.5%). We should be able to solve for the missing rate by setting the interpolation table:

		Yield Rate (%)	PurchasePrice ($)	
		2.25	4412.16	4412.16 – $4350.00
2.25 – 2.25	2.25 – x	x	4350.00	
				4412.16 – $4199.65
		2.5	4199.65	

$$\frac{2.25 - x}{2.25 - 2.5} = \frac{\$4,412.16 - \$4,350.06}{\$4,412.16 - \$4,199.65}$$
$$\frac{2.25 - x}{-.25} = \frac{62.16}{212.51}$$

$$\$478.15 - \$212.51x = -\$15.54$$

$$493.69 = 212.51x$$

$$\frac{493.69}{212.51} = x$$

$$\boxed{2.32 = x}$$

So the approximate yield is 2.32%. In fact, if we use it to get the purchase price, and if we use a table value that is in between the two readings above, we get a price very close to $4,350. With a sophisticated computer program, the exact yield can be found by the touch of a button.

The Current Yield Method

The **current yield method** approximates the yield rate based on a calculated rate of interest (Cr) which is obtained by dividing the interest (I) by the quoted price (B_q), then using Cr to get the approximate yield rate (YR):

$$\boxed{YR = Cr(2 + Cr)}$$

where Cr is

$$Cr = \frac{I}{B_q}$$

Note that in case of the premium price, the average periodic premium P_m/n has to be subtracted from the interest.

$$Cr = \frac{I - P_m/n}{B_q}$$

Example 2.6.3 If we run this method over Example 2.6.2, where the premium price was $4,350, the face value was $4,000, and the bond rate was 5.5% for 14 years, we should obtain the average premium first:

$$P_m = \$350$$

$$\frac{P_m}{n} = \frac{\$350}{28} = \$12.50$$

which has to be taken off the interest I:

$$I = \$4,000(.055) = \$220$$

$$I - \frac{P_m}{n} = \$220 - \$12.50 = \$207.50$$

$$Cr = \frac{I}{B_q}$$

$$= \frac{\$207.50}{\$4,350} = .048$$

$$YR = Cr(2 + Cr)$$

$$= .048(2 + .048)$$

$$= .098 \quad \text{annual rate}$$

and the semiannual rate is .049.

2.7. DURATION

Duration is the average maturity of cash flows on an investment weighted by the present value of the cash flows. It is a measure of the reinvestment rate risk, specifically gauging how sensitive bond prices are to changes in interest rates. It is expressed by

$$D = \frac{I \sum_{t=1}^{n} \left[\dfrac{t}{(1+i)^t} \right] + \dfrac{nM}{(1+i)^n}}{I \sum_{t=1}^{n} \left[\dfrac{1}{(1+i)^t} \right] + \dfrac{M}{(1+i)^n}}$$

where I is the annual interest on a bond, t is the payment term, n is the number of years to redemption, i is the yield, and M is the par value of the bond.

Example 2.7.1   Consider a bond with face value $1,000 at a bond rate of 9% paying annually and redeemable in 5 years. Given that the market yield is 12%, calculate the duration of this bond and explain its effect.

$$t = 1, 2, 3, 4, 5; \, n = 5; \, i = .12; \, M = \$1,000.$$

$$I = M(r)$$

$$= \$1,000(.09) = \$90$$

$$D = \frac{I \sum_{t=1}^{n} \dfrac{t}{(1+i)^t} + \dfrac{nM}{(1+i)^n}}{I \sum_{t=1}^{n} \dfrac{1}{(1+i)^t} + \dfrac{M}{(1+i)^n}}$$

$$= \frac{\$90(10) + (5 \times 1,000)/(1+.12)^5}{\$90(3.61) + (1,000)/(1+.12)^5}$$

TABLE E2.7.1

Terms of Payments t	$\dfrac{t}{(1+i)^t}$	$\dfrac{1}{(1+i)^t}$
1	.893	.893
2	1.595	.797
3	2.135	.712
4	2.542	.636
5	2.837	.567
	10.00	3.61

$$D = \frac{\$3,737.13}{\$892.33}$$

$$= \$4.19$$

Duration is an element in gauging the sensitivity of bond price to any change in the yield rate.

$$\%\Delta B = \frac{\Delta i \cdot D}{1+i}$$

Let's suppose that the yield has increased from 12% to 12.1%. The percentage change in bond price ($\%\Delta B$) would be

$$\%\Delta B = \frac{.1(\$4.19)}{1+.12}$$

$$= \$.37$$

which means that as the yield changes by .1, the bond price changes by \$.37.

The term $D/(1+i)$ called the **volatility** (VL), which leads us to consider the last formula and rewrite it as

$$\%\Delta B = \Delta i \cdot V L$$

where volatility is defined as a measure of how briskly the present value of cash flows respond to a change in the market interest rate.

3 Mutual Funds

Mutual funds are not specific securities such as stocks or bonds. They are actually a financial intermediary that pools funds of investors and makes them available for various investment opportunities by businesses and governments. Most common is the purchase of large blocks of stocks and bonds and other investment instruments. The most striking feature of mutual funds is the creation of a diversified portfolio of investment securities that is professionally managed and monitored. Each fund has its own specific investment goals and risk tolerance to reflect the individual objectives and character of the investors. Mutual fund companies claim a unique role in providing high-quality services to their investors, characterized by:

- Minimizing the levels of unsystematic risk by instituting the formal disciplined diversification of investment instruments in the face of transaction costs.
- Providing a high-standard around-the-clock level of professional management that has been able to deliver high rates of return and successful predictions of future trends and price charges.

Mutual funds are tailored to fit the need, nature, and specific objectives of various investors. They come into four major categories:

1. *Funds for income*. Their major objective is to provide a stable level of income. They are focused primarily on corporate and government bonds.
2. *Funds for growth*. The major objective of these funds is capital appreciation. They focus on the common stock of publicly held corporations. They vary in their aggressiveness and risk levels.
3. *Balanced funds*. These funds provide a balance of income and appreciation. They provide a fixed income as well as seeking capital growth opportunities. They invest in both stocks and bonds and are very popular with the large sector of investors who have a moderate level of risk tolerance.
4. *Global funds*. These are basically balanced funds, but they focus on investment opportunities abroad.

Mathematical Finance, First Edition. M. J. Alhabeeb.
© 2012 John Wiley & Sons, Inc. Published 2012 by John Wiley & Sons, Inc.

3.1. FUND EVALUATION

Unlike stocks, which are traded throughout the entire business day, a mutual fund's value is determined only at the end of each trading day. They are priced based on what is called **net asset value** (NAV), which would settle after the market has been closed. The net asset value is the mutual fund's equivalent of share price. It usually stands for the ability of the fund's management to deliver continuous and consistent profits. It is also a direct indicator of the market value. Net asset value consists of the total value of the holdings of the fund (market value and cash) after subtracting any liabilities or obligations (O). Dividing by the number of shares outstanding, we obtain the net asset value per share:

$$NAV = \frac{1}{S}[(MV + C) - O]$$

where NAV is the net asset value per share, MV is the market value of assets invested, C is the cash on hand, S is the number of shares outstanding, and O is the obligations or liabilities.

Example 3.1.1   The Bright Future Mutual Funds Company owns the following shares in four major securities:

Security 1: 333,200 shares

Security 2: 298,513 shares

Security 3: 197,814 shares

Security 4: 88,500 shares

The company also holds $244,000 in cash, is responsible for $153,000 in liabilities, and has a total of 395,667 shares outstanding. Calculate the fund's net asset value on a day where the share price for each security is $12, $16, $20, and $22, respectively.

First we calculate the market value for four securities at four prices:

$$MV = (333,210 \times 12) + (298,513 \times 16) + (197,814 \times 20)$$

$$+ (88,500 \times 22)$$

$$= 14,678,008$$

$$NAV = \frac{1}{S}[(MV + C) - L]$$

$$= \frac{1}{395,667}[(14,678,008 + 244,000) - 153,000]$$

$$= \$37.33$$

3.2. LOADS

Loads are the commission or transaction fees imposed on the purchase and sale of mutual funds. Not all mutual funds carry these loads. In fact, they are named based on their inclusion as fees. A **load fund** is a fund that requires investors, buyers, and sellers of funds to pay these charges either per transaction, or as a percentage of return, or both. A **front-end load** is a charge imposed on the purchase transaction and a **back-end load** is a charge imposed on the sale transaction. It is also called a **deferred sales charge** or **contingent commission**. Loads range in value between 1% and 10% (at most) on the investment amount. **No-load funds** do not require paying these transaction charges but may include other types of charges. Factoring those charges in, we obtain the purchase price (PP) and the selling price (SP), which comprise the net asset value adjusted for the front-end load (L_F) and the back-end load (L_B).

$$PP = NAV \left(\frac{1}{1 - L_F} \right)$$

$$SP = NAV(1 - L_B)$$

Example 3.2.1   If the front-end load is $6\frac{1}{2}$ % and the back-end load is $5\frac{1}{4}$ %, what is the previous net asset value as to the offering and selling prices?

$$PP = NAV \left(\frac{1}{1 - L_F} \right)$$

$$= \$37.33 \left(\frac{1}{1 - .065} \right)$$

$$= \$39.93$$

$$SP = NAV(1 - L_B)$$

$$= \$37.33(1 - .0525)$$

$$= \$35.37$$

3.3. PERFORMANCE MEASURES

The Expense Ratio (ER)

The **expense ratio** is a performance measure especially useful for comparing the cost of investing between two or more funds. It reflects a fund's operating expenses as they are relative to the average asset throughout the year. It is therefore obtained by dividing the total expenses charged by the average of net assets of the fund.

$$ER = Exp\left[\frac{1}{(A_1 + A_2)/2}\right]$$

Example 3.3.1 Suppose that a mutual fund company shows its total assets at the start of the year as \$3,950,000, and at the end of the year as \$3,873,150. Suppose that the fund's total operating expenses have reached \$35,000. What would the expense ratio be?

$$ER = \$35,000\left[\frac{1}{(\$3,950,000 + \$3,873,150)/2}\right]$$
$$= .009 \text{ or } .09\%$$

An expense ratio of .09% is reasonable since the range is .4 to 1.5%. Of course, the higher the expense ratio, the more it takes away from the return on investment.

The Total Investment Expense (TIE)

The **total investment expense** is an expansion of the expense ratio. It adds to it the load expenses as they are adjusted to the holding period.

$$TIE = \frac{1}{n}(L_f + L_B) + ER$$

Example 3.3.2 Let's suppose that the front-end load of a mutual fund was 4% and the back-end load was 3.5% and the fund was held for 3 years. The total investment expense would be

$$TIE = \tfrac{1}{3}(.04 + .035) + .009$$
$$= .034 \text{ or } 3.4\%$$

The Reward-to-Variability Ratio (RVR)

The **reward–variability ratio** was developed by W. F. Sharpe in 1966 (*Journal of Business*, January, pp. 119–138). It assesses the mutual fund beyond the risk-free return for every unit of total risk that may face the fund. It compares the fund return to the risk-free return of 91-day U.S. Treasury bills and assesses the difference between the rates against the fund's standard deviation of returns.

$$\text{RVR} = \frac{1}{\sigma_j}(R_{jt} - R_{ft})$$

where R_{jt} is the return on the jth fund for time t and R_{ft} is the return on a risk-free asset, usually Treasury bills, and σ_j is the standard deviation of return on the jth fund.

Example 3.3.3   Suppose that at a time when a mutual fund return was 9%, the risk-free return was 5%, and the standard deviation of returns in the same time was 18.53. The reward-to-variability ratio would be

$$\text{RVR} = \frac{1}{.1853}(.09 - .05)$$
$$= 21.6\%$$

which means that the fund is providing 21.6% return beyond the risk-free rate.

Note that if the fund rate of return has to be calculated, we would calculate it using the following:

$$R = \frac{S}{\text{IF}}(\Delta P + D + G)$$

where R is the rate of return on mutual funds, S is the number of shares owned, IF is the amount invested in funds, ΔP is the change in the price of the fund, D is the dividend received per share, and G is the capital gain per share.

Example 3.3.4   Suppose that we invested $2,500 in purchasing 200 shares of a mutual fund at $12.50 a share and after a year this price increased to $13.70 and the company gave out 30 cents in dividends and 45 cents in capital gain. What would the rate of return be?

$$R = \frac{S}{\text{IF}}(\Delta P + D + G)$$
$$= \frac{200}{\$2,500}[(13.70 - 12.50) + .30 + .45]$$
$$= 15.65\%$$

Also, if the standard deviation of returns has to be calculated, we would calculate it by the normal statistical formula of the standard deviation:

$$\sigma = \sqrt{\frac{\sum(x_t - \overline{x})^2}{n - 1}}$$

where x_t is the return for period t, $\overline{x}$ is the mean return, and n is the number of periods. Standard deviation is often used to estimate the extent of risk in surrounding the flow of fund returns. A large standard deviation usually indicates a higher level of risk.

Example 3.3.5   Suppose that we track down the 7-year returns of a specific mutual fund and find them to be as follows:

$$5, \quad 5\tfrac{3}{4}, \quad 6\tfrac{1}{2}, \quad 6, \quad 7\tfrac{1}{3}, \quad 8, \quad 9\tfrac{1}{2}$$

We can calculate the standard deviation by first finding the mean, which would be 6.87, and then arranging the data we need in Table E3.3.5.

$$\sigma = \sqrt{\frac{14.06}{7 - 1}}$$
$$= \sqrt{2.34}$$
$$= 1.53$$

This is a relatively small standard deviation, indicating a lower level of risk. Similarly, we can follow other regular statistical measures when we need their aid for analysis, such as in the case of testing the degree of diversification of funds in an investor's portfolio. We can see how these funds are tied to each other and to what degree by employing a correlation index such as Pearson's

TABLE E3.3.5

t	x_t	$\overline{x}$	$x_t - \overline{x}$	$(x_t - \overline{x})^2$
1	5	6.78	−1.87	3.5
2	5.75	6.78	−1.12	1.25
3	6.5	6.78	−.37	.137
4	6	6.78	−.87	.757
5	7.34	6.78	.47	.221
6	8	6.78	1.13	1.28
7	9.5	6.78	2.63	6.92
			$\sum(x_t - \overline{x})^2$	14.06

correlation coefficient (r_p):

$$r_p = \frac{1}{\sigma_i \sigma_j} \left[\frac{\sum (R_i - \overline{R}_i)(R_j - \overline{R}_j)}{n - 1} \right]$$

where σ_i and σ_j are the standard deviations of returns for funds i and j, R_i and R_j are the series of returns of funds i and j, $\overline{R}_i$ and $\overline{R}_j$ are the means of those returns, and n is the number of return periods or observations.

Example 3.3.6 Suppose that we want to test how diversified Jim's investment portfolio of mutual funds is by looking at only two funds and observing their returns in the last six years (Table E3.3.6).

$$\boxed{\sigma_i = \sqrt{\frac{\sum (R_i - \overline{R}_i)^2}{n - 1}}}$$

$$= \sqrt{\frac{18.05}{6 - 1}}$$

$$= 1.9$$

$$= \sqrt{\frac{26.78}{6 - 1}}$$

$$= 2.31$$

$$r_p = \frac{1}{\sigma_i \sigma_j} \left[\frac{\sum (R_i - \overline{R}_i)(R_j - \overline{R}_j)}{n - 1} \right]$$

$$= \frac{1}{1.9(2.31)} \left(\frac{-20.26}{6 - 1} \right)$$

$$= .923 \text{ or } -92.3\%$$

This very high and negative correlation means that the funds in the portfolio are highly diversified. They would move in a way strongly opposite to each other,

TABLE E3.3.6

n	R_i	$\overline{R}_i$	$R_i - \overline{R}_i$	$(R_i - \overline{R}_i)^2$	R_j	$\overline{R}_j$	$R_j - \overline{R}_j$	$(R_j - \overline{R}_j)^2$	$(R_i - \overline{R}_i)$ $(R_j - \overline{R}_j)$
1	5	7.04	−2.04	4.16	12	8.29	3.71	13.76	−7.57
2	5.5	7.04	−1.54	2.37	9	8.29	.71	.50	−1.09
3	6.25	7.04	−.79	.624	8.25	8.29	−.04	.0016	.032
4	7	7.04	−.04	.0016	8.5	8.29	.21	.044	−.0084
5	8.5	7.04	1.46	2.13	7	8.29	−1.29	1.66	−1.88
6	10	7.04	2.96	8.76	5	8.29	−3.29	10.82	−9.74
				18.05				26.78	−20.26

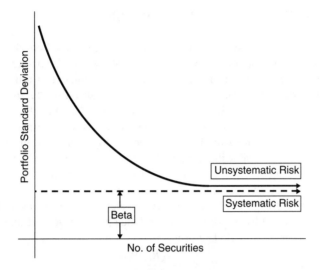

FIGURE 3.1

which is the aim of diversification. While one fund does not do well, the other does extremely well, and a nice balance is achieved.

Treynor's Index

Treynor's index is named after J. L. Treynor (*Harvard Business Review*, January–February 1965, pp. 63–75). It is similar to the share's reward-to-variability ratio (RVR) except that the difference between the rate of return generated by the mutual fund and the risk-free rate of U.S. Treasury bills is divided by beta (β_j) instead of σ_j where beta is a coefficient representing the estimated systematic risk that the fund in question may face.

$$\text{TI} = \frac{1}{\beta_j}(R_{jt} - R_{ft})$$

Example 3.3.7 If the coefficient estimated for the market systematic risk (β) at the time of calculating the previous rate of return is estimated at 1.07, the Treynor index would be

$$\text{TI} = \frac{1}{1.07}(.09 - .05)$$
$$= 3.7\%$$

which says that this fund is delivering a 3.7% return beyond the risk-free return for every level of the systematic risk that the fund was expected to return.

3.4. THE EFFECT OF SYSTEMATIC RISK (β)

Systematic risk (β) is also called **market risk** or **undiversified risk**. It refers to the unavoidable type and level of risk that is inherent in the nature of the free market and tied inextricably to its fluctuations. In the context of a mutual funds portfolio, this type of risk cannot be eliminated or reduced by the usual remedy of diversifications of funds. It is associated with the fact that there are other economy-wide perils that continue to threaten all business performance, and it is the reason that stocks, for example, tend to move together. It is an inevitability that investors have to deal with. Figure 3.1 shows how the level of systematic risk is determined independently with regard to the unsystematic or specifically unique level of risk to which securities are individually exposed.

Systematic risk for a specified fund return such as R_i is measured by beta (β_i), which is obtained by dividing the covariance between R_i and market return (R_m) by the variance of market returns:

$$\beta_i = \frac{\text{Cov}(R_i, R_m)}{\sigma_m^2}$$

Keeping in mind that for a risk-free return (R_f), beta would be zero because its covariance with the market return is zero, and for the collective market returns, beta would be equal to 1 because the covariance of market return with itself is the variance of itself:

$$\beta_m = \frac{\text{Cov}(R_m, R_m)}{\sigma_m^2} = \frac{\sigma_m^2}{\sigma_m^2} = 1$$

As for many individual funds in a portfolio, beta would be the weighted average of all individual betas of the returns of the portfolio funds:

$$\beta_p = \sum_{j=1}^{k} \beta_j w_j \qquad j = 1, 2, \ldots, k$$

where β_p is the beta for the portfolio, β_j is the individual beta for each fund in the portfolio, and w_j is the weight of each fund in the portfolio, which is simply the proportion of the fund market value to the entire value of the portfolio.

Example 3.4.1 Suppose that we have a portfolio of five funds, the market values of which are $250,000, $370,000, $588,000, $610,000, and $833,000, with estimated betas of .8, .75, 1.2, 1.35, and .97 (see Table E3.4.1). Calculate the portfolio beta, β_p.

First, we obtain the market value of the portfolio MV$_p$ and then we calculate the funds weights (w$_j$) as the individual fund percentages of the entire portfolio:

$$\text{MV}_p = \sum_{j=1}^{5} \text{MV}_j = \$250.000 + \$370,000 + \$588,000 + \$610,000 + \$833,000$$

$$= \$2,651,000$$

TABLE E3.4.1

Fund	MV_j	W_j	B_j	$B_j W_j$
1	250,000	.094	.8	.075
2	370,000	.139	.75	.104
3	588,000	.222	1.2	.2666
4	610,000	.231	1.35	.312
5	833,000	.314	.97	.305
Portfolio	2,651,000			$\beta_p = 1.06$

$$W_1 = \frac{MV_1}{MV_p} = \frac{\$370,000}{\$2,651,000} = 9.4\%$$

$$W_2 = \frac{MV_2}{MV_p} = \frac{\$250,000}{\$2,651,000} = 13.9\%$$

$$W_3 = \frac{MV_3}{MV_p} = \frac{\$588,000}{\$2,651,000} = 22.2\%$$

$$W_4 = \frac{MV_4}{MV_p} = \frac{\$610,000}{\$2,651,000} = 23.1\%$$

$$W_5 = \frac{MV_5}{MV_p} = \frac{\$833,000}{\$2,651,000} = 31.4\%$$

The Abnormal Performance Alpha (α_{jt})

The **abnormal performance alpha** measure is also called **Jensen's index** after M. Jensen (*Journal of Finance*, May 1968, pp. 389–416), who used it to test the abnormality in mutual fund performance by calculating the coefficient alpha (α), which compares two rate differences:

1. The difference between the returns of the performing fund and a risk-free asset such as U.S. Treasury bills.
2. The difference between the market (R_{mt}) return and the risk-free asset return, adjusted for systematic risk.

$$\alpha_{it} = (R_{jt} - R_{ft}) - [\beta_j(R_{mt} - R_{ft})]$$

Jensen explained that a positive alpha means that after adjusting for risk and movements in the market index, the abnormal performance of a portfolio stays on, and that the fund is able to handle its own expenses. A negative alpha suggests that the fund is not able to forecast future security prices well enough to cover expenses. Jensen measured primarily net costs such as research cost, management fees, and brokerage commissions.

Example 3.4.2   If we keep the same data for the rate we used before and assume that the market rate is 10%, we can calculate alpha as

$$\alpha_{it} = (.09 - .05) - [1.07(.10 - .05)]$$
$$= 1.35\%$$

3.5. DOLLAR-COST AVERAGING

One of the simplest and most popular techniques in the mutual fund trust is **dollar-cost averaging**, which minimizes the cost and increases the return over the long run. It is used basically to maintain a regular periodic investment that over time would automatically purchase more of low-priced shares and fewer high-priced shares, striking a natural balance. The result is that the average cost per share will always be less than the average price. There are two ways to achieve regularity in investment:

1. Buying the same number of shares regardless of the cost per share.
2. Buying for the same dollar amount regardless of the cost per share.

Either way should result in an average cost that is less than the average price per share.

Example 3.5.1   Suppose that we invest $250 a month to buy shares in a specific fund throughout its price changes over the next six months: July, 14.25; August, 13.35; September, 12.20; October, 11.92; November, 13.10; and December, 12.50 (Table E3.5.1). Calculate the average cost and compare it to the average price.

$$\text{average cost} = \frac{\sum_{i=1}^{6} I_i}{\sum_{i=1}^{6} S_i} = \frac{1,500}{116.82} = 12.84$$

$$\text{average price} = \frac{\sum p_i}{n} = \frac{77.32}{6} = 12.89$$

TABLE E3.5.1

Month, n	Share Price, P_i ($)	Amount Invested, I_i ($)	No. Shares, S_i
July	14.25	250	17.54
August	13.35	250	18.73
September	12.20	250	20.50
October	11.92	250	20.97
November	13.10	250	19.08
December	12.50	250	20.00
Total	77.32	1,500	116.82

4 Options

As a form of investment, the organized options market has been advancing rapidly in the last four decades. It began in 1973 when the Chicago Board of Options Exchange (CBOE) ushered in the era of trading in standardized option contracts. Since then it has became a significant part of the investment scene, has attracted its own zealous investors, and has introduced its own culture and language, which we need to explore. **Options** are financial instruments that act like a standardized contract and provide their holders with the opportunity and right, but not the obligation, to purchase or sell a certain asset, at a stated price, on or before a short-term expiration date, usually set as within a year.

There are three basic forms of options: rights, warrants, and calls and puts. We focus on the most common: call options and put options. A **call option** gives its holder the right to purchase a specific number of shares (usually, 100 shares of common stock) at a price called a **strike** or **exercise price**, on or prior to a specific expiration date. The strike price is often set at or near the prevailing market price of a stock at the time the option is issued. A **put option** is similar in every way to a call option except that it gives the right to sell instead of to purchase. Although the most common underlying assets for options are common stocks, they can also be based on stock indices, foreign currencies, debt instruments, and commodities. Those underlying assets are why options are considered derivatives: because the option value depends on the value of its basis asset. This is also what justifies the existence of options. It is because investors expect the market price of the underlying asset to rise high enough to cover the cost of the option and still leave room to make a profit.

The option trading cycle begins when an option contract is generated by an **option writer**, who would be paid a premium by the asset owner to create such an option, and grants its selling and buying rights that can be exercised during the set life of that option. An **option holder** can monitor his or her action based on the market prices of the underlying assets and the level of risk involved. Generally, the option holder may follow one of the following actions:

1. *To exercise the option:* to exercise the right to buy or sell an option. When the market price of the underlying asset, such as a common stock, is higher than the strike price of the option, a holder of a call option would exercise his or her right to buy a call and make a profit. Similarly, when the market

Mathematical Finance, First Edition. M. J. Alhabeeb.
© 2012 John Wiley & Sons, Inc. Published 2012 by John Wiley & Sons, Inc.

price is lower than the strike price of the put option, the put holder would exercise his or her right to sell and make a profit. This is the case called *in-the-money*.

$$\text{In-the-money:} \quad \begin{cases} \text{MP}_c > \text{SP} \\ \text{MP}_p < \text{SP} \end{cases}$$

MP_c is the market price of a call, MP_p is the market price of a put, and SP is the strike price.

2. *Not to exercise the option:* when the investor does not see any opportunity to make a profit. It is usually the case when the market price of the underlying asset is either equal to or lower than the strike price of a call and either equal or higher for a put. This case is called *out-of-the-money* when prices are different and *at-the-money* when prices are equal.

Out-of-the-money:

$$\text{MP}_c < \text{SP}$$

$$\text{MP}_p > \text{SP}$$

At-the-money:

$$\text{MP}_c = \text{SP}$$

$$\text{MP}_p = \text{SP}$$

3. *To let the option expire:* when investors keep waiting for the prices of the underlying assets to change in their favor so that they can exercise the option. Sometimes, no change in price of this sort occurs and investors cannot take any action until time runs out, the expiration date arrives, and their options are deemed worthless. In this case, investors suffer a loss. There are strategies used by financial institutions to protect investors against this loss and other types of loss in the investment market. One very common strategy, **hedging**, is an action taken with one security to protect another security against risks such as buying on one side and selling on the other. In option investment there are three common types of strategies: spreads, straddles, and a combination of both. A **spread** is a simultaneous purchase and sale of calls and puts on the same underlying asset, such as stock, which are written with either different strike prices, different expiration dates or both. Three spreads can be identified:

   a. A **horizontal spread** is characterized by buying and selling an option with an identical strike price but different expiration dates.

   b. A **vertical spread** involves the purchase and sale of an option with identical expiration dates but a different strike price.

   c. A **butterfly** represents many options purchased at different strike prices.

A **straddle** involves a combination of the same number of calls and puts purchased simultaneously at an identical strike price for the same expiration date.

4.1. DYNAMICS OF MAKING PROFITS WITH OPTIONS

Investors make profits by buying and selling calls and puts and through the differences between the market price of the underlying assets and the strike price of those options exercised within the proper timing.

Buying and Selling Calls

An investor who buys calls has the right to purchase 100 shares of the underlying asset (let's say stock here) per call at a strike price that would stay valid for a certain maturity time. This investor would watch the fluctuations of the market price of that stock and hope that it would rise above the strike price that she paid so that she can sell and profit from the difference between the two prices, all before the expiration day, which is the last day of the maturity. This type of profit can be substantial and achieved in a short period of time, but comes with a high risk. The basic risk is that the maturity time will expire without the investor being able to sell at a profit. It is very possible that the market price of such stocks would not rise within the time frame of maturity, rendering those calls worthless. Some investors try to sell at attractive discounts before the expiration date, to recover at least a portion of their investment.

There are two common ways of selling a call: the risky and the conservative. In the risky way, called an **uncovered sale**, the seller grants to the buyer the contractual right that 100 shares will be delivered at the strike price no matter how high the market value of the stock. Suppose that a seller grants such a right to a buyer to deliver 100 shares of a stock at a strike price of $30, and suppose by the expiration date the stock market price rises to $59. The seller would suffer a significant loss: $2,900. The less risky type of call sale, the **covered sale**, involves selling shares when the seller already owns those shares instead of having to buy them, even when the market value is substantially high. It is most likely that they were purchased at a lower price, so that this sale is significantly less risky than an uncovered sale of calls.

Buying and Selling Puts

In contract to a call investor, the buyer of a put waits for the market price of a stock to fall below the strike price so that she can make a profit by buying more cheaply. But just like calls, puts can easily lose their value if a drop in the market value of the underlying stock does not occur by the expiration date. In selling puts, there are no covered and uncovered sales as in the case of selling calls. The seller would wish for the market price of the underlying stock to rise above the strike price so that he can make a profit when exercising his option by selling at a higher value than that at which he purchased the put.

4.2. INTRINSIC VALUE OF CALLS AND PUTS

The intrinsic value of a call option to the buyer is what he would gain through the difference between the market price of the underlying stock and the strike price. He would gain only when the market price of the underlying stock exceeds the strike price when he chooses to exercise his option right to buy at the strike price and sell at the market price. However, he would not lose anything if the market price of the stock decreases, because he has the choice not to exercise his option but wait for a better opportunity. The intrinsic value for the call writer is what he loses when the market price of the underlying stock increases at the time he issued the option at a strike price less than the current market price of the underlying stock.

The value in both cases, buying a call and writing a call, is determined dollar by dollar by how much the market price of a stock increases over the strike price, but it is determined inversely for both of them (i.e., what the buyer gains, the writer loses). Mathematically, the intrinsic value of a call to a call buyer (IVC_B) and for a call writer (IVC_W) can be expressed by

$$IVC_B = \max[(MP - SP), 0]$$
$$IVC_W = \min[(SP - MP), 0]$$

where MP is the market price of the stock and SP is the strike price for the call option.

Example 4.2.1 Suppose that a call option is written for a strike price of $25 per share of a certain stock, and suppose that after some time, the market price for that stock goes up to $37. What is the intrinsic value for the buyer and the writer?

$$
\begin{aligned}
IVC_B &= \max[(MP - SP), 0] \\
&= \max[(\$37 - \$25), 0] \\
&= \max[\$12, 0] \\
&= \$12 \\
IVC_W &= \min[(SP - MP), 0] \\
&= \min[(\$25 - \$37), 0] \\
&= \min[-\$12, 0] \\
&= -\$12
\end{aligned}
$$

So the value to the buyer is $12 and to the writer is −$12, which means that the writer loses as much as the buyer gains because the writer had to stay on her contractual obligation to deliver the option at the strike price of $25 even if she had to buy the stock at $37 in order to make it available to the buyer.

In Figure E4.2.1 we see that everything was flat before the $25 price, but after the market price increased to $37, each dollar of that increase raised the potential

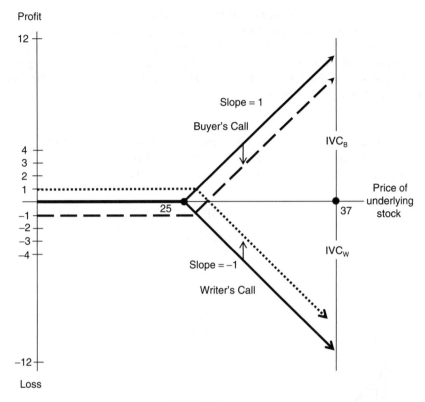

FIGURE E4.2.1

profit by the same dollar amount, which is why the slope is 1 and the line is at 45°. The increase was $12 and the gain was also $12. It is exactly the opposite for the writer, whose curve dropped by each dollar increase in the stock price and ultimately became a mirror image of the buyer's call curve. The entire loss was also $12.

Note that in the calculation we did not account for the fee paid to the call writer, called the **option premium**. In the figure, the fee is shown to reduce the gain for the buyer by shifting the entire curve down to the dashed line. At the same time, this fee is received by the writer, reducing her loss by shifting her curve up to the dotted line.

From the perspective that puts are the opposite of calls, we can see that the intrinsic value equation for a put would be the same as the call value equation except in switching the order of the prices. Therefore, we can write the intrinsic value of a put to a buyer (IVP_B) and to a put writer (IVP_W) as

$$IVP_B = \max[(SP - MP), 0]$$
$$IVP_W = \min[(MP - SP), 0]$$

In this case, the put buyer seeks a drop in the market price of the underlying stock so that he can make a potential profit through buying cheap. The put writer, on the other hand, would make a loss by delivering an option at a strike price higher than what the market sells. Again, in both cases, what the buyer gains, the writer loses, and both gain and loss are dollar for dollar the same as the drop in the market price of the stock.

Example 4.2.2   An investor has a put option with a strike price of $70, but the market price of its underlying stock goes down to $65 (see Figure E4.2.2). What is the intrinsic value of the put for this buyer and for the writer?

$$IVP_B = \max[(SP - MP), 0]$$
$$= \max[(\$70 - \$65), 0]$$
$$= \max(\$5, 0)$$
$$= 5$$

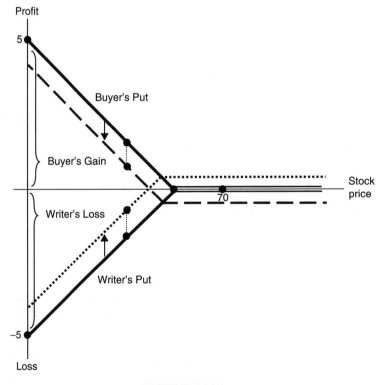

FIGURE E4.2.2

$$IVP_W = \min[(MP - SP), 0]$$
$$= \min[(\$65 - \$70), 0]$$
$$= \min(-\$5 - 0)$$
$$= -\$5$$

Similar to the call case, what the buyer gains, which is $5 per share, the writer loses, and both gain and loss are dollar for dollar as much as the drop in the market price of the stock. The reason behind the put buyer's gain is that he pays $70 as a contracted strike price, but the writer is committed to deliver a $65 per share, although she has to carry the loss of $5 per share.

As expected, the graphic presentation of the put payoff would be very similar to the call payoff except that it is switched to the left side. The reason for the switch is that the entire sequence follows a drop in the market price of the stock, not an increase, which warrants the movement from right to left. Again, the actual buyer's and writer's curves are the dashed line for the buyer, which is shifted down since it reflects the reduction caused by paying the option premium. The dotted line represents the actual writer's curve, which is shifted up to reflect the gain in receiving the option premium.

4.3. TIME VALUE OF CALLS AND PUTS

The time value of an option, a call (TV_c) or a put (TV_p), is the difference between the option price (OP) and the intrinsic value of the option, whether it is for the call (IVC) or for the put (IVP). It is also defined as the portion of the premium above any in-the-money premium.

$$TV_c = OP - IVC$$

and

$$TV_p = OP - IVP$$

Example 4.3.1 A call option is currently worth $600 for a 100 share, and its strike price is $40. Find the call time value for its buyer if the market price of the stock rises to $45.

$$IVC_B = \max[(MP - SP), 0]$$
$$= \max[(\$45 - \$40), 0]$$
$$= \max[\$5, 0]$$
$$= \$5$$
$$OP = \frac{\$600}{100} = \$6 \text{ per share}$$
$$TV_C = OP - IVC$$
$$= \$6 - \$5 = \$1.00$$

Example 4.3.2   What is the intrinsic value of an option if it costs $500 (for 100 shares), and its time value is $2?

$$OP = \frac{\$500}{100} = \$5$$

$$IV = OP - TV$$

$$= \$5 - \$2 = \$3$$

Example 4.3.3   If we are told that the option in Example 4.3.2 was a call and the market price of its underlying stock was $75, what would the strike price be that has been granted to the buyer?

The intrinsic value of a call for the buyer is ultimately equal to the difference between the market price of the stock and the strike price of the call:

$$IVC_B = MP - SP$$

$$SP = MP - IVC_B$$

$$= \$75 - \$3 = \$72$$

Note that the time value would be equal to the premium or the option price if the option is either at-the-money or out-of-the-money. In the case of at-the-money, the market price of the stock and the strike price would be equal and the intrinsic value would be zero. The time value would be the premium minus zero, which would end at the equality between the time value and the premium. In the case of out-of-the-money, the difference between the market price of the stock and the strike price would be negative, lending to a consideration of zero as the intrinsic value to signify that such an option would have no intrinsic value. Here, too, the time value would be the result of taking away zero from the premium, which means keeping the premium value as it is. Table 4.1 shows how this works.

As an example, when we look at calls 3 and 10 and puts 1 and 6 (marked by arrows), we observe that the time value equals the premium $OP = TV$ (4's; 10's; and 1's; and 3's) simply because the intrinsic values were zero. Also, we can observe that the intrinsic value cannot exist ($= 0$) if the market price is lower than the strike price in the case of calls or higher than the strike price in the case of puts (cases marked by stars). Therefore, zero is assigned to the intrinsic value in these cases. Naturally, the intrinsic value would be zero when the market price equals the strike price (cases marked by x).

4.4. THE DELTA RATIO

Delta (D) describes the relationship between the change in the market price of the underlying stock (ΔMP) and the change in option price or the premium

(ΔOP). It is an index that tells investors about the dynamics of their profits and losses out of option trading. Mathematically, it is a ratio of the changes in these two prices:

$$D = \frac{\Delta OP}{\Delta MP}$$

If delta is equal to 1, it means that the option price follows the market price dollar for dollar. If it is more than 1, the option price would respond faster to the change in market price, and if it is less than 1, we know that the option price would lag behind in its response to the change in market price. Table 4.1 shows that cases of calls 1 and puts 2, 3, and 5 are where the option price and market price go together. Calls 4, 5, and 8 and puts 9 and 10 are where the option prices were ahead of the market prices, and finally, calls 6, 9, and 10 and puts 8 are where the option prices stayed behind in their response to the market change.

TABLE 4.1

		(1)	(2)	(3)	(4)	(5)	(6)	(7)	(8)
				IV		TV			D
Option		MP	SP	[(1) − (2)]	OP	[(4) − (3)]	ΔOP	ΔMP	[(6) ÷ (7)]
Calls									
x	1	17	17	0	2	2			
	2	19	18	1	4	3	2	2	1
→	3	20	20	**0**	4	**4**	0	1	0
	4	22	20	2	7	5	3	2	1.5
*	5	21	23	0	5	5	−2	−1	2
	6	23	22	1	6	5	1	2	.5
*	7	24	25	0	6	6	0	1	0
	8	28	25	3	11	8	5	4	1.25
*	9	25	28	0	9	9	−2	−3	.66
→	10	27	27	**0**	**10**	**10**	1	2	.5

				(3) =		(5) =			
		(1)	(2)	(2) − (1)	(4)	(4) − (3)	(6)	(7)	(6) ÷ (7)
Puts									
→	1	17	15	**0**	**1**	**1**			
	2	19	20	1	3	2	2	2	1
x	3	20	20	0	4	4	1	1	1
*	4	22	20	0	4	4	0	2	0
	5	21	23	2	3	1	−1	−1	1
→	6	23	22	**0**	3	**3**	0	2	0
*	7	24	23	0	3	3	0	1	0
*	8	28	25	0	6	6	3	4	.75
x	9	25	25	0	2	2	−4	−3	1.3
	10	27	29	2	5	3	3	2	1.5

4.5. DETERMINANTS OF OPTION VALUE

Five major factors determine the market value of an option, especially a call option, since it is the most common and popular option in the exchange market. The following equation describes the value of a call option (VC) as a function of those factors:

$$VC = f[MP, \ SP, \ T, r_f, \sigma_{mp}^2]$$

$$\frac{\partial VC}{\partial MP} > 0 \qquad \frac{\partial VC}{\partial T} > 0 \qquad \frac{\partial VC}{\partial r_f} > 0 \qquad \frac{\partial VC}{\partial \sigma_{mp}^2} > 0$$

but

$$\frac{\partial VC}{\partial SP} < 0$$

where MP is the market price of the underlying stock that would affect the call value positively. The higher the market price of stock, the greater the call value, *ceteris paribus*; SP is the strike price of the call option, which affects the value of option negatively. The lower the strike price, the greater the call value, *ceteris paribus*. T is the length of maturity time, r_f is the risk-free rate, and σ_{mp}^2 is the variance in the market price of the underlying stock. All of these last three factors affect the value of the call option in a positive manner.

Figure 4.1 shows two hypothetical distributions of the market price of the underlying stock. Distribution 2 has a higher variance, and therefore the price movement is harder to predict and more volatile than distribution 1. Both distributions are assumed to have the same expected stock price and same strike price. Since the call option is characterized by being a contingent claim (i.e., the

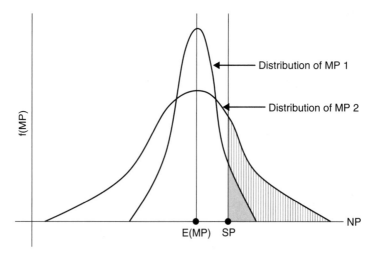

FIGURE 4.1

call holder would make a profit only when the stock price is greater than the strike price), the second distribution with the larger variance would offer much higher probability for the market price to exceed the strike price compared with the limited probability offered by the first distribution.

Consequently, we can conclude that the same set of determinants also affects the value of the put (VP):

$$VP = f[MP, SP, T, r_f, \sigma_{mp}{}^2]$$

$$\frac{\partial VP}{\partial MP} < 0 \qquad \frac{\partial VP}{\partial r_f} < 0$$

$$\frac{\partial VP}{\partial SP} > 0 \qquad \frac{\partial VP}{\partial T} > 0 \qquad \frac{\partial VP}{\partial \sigma_{mp}^2} > 0$$

But the dynamics of the factor changes are different. The market price of the stock and the risk-free rate affect the value of puts negatively, but the rest of the factors—the strike price, the time of maturity, and the variance of the market price—affect the value of puts positively.

4.6. OPTION VALUATION

The 1973 study by F. Black and M. Scholes (*Journal of Political Economy*, **81**, 637–654) on option pricing became a classic reference in the option evaluation calculation. The call option value (VC) is determined by

$$VC = MP[N(d_1)] - (e)^{-rT} \cdot SP[N(d_2)]$$

where MP is the market price of the underlying stock, SP is the strike price of the call, r is the risk-free rate, T is the length of maturity time, and $N(d)$ is the cumulative normal probability density function. VC stands for the probability that a normally distributed random variable will be less than or equal to the area d, where

$$d_1 = \frac{\ln(MP/SP) + T(r + \sigma^2/2)}{\sigma\sqrt{T}}$$

and

$$d_2 = d_1 - \sigma\sqrt{T}$$

where σ is the standard deviation per period of the continuously compounded rate of return on stock. As for the value of a put (VP), it can be obtained using the calculated value of the call (VC), the current value of the strike price CV(SP), and the market price of the stock:

$$VP = [VC + CV(SP)] - MP$$

$$VP = [VC + SP(e)^{-rT}] - MP$$

Although these formulas may sound tedious in application and require obtaining two table values, $N(d_1)$ and $N(d_2)$, Black and Scholes' formulas were simplified for the hand calculation even though computerized programs can handle much more complex values in a split second. The following is the simplified way to get the call and put values:

$$VC = MP(PSP\ value)$$

where the value of a call is a certain percentage of the market price of stock. This percentage is determined by a table value called the **percentage of share price** (PSP). Look at Table 9 in the Appendix. PSP is obtained according to two calculated values:

1. *Vertical value:* calculated as the product of the standard deviation (σ) and the square root of maturity $\sqrt{T}$:

$$\sigma\sqrt{T}$$

2. *Horizontal value:* calculated by dividing the market price of stock by the current value of the strike price using the risk-free rate and the actual maturity time in annual format.

$$\frac{MP}{CV(SP)} = \frac{MP}{SP/(1 + r_f)^T}$$

Example 4.6.1 Calculate the value of a call option with a strike price of $90 and a maturity of 9 months. Given that the current stock price is $108.50, the risk-free rate is 5% and the standard deviation of the rate of return on stock is .75.

First we have to look up the PSP value in Table 9 in the Appendix, but we need to calculate the vertical and horizontal values first:

- Vertical value: $\sigma\sqrt{T} = (.75)\sqrt{.75}$ considering the nine-month maturity as three-fourths of the year.

$$\sigma\sqrt{T} = .65$$

- Horizontal value:

$$\frac{MP}{SP/(1+r_f)^T} = \frac{\$108.50}{\$90/(1+.05)^{.75}} = \$1.25$$

Next, we look at the value in Table 9 corresponding to the vertical value of .65 and the horizontal value of $1.25. The value is 34.2 and that is the PSP value we need (it is in a percentage format).

$$VC = MP(PSP)$$
$$= \$108.50(.342)$$
$$= \$37.11$$

Example 4.6.2   What would be the value of the put associated with the Example 4.6.1 call option?

$$VP = [VC + CV(SP)] - MP$$

Recall that the current value of the strike price was the denominator of the horizontal value calculated above, which was

$$CV(SP) = \frac{SP}{(1+r_f)^T}$$
$$= \frac{\$90}{(1+.05)^{.75}} = \$86.77$$
$$VP = (\$37.11 + \$86.77) - \$108.50$$
$$= \$15.38$$

4.7. COMBINED INTRINSIC VALUES OF OPTIONS

One of the best known business strategies to diversify assets and protect against risks is **hedging**. It is used to combine options and create a mix of possibilities and returns through utilizing a variety of elements, such as different maturities, strike prices, and market prices. We have briefly described the types of option mixes in the early pages of this chapter. Below we calculate the combined intrinsic values of options in two forms of those combinations: a straddle and a butterfly spread.

Example 4.7.1   Calculate the intrinsic value of a combined option consisting of buying a call at a strike price of $55 and buying a put at the same strike price when the market price of the stock is $68. Calculate the value when the stock price goes down to $50.

combined intrinsic value = intrinsic value of call + intrinsic value of put

$$= IVC_B = IVP_B$$

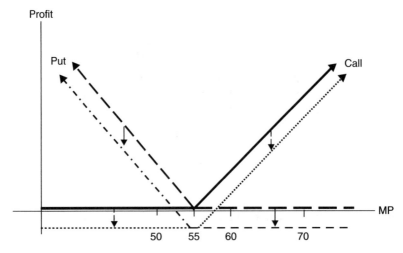

FIGURE E4.7.1 A straddle payoff.

$$
\begin{aligned}
\text{1.} \quad &= \max[(MP - SP), 0] + \max[(SP - MP), 0] \\
&= \max[(\$68 - \$55), 0] + \max[(\$55 - \$68), 0] \\
&= \max(\$13, 0) + \max(-\$13, 0) \\
&= \$13 \\
\text{2.} \quad &= \max[(\$50 - \$55), 0] + \max[(\$55 - \$50), 0] \\
&= \max(-\$5, 0) + \max(\$5, 0) \\
&= \$5
\end{aligned}
$$

This combination, called a **straddle**, is illustrated in Figure E4.7.1.

Example 4.7.2 Suppose that a small business investing in options buys two calls, whose strike prices are $20 and $30, and writes two puts, whose strike prices are $25 and $15. Calculate the intrinsic value of this combination when the stock price is $40 and when it goes down to $30.

When MP = $40:

$$
\begin{aligned}
\text{IVC}_B^1 &= \max[(MP - SP), 0] \\
&= \max[(\$40 - \$20), 0] \\
&= \max(\$20, 0) = \$20
\end{aligned}
$$

$$IVC_B^2 = \max[(\$40 - \$30), 0]$$
$$= \max(\$10, 0) = \$10$$
$$IVP_W^1 = \min[(MP - SP), 0]$$
$$= \min[(\$40 - \$25), 0]$$
$$= \min(\$15, 0) = \$0$$
$$IVP_W^2 = \min[(\$40 - \$15), 0]$$
$$= \min(\$25, 0) = \$0$$

The combined intrinsic value $= \$20 + \$10 + \$0 + \$0 = \$30$.
When MP $= \$30$:

$$IVC_B^1 = \max[(MP - SP), 0]$$
$$= \max[(\$30 - \$20), 0]$$
$$= \max(\$10, 0) = \$10$$

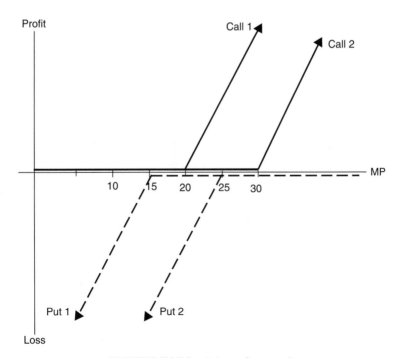

FIGURE E4.7.2   A butterfly spread.

$$\text{IVC}_B^2 = \max[(\$30 - \$30), 0]$$
$$= \max(\$0, 0) = \$0$$
$$\text{IVP}_W^1 = \min[(\text{MP} - \text{SP}), 0]$$
$$= \min[(\$30 - \$25), 0]$$
$$= \min(\$5, 0) = \$0$$
$$\text{IVP}_W^2 = \min[(\$30 - \$15), 0]$$
$$= \min(\$15, 0) = \$0$$

The combined value $= \$10 + \$0 + \$0 + \$0 = \$10$.

This type of combination of options with different strike prices, called a **butterfly spread**, is illustrated in Figure E4.7.2.

5 Cost of Capital and Ratio Analysis

The **cost of capital** is a crucial concept in the context of financial decision making, especially in terms of being the accepted criteria by which a firm would decide whether an investment can or cannot potentially increase the firm's stock price. It is defined as:

1. The rate of return that the firm must earn on its investment to maintain a proper market value for its stock.
2. The rate of return that the investor must require to make its capital attractive for rewarding investment opportunities.

5.1. BEFORE- AND AFTER-TAX COST OF CAPITAL

If a firm uses a long-term debt such as selling bonds to finance its operation, a before-tax cost of debt can be calculated as

$$CC_b = \frac{I + [(M - NP)/n]}{(NP + M)/2}$$

where CC_b is the cost of capital for bonds, I is the annual interest, M is the face value of bonds, NP is the net proceeds, which is the face value adjusted to the flotation cost, and n is the number of years to redemption.

Example 5.1.1 Suppose that a corporation is planning to collect a capital of $5 million by selling its bonds of $1,000, $8\frac{1}{2}$ % coupon rate. Given that the firm is selling at a discounts of $30 per bond and that the flotation cost is 2% per bond, calculate the 20-year and the before-tax cost of capital.

$$I = \$1,000(.085) = 85$$

$$B_d = M - D$$

$$= \$1,000 - \$30 = \$970$$

$$NP = \$970 - (\$970 \times .02) = \$950.60$$

Mathematical Finance, First Edition. M. J. Alhabeeb.
© 2012 John Wiley & Sons, Inc. Published 2012 by John Wiley & Sons, Inc.

$$CC_b = \frac{I + [(M - NP)/n]}{(NP + N)/2}$$

$$= \frac{85 + [(\$1,000 - \$950.60)/20]}{(\$950.60 + \$1,000)/2}$$

$$= \frac{87.47}{\$975.30} = .09 \qquad \text{before-tax cost of capital is } 9\%$$

To get the after-tax cost of capital (CC_a), we use

$$CC_a = CC_b(1 - T)$$

where T is the corporate tax rate. Suppose that the corporate tax rate is 39%; then the after-tax cost of capital would be

$$CC_a = .09(1 - .39) = .055 \text{ or } 5.5\%$$

5.2. WEIGHTED-AVERAGE COST OF CAPITAL

A firms **capital structure** is the mix of debt and equity used to finance the firm's operation. The cost of many basic long-term sources of capital, such as stocks and bonds, have been detailed before. What remains is how these types of capital relate to the firm's capital structure and that is what the overall weighted-average cost of capital does. It is a method to determine the cohesiveness of the firm's capital structure by weighting the cost of each capital component based on its proportion as measured by the market value or book value.

$$CC_{wa} = \sum_{i=1}^{n} w_i k_i$$

where CC_{wa} is the cost of capital weighted average, w_i is the proportion of any type of capital in the firm's capital structure, and k_i is the cost of any type of capital.

Example 5.2.1   The components of a corporation's capital structure and their individual costs are ontlined below and in Table E5.2.1.

1. Long-term debt takes 38% of capital structure and costs 5.59%.
2. Preferred stock represents 14% and costs 9.62%.
3. Common stock takes the remaining 48% and costs 12.35%.

Calculate the corporation's weighted-average cost of capital and explain what it means.

TABLE E5.2.1

Source of Capital	% of Capital Structure, w_i	Cost of Capital, k_i	$w_i k_i$
Long-term debt	.38	.0559	.0212
Preferred stock	.14	.0962	.0135
Common stock	.48	.1235	.0593
	100.00		$\sum_1^3 w_i k_i = .094$

$$CC_{wa} = \sum_{i=1}^{3} w_i k_i = 9.4\%$$

The weighted-average cost of capital is 9.4%, which means that this corporation would be able to accept all investment projects that would potentially earn returns greater than or at least equal to 9.4%.

5.3. RATIO ANALYSIS

Shareholders, creditors, and managers are all very interested in a firm's performance as expressed through its financial statements. Prospective investors as well as current stockholders wish to know more about the firm's trends in returns and potential risks, and ultimately, to have a better understanding of what affects the share price and their share of the firm's profits. Managers' main interest is in their capacity to control and monitor a firm's performance and to take it to the best possible level. All of this would establish the need to analyze the firm's financial statements by the way of constructing and calculating a variety of ratios that would serve as general indicators to assess the firm's performance. Ratio analysis also considers two points of view: the **cross-sectional**, where a comparison of those financial indicators is made at the same point in time, and the **time series**, where the indicators are analyzed as trends extending over a period of time. Most financial ratios are related to the investment, as they are related to the way in which the firm employs and manages its capital. Regarding the term of analysis, most financial ratios are related to short-run analysis, as they address specific aspects of performance, such as the ratios of profitability, liquidity, and operations. In long-run analysis, ratios of debt would be a typical example.

Profitability Ratios

Profitability ratios are also called **efficiency ratios** since the major objective is to assess how efficiently firms utilize their assets and, ultimately, how they are able to attract investers and their capital.

Gross Profit Margin Ratio (GPMR)   This ratio shows how much gross profit, GP (sales after paying for the cost of goods sold) is generated by each dollar of net sales, NS (gross sales minus all goods returned).

$$GPMR = \frac{GP}{NS}$$

Example 5.3.1   If gross profit is $83,420 and net sales is $185,377, the GPMR would be

$$GPMR = \frac{\$83,420}{\$185,377} = .45$$

A gross profit margin of 45% means that out of each dollar of net sales, 45 cents would be gross profit.

Operating Profit Margin Ratis (OPMR)   Instead of the gross profit in the GPMR, the OPMR shows the operating profits as they are related to net sales. Operating profit is another term for operating income, which is the same as EBIT (earnings before income and taxes).

$$OPMR = \frac{OY}{NS}$$

Example 5.3.2   If the operating income is $45,000 the and net sales is $300,000, the OPMR is

$$OPMR = \frac{\$45,000}{\$300,000} = .15$$

which means that 19 cents out of each dollar of net sales in this firm goes to the operating income budget.

Net Profit Margin Ratio (NPMR)   This time, the net profit is related to the net sales. The NPMR states how much net profit the firm earns from its volume of sales.

$$NPMR = \frac{NP}{NS}$$

Example 5.3.3   Suppose that the net profit in one firm is $35,287 and the net sales are $298,971. The NPMR would be

$$NPMR = \frac{\$35,287}{\$298,971} = .118$$

or 11.8%, meaning that of each dollar of net sales this firm would have a little less than 12 cents as net profit. This measure is important especially because it paints a picture of the profit after all expenses, including interest and taxes, have been paid for.

Return on Investment Ratio (ROIR)   The ROIR is also known as the ROA (return on assets). It relates net profit (i.e., after interest and taxes) to total assets (TA) of a firm.

$$ROIR = \frac{NP}{TA}$$

Example 5.3.4   Let's use the previous net profit figure of $35,287 against a total asset of $250,000. The ROIR would be

$$ROIR = \frac{\$35,287}{\$250,000} = 14\%$$

which says that each dollar of the total asset value would give 14 cents in net profit.

Return on Equity Ratio (ROER)   In this ratio, the net profit (NP) is related to the owner's equity (OE) in its format of both preferred and common stock. It basically tells stockholders a crucial piece of information: how much of their money a firm would turn into net profit:

$$ROER = \frac{NP}{OE}$$

Example 5.3.5   Suppose that the owner's equity is valued at $79,500. The net profit of $35,287 would be forming an ROER as

$$ROER = \frac{\$32,287}{\$79,500} = 44\%$$

which tells shareholders that this firm is able to turn 44 cents of each dollar of their investment into a net profit.

Sales–Asset Ratio (SAR)   The SAR is another efficiency ratio. It shows how efficient the use of resources is, as an important aspect of a firm's performance and its ability to generate profits. It relates sales to total assets.

$$SAR = \frac{S}{TA}$$

Example 5.3.6   Suppose that the volume of sales for a firm reached $46,890 and its records indicate that the value of its total assets at the beginning of the year was $60,522 and at the end of the year was $50,177. The firm's SAR would consist of dividing the sales (S) by the average value of assets since we have two readings:

$$SAR = \frac{\$46,890}{(\$60,522 + \$50,177)/2}$$

$$= .85$$

This ratio says that the firm is working hard to put its assets to use in producing and selling its products.

Sales to Net Working Capital Ratio (SNWCR)   This time we relate sales to the net working capital, which is basically the firm's short-run net worth or the difference between the current assets and the current liabilities. That current sense of measure is what gives this ratio its more important meanings:

$$\boxed{SNWCR = \frac{S}{NWC}}$$

Example 5.3.7   Suppose that the working capital in the firm of Example 5.3.6 is \$3,590. Its SNWCR would be

$$SNWCR = \frac{\$46,890}{\$3,590} = 13.1$$

This ratio reflects how the volume of sales relates to the firm's current net worth: in other words, how the net working capital has been put to use.

Market-Based Ratios

Market-based ratios reflect a firm's performance as it is associated with the related market, and therefore the ratios would be looked at with great interest by current investors, potential investors, and managers.

Price–Earnings (P/E) Ratio   The P/E ratio is one of the most important and commonly used ratios. It relates the market price of a firm's common stock (MPS) to its earnings per share (EPS).

$$\boxed{P/E = \frac{MPS}{EPS}}$$

This ratio reflects investors' confidence in a firm's financial performance; therefore, the higher the P/E, the higher the appraisal given by the stock market.

Example 5.3.8 If a firm's market price per share of common stock is $65 and the firm has a $7.45 earnings per share, the firm's P/E is

$$P/E = \frac{\$65}{\$7.45} = 8.72$$

which means that this firm's common stock is selling in the stock market for nearly nine times its earnings.

Price–Earnings–Growth Ratio (PEGR)   This ratio employs the P/E ratio and relates it to a firm's expected growth rate per year (EGR). It reflects the firm's potential value of a share of stock.

$$\boxed{PEGR = \frac{P/E}{EGR}}$$

Example 5.3.9   Suppose that a firm with a P/E of 8.72 expects an annual growth rate of 8%. Its PEGR would be

$$PEGR = \frac{8.72}{8} = 1.09$$

It is theorized that PEGRs represent the following:

- If PEGR $= 1$ to 2: The firm's stock is in the normal range of value.
- If PEGR < 1: The firm's stock is undervalued.
- If PEGR > 2: The firm's stock is overvalued.

Earnings per Share (EPS)   The EPS is more important to common stockholders in particular because it is calculated by dividing the net profit (after subtracting the dividends for preferred stock) by the outstanding number of shares of common stock.

$$\boxed{EPS = \frac{NP - D_p}{\text{no. shares}}}$$

Example 5.3.10   Suppose that a firm has a net profit of $600,000. It pays 7% of its dividends to preferred stockholders and distributes the remainder among the 40,000 shares of common stock. Its EPS would be

dividends for preferred stocks, $D_p = \$600,000 \times .07 = \$42,000$

$$EPS = \frac{\$600,000 - \$42,000}{40,000} = \$13.95$$

This means that for each share of common stock that investors own, they earn $13.95.

Dividend Yield (DY)   The DY is obtained by dividing dividends of common stock per share (DPS) by the market price of stock (MPS):

$$DY = \frac{DPS}{MPS}$$

If the dividend per share is $1.95 and the stock price is $35, the dividend yield is

$$DY = \frac{\$1.95}{\$35} = 5.6\%$$

which says that common stockholders receive only 5.6% as a dividend for the price that each share of stock sells for in the market.

Cash Flow per Share (CFPS)   The CFPS is just like earnings per share (EPS) except that it uses cash flow instead of net profit. Some financial analysts believe that real operating cash flow (OCF) is a much more reliable measure than net profit, which includes a lot of accounts receivable. A measure of the cash available as related to the number of shares of common stock is a good indicator of a firm's financial health.

$$CFPS = \frac{OCF}{no.\ shares}$$

Example 5.3.11   Suppose that a firm has an operating cash flow of $65,000 and its shares of common stock reach 500,000 shares outstanding. Its CFPS would be

$$CFPS = \frac{\$65,000}{500,000} = \$.13$$

which means that the cash flow per share in this firm is 13 cents.

Payout Ratio (PYOR)   This ratio shows how much earnings per share would be paid out as cash dividends for common stockholders (D_c).

$$PYOR = \frac{D_c}{EPS}$$

Example 5.3.12   Let's suppose that for a firm with an EPS of $13.95, $3.10 is paid out as a cash dividend per share. The payout ratio would then be

$$PYOR = \frac{\$3.10}{\$13.95} = \$.22$$

which means that 22 cents out of each dollar earned per share is being paid out as a dividend.

Book Value per Share (BVPS) This ratio shows the stockholder's equity or net worth (NW) for each share held.

$$BVPS = \frac{NW}{no.\ shares}$$

Example 5.3.13 Suppose that a firm's total assets are $747,000, and total liabilities are $517,000 and there are 20,000 shares outstanding. What would be the book value per share?

$$NW = A - L$$
$$= \$747,000 - \$517,000$$
$$= \$230,000$$
$$BVPS = \frac{230,000}{20,500} = \$11.50$$

This means that each share is worth $11.50 of the firm's net worth.

Price–Book Value Ratio (PBVR) This ratio shows how the market price of a stock (MPS) is related to the book value per share (BVPS):

$$PBVR = \frac{MPS}{BVPS}$$

Example 5.3.14 Suppose that the stock of the firm in Example 5.3.14 is sold in the market for $20. The price–book value ratio would be

$$PBVR = \frac{\$20}{\$11.50} = \$1.74$$

A PBVR of $1.74 means that this firm is worth 74% more than the shareholders put into it.

Generally, the PBVR can be read like this:

- If PBVR > 1: The firm is utilizing assets efficiently.
- If PBVR < 1: The firm is utilizing assets inefficiently.
- If PBVR = 1: The firm is utilizing on the margin.

Price/Sales (P/S) Ratio   This ratio shows how many dollars it takes to buy a dollar's worth of a firm's revenue. It is calculated by dividing the market capitalization (MC), which is (stock price × no. shares) by the firm's revenue for the last year (TR).

$$MC = MPS \times \text{no. shares}$$

$$\boxed{P/S = \frac{MC}{TR}}$$

Example 5.3.15   If we take the stock price and the number of shares from the preceding examples: MPS = $20 and the number of shares = 20,000, and if we suppose that the revenue of this firm last year was $650,000, the market capitalization (MC) would be

$$MC = \$20 \times 20,000 = \$400,000$$

$$P/S = \frac{\$400,000}{\$650,000} = .62$$

A price/sales ratio of 62% is good; it describes a case in which the investors get more than they invest. Generally, market analysts have come up with the criterion that the P/S ratio should be less than if not equal to 75%:

$$\boxed{P/S \leq .75}$$

and investors should avoid firms with price/sales ratios above 150%.

Tobin's Q-Ratio   Tobin's Q-ratio is named after the economist James Tobin, who came up with this ratio as an improvement over the traditional price–book value ratio (PBVR). Tobin believes that both the debt and equity of a firm should be included in the top of the ratio, and instead of depending on the firm's book value the bottom should be the firm's entire assets in their replacement cost, which is adjusted for inflation. In this case, Tobin's Q would reflect accurately where the firm stands.

$$\boxed{\text{Tobin's } Q = \frac{TA_{mv}}{TA_{rv}}}$$

where TA_{mv} is the market value of the firm's total assets and TA_{rv} is the replacement value of total assets.

Example 5.3.16   If the market value of total assets of a firm is $127 million and its replacement cost is $150 million, its Tobin's Q value would be

$$\text{Tobin's } Q = \frac{127}{150} = 84.6\%$$

Tobin referred to the rule of thumb for this ratio:

- If Tobin's $Q > 1$: The firm would have the capacity and incentive to invest more.
- If Tobin's $Q < 1$: The firm cannot invest and may acquire assets through merger.

Operational Ratios

Members of this group of ratios are also called **activity ratios**. They deal with the extent to which the firm is able to convert various accounts into cash or sales. These accounts include inventory, accounts receivable, accounts payable, fixed assets, and total asset turnover.

Inventory Turnover Ratio (ITR)   This ratio relates the cost of goods sold (COGS) to the value of inventory (INY):

$$\boxed{ITR = \frac{COGS}{INY}}$$

Often, inventory is calculated as an average of the inventory at the beginning of the year and at the end of the year.

Example 5.3.17   If the cost of goods sold is \$130,000 and the average value of inventory is \$53,560, the ITR would be

$$ITR = \frac{\$130,000}{\$53,560} = 2.43$$

An ITR of 2.43 means that the firm moves its inventory 2.43 times a year. The ITR can also be expressed as the **average age of inventory** (AAINY), which is a measure of how many days the average inventory stays in stock. That would be done by dividing the number of days a year (365) by the ITR.

$$\boxed{AAINY = \frac{365}{ITR}}$$

So if we divide 365, by 2.43 we get

$$AAINY = \frac{365}{2.43} = 150$$

which means that it would take 150 days for this firm to carry its inventory.

Accounts Receivable Turnover (ART)   The ART is also called the **average collection period**, which shows the extent to which customers pay their credit bills. It is the account receivable (AR) divided by the average daily sales (DS):

$$ART = \frac{AR}{DS}$$

Example 5.3.18   If the account receivable has $550,000 and the annual sales are $3,650,000, we can get ART by first getting the daily sales by dividing the annual sales by 365:

$$\frac{\$3,650,000}{365} = 10,000$$

$$ART = \frac{\$550,000}{10,000} = 55$$

which means that it would take the firm 55 days to collect its bills. This is not good unless the firm has a 60-day collection standard, but it is usually 30 days.

Account Payable Turnover (APT)   The APT is also called the **average payment period**. It is similar to the accounts receivable turnover in that it divides the account payable (APY) by the average daily purchase (DP).

$$APT = \frac{APY}{DP}$$

Example 5.3.19   Suppose that a firm's accounts payable shows $480,000 and its daily purchases are estimated by $15,517. Its APT would be

$$APT = \frac{\$480,000}{\$15,517} = 31 \text{ days}$$

That means that on average the firm has 31 days to pay its bills, which would be a very good standard.

Fixed Asset Turnover (FAT)   The FAT relates the volume of sales (NS) to the firm's fixed assets (FA).

$$FAT = \frac{NS}{FA}$$

Example 5.3.20 Suppose that a firm has a total value of fixed assets of $79,365 and its net sales are estimated at $133,773. Its FAT value would be

$$FAT = \frac{\$133,773}{\$79,365} = 1.7$$

which means that this firm is able to generate a sales value 1.7 times that of the value of its fixed assets.

Total Asset Turnover (TAT) The TAT is just like the FAT except that this time the net sales value is related to all assets in the firm instead of only the fixed assets.

$$TAT = \frac{NS}{TA}$$

Example 5.3.21 Suppose that all assets in Example 5.3.20 are $140,593; the total asset turnover TAT would then be

$$TAT = \frac{\$133,773}{\$140,593} = .95$$

A total asset turnover of 95% means that a firm is able to turn over 95% of its asset value in net sales.

Liquidity Ratios

Liquidity ratios show a firm's ability to handle and pay its short-term liabilities and obligations. The more liquid assets the firm can lay its hands on, the easier and smoother the entire performance will be. Liquidity ratios include the current ratio, the quick ratio, the net working capital ratio, and the cash ratio.

Current Ratio (CR) This ratio is probably the most popular among financial ratios for its direct relevance. It simply describes how current assets (CA) are related to current liabilities (CL):

$$CR = \frac{CA}{CL}$$

Example 5.3.22 Suppose that a firm's current assets are valued at $1.5 million and its current liabilities are estimated at $980,711. The firm's current ratio would be

$$CR = \frac{\$1,500,000}{\$980,711} = \$1.53$$

A current ratio of 1.53 means that this firm has 1 dollar and 53 cents in its current asset value to meet each dollar of its current obligations. Generally, the current ratio is recommended by most financial analysts to be 2 or more, which means that for a firm to be robust, it has to own at least twice as much as it owes.

$$CR \geq 2$$

The firm here has to dedicate 65 cents out of each dollar of its current assets to pay its current creditor's claims $(1/1.53) = .65$.

Acid-Test Ratio (QR)   This ratio is also called the **quick ratio**. It is similar to the current ratio above except that the value of inventory is taken away from the current assets.

$$QR = \frac{CA - INY}{CL}$$

Example 5.3.23   If the entire inventory in the firm of Example 5.3.22 was estimated at $380,664, the quick ratio would be

$$QR = \frac{\$1,500,000 - \$380,664}{\$980,711} = 1.14$$

which means that the firm has a dollar and 14 cents for each dollar of its creditor's claims. It is noteworthy to mention here that if there are any prepaid items, they would also be subtracted from the current assets along the inventory value.

Net Working Capital Ratio (NWCR)   The net working capital is the short-run net worth of a firm. It is the difference between the current assets and the current liabilities. If we divide the net working capital (NWC) by the available total assets (TA), we get the net working capital ratio (NWCR), which shows the firm's potential cash capacity.

$$NWCR = \frac{NWC}{TA}$$

Example 5.3.24   Suppose that the total assets for the firm in Example 5.3.23 is $49,950,592 and that the current assets and liabilities stay at $1,500,000 and $980,771, respectively. The firm's net working capital would be

$$NWC = CA - CL$$
$$= \$1,500,000 - \$980,711$$
$$= \$519.289$$

and its net working capital ratio would be

$$\text{NWCR} = \frac{\$519,289}{\$4,950,592} = .10$$

which means that this firm has 10 cents in current net worth in each dollar of its total assets.

Cash–Current Liabilities Ratio (CCLR)   This ratio tracks down cash and marketable securities (C + MS) that are at hand and weighs them against the due current obligations and liabilities (CL).

$$\boxed{\text{CCLR} = \frac{C + MS}{CL}}$$

Example 5.3.25   If we keep the current liabilities of the last firm at $980,711 and assume that cash is counted as $27,500 and marketable securities estimated at $31,342, the firm's cash-to-current liabilities ratio (CCLR) would be

$$\text{CCLR} = \frac{\$27,500 + \$31,342}{\$980,711} = .06$$

which says that this firm holds some liquid assets in terms of cash and marketable securities equal to 6 cents to meet each dollar of its current liabilities.

Interval Ratio (InR)   This ratio is another expression of the cash–current liabilities ratio but in terms of time. It reveals how many days a firm is able to meet its short-term obligations. It is obtained by dividing not only cash and marketable securities, but also accounts receivable (AR), by the daily expenditures on current liabilities or current liabilities per day (CLPD).

$$\boxed{\text{InR} = \frac{C + MS + AR}{CLPD}}$$

Example 5.3.26   Let's consider $18,500 in accounts receivable in Example 5.3.25. Also consider that the average daily expenditures on obligations is calculated at $1,250.

$$\text{InR} = \frac{\$27,500 + \$31,342 + \$18,500}{\$1,250} = 62 \text{ days}$$

An interval ratio of 62 days means that the firm can continue to meet its average daily spending of $1,250 on obligations for 2 months, tapping its reserve of cash, marketable securities, and accounts receivable.

Debt Ratios

Debt ratios are also called **leverage ratios**. Because of the increased financial leverage and risk that comes with using more debt in a firm's financing, debt ratios have more importance. These ratios indicate the extent to which a firm's assets are tied to a creditor's claims and therefore the firm's ability to meet the fixed payments that are due to pay off debt.

Debt–Asset (D/A) Ratio   This is a direct measure of the percentage of a firm's total assets that belong to creditors: in other words, how much of other people's money is used to generate business profits. It is obtained simply by dividing total liabilities or debt (TD) by total assets (TA).

$$D/A = \frac{TD}{TA}$$

Example 5.3.27   If a firm has a total debt of $734,000 and its total assets are estimated at $1,930,570, its debt–asset ratio would be

$$D/A = \frac{\$734,000}{\$1,930,570} = .38$$

which means that 38% of the firm's assets is financed with debt.

Debt–Equity (D/E) Ratio   This ratio weighs a firm's total debt (TD) to its owner's equity (E). It shows the percentage of owner's equity that is generated by debt.

$$D/E = \frac{TD}{E}$$

Example 5.3.28   If the firm in Example 5.3.27 has an equity estimated at $1,200,000, its D/E ratio would be

$$D/E = \frac{\$734,000}{\$1,200,000} = .61$$

A D/E of 61% means that for every dollar of owner's equity in the firm, 61 cents is owed to creditors.

Solvency Ratio (Sol)   The solvency ratio is the opposite of the D/A ratio: It is obtained by dividing total assets by total debt. It shows to what extent a firm's total assets can handle its total liabilities or debt.

$$\text{Sol} = \frac{\text{TA}}{\text{TD}}$$

Example 5.3.29   Let's reverse the D/A ratio in Example 5.3.27, and see what kind of solvency ratio we get:

$$\text{Sol} = \frac{\$1,930,570}{\$734,000} = 2.6$$

This solvency ratio means that the firm actually owns 2.6 times more than it owes, and therefore it is solvent. Solvency criteria are:

- If Sol > 1: The firm is solvent.
- If Sol < 1: The firm is insolvent.
- If Sol = 1: The firm is on the margin when its total debt is equal its total assets.

Times Interest Earned Ratio (TIER)   This ratio measures the extent to which a firm is able to make its interest payments. It is obtained by dividing the firm's operating income (OY) (earnings before interest and taxes) by the annual amount of interest due to creditors.

$$\text{TIER} = \frac{\text{OY}}{I}$$

Example 5.3.30   Suppose that a firm's operating income is $170,000 and its total annual interest payment is $35,000. The times–interest earned ratio would be

$$\text{TIFR} = \frac{\$170,000}{\$35,000} = 4.9$$

This means that this firm has an operating income almost five times larger than the interest payment due. We can also say that for every dollar of interest the firm pays to creditors, it has almost $5 in the form of operating income. Still yet, we can say that the firm has enough of a paying capacity to be able to service its debt for about five years.

Operating Income–Fixed Payments Ratio (OYFPR)   This ratio is an expanded TIER. Instead of interest payments only in the denominator, all other fixed payments are added to the interest payments, such as payment for principal (P), payments for preferred stocks as dividends (D_{ps}), and scheduled lease payments (L).

$$\text{OYFPR} = \frac{\text{OY}}{I + P + D_{ps} + L}$$

Example 5.3.31 Consider the following fixed payments as additions to the interest payment in Example 5.3.30: P, \$22,000; D_{ps}, \$51,000; L, \$11,000. Then the (OYFPR) would be

$$\text{OYFPR} = \frac{\$170,000}{\$35,000 + \$22,000 + \$51,000 + \$11,000}$$

$$= 1.43$$

Still, this firm's operating income is 1.43 times more than all the fixed payments due.

Note that the principal payment, lease payment, and preferred stock payment have to be in before-tax status. If they are in after-tax status, they have to be converted to before-tax status by dividing them by $(I - T)$.

$$(P + D_{ps} + L)_b = \frac{(P + D_{ps} + L)_a}{(I - T)}$$

where T is the corporate tax rate.

5.4. THE DuPONT MODEL

The **DuPont model** is a system of financial analysis that has been used by financial managers since its invention by financial analysts working at the DuPont Corporation in the 1920s. It can be described as a collective method of financial analysis, although it has been characterized by some analysts as a complete system of financial ratio utilization. The basic premise of this model is to combine the firm's two financial statements:

1. The income–expense statement
2. The balance sheet

and to incorporate the impact of three important elements:

- The profits on sales, represented by the net profit margin ratio
- The efficiency of asset utilization, represented by the total asset turnover ratio
- The leverage impact, represented by the equity multiplier

The model has two major objectives:

1. To analyze what determines the size of return that investors look forward to receiving from firms in which they invest. This objective is achieved by breaking the return on equity (ROE) into two components: the return on investment (ROI) and the equity multiplier (EM).

$$\boxed{\text{ROE} = \text{ROI} \cdot \text{EM}}$$

2. To break the elements of ROE into subelements: The return on investment is obtained by multiplying the net profit margin (NPM) by the total assets turnover (TAT):

$$\boxed{\text{ROI} = \text{NPM} \cdot \text{TAT}}$$

Furthermore, the net profit margin is obtained by dividing net profits by net sales, and total asset turnover is obtained by dividing net sales by total assets:

$$\text{NPM} = \frac{\text{NP}}{\text{NS}} \quad \text{and} \quad \text{TAT} = \frac{\text{NS}}{\text{TA}}$$

The equity multiplier (EM) is the ratio of total assets to owner's equity:

$$\text{EM} = \frac{\text{TA}}{\text{OE}}$$

To substitute all of the elements, we get

$$\text{ROE} = \frac{\text{NP}}{\text{NS}} \cdot \frac{\text{NS}}{\text{TA}} \cdot \frac{\text{TA}}{\text{OE}}$$

Canceling out NS and TA, we get:

$$\boxed{\text{ROE} = \frac{\text{NP}}{\text{OE}}}$$

Figure 5.1 shows how the various elements are taken from two financial statements, the balance sheet and the income–expense statement, to establish the return on equity.

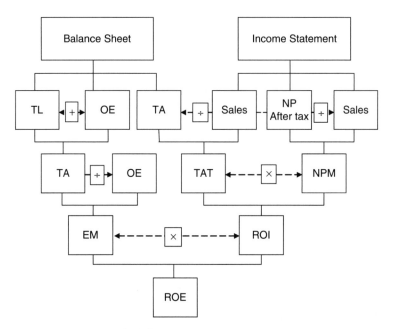

FIGURE 5.1 The DuPont Model.

5.5. A FINAL WORD ABOUT RATIOS

We have discussed a large number of ratios over five categories covering almost every possible aspect of business performance. These ratios are not to be memorized but to be understood and used and interpreted well. They are mathematical terms of one amount divided by another and therefore must be understood as such. The interpretation simply has to be focused on reading the numerator as part of the denominator or the denominator as the whole, including the numerator: simply how the part above the division line relates to the part below the line. Business performance has many aspects, and this is a reason to say that an analyst would be much wiser to use many ratios than only one or two. A comparison has to be consistent in terms of the time period, and requires consistency in size and line of product, among many other factors. A comparison can be made horizontally by the cross-section approach to compare the same ratio across firms, and vertically by the time-series approach to compare ratios of the same firm over the years. Data have to be from sources that were already checked and approved and should be from audited statements. Because of many overlaps, ratios for particular purposes of financial analysis have to be chosen carefully to prevent redundancy.

Unit VI Summary

In this unit four major financial securities were discussed in detail as to matters related to their calculations. Stocks, bonds, mutual funds, and options are the fundamental tools of investment, in addition to an integrating topic as to the cost of capital and ratio analysis.

Types of stocks were discussed briefly, and the process of buying and selling stocks was illustrated with examples. The focus was on common stocks as to their evaluation, the cost of new issues, their value with two-stage dividend growth, their cost through the CAMP model, and other methods of evaluation, such as the P/E multiples method and the liquidation value per share method. Valuation and cost of preferred stocks followed.

The other important security was bonds. The discussion followed a similar pattern, starting with bond evaluation, premium, and discount prices, followed by illustrations of premium amortization and discount accumulation. Several examples explained what happened when bonds were purchased between interest days and how the yield rate was established by the average method as well as by the interpolation and current yield methods. This issue of duration concluded the discussion of bonds.

The third major security is the mutual fund, a combination of stocks and bonds but with their own character and therefore worthy of separate consideration. Discussed were fund evaluation and loads, which are commissions on the purchase and sale of funds. Also discussed was the performance measures of four major criteria: the expense ratio, the total investment expense, the reward–variability ratio, and Treynor's index. The bonds chapter concluded with a related subject on systematic risk, where beta and alpha were explained.

The last major security discussed in this unit was options. We started with a discussion of the choices available to an option holder: whether he would or would not exercise his rights to buy or sell, or even choose to skip action and let the time run out and the rights expire. This discussion was followed by a detailed explanation of the dynamics of making profits with options through buying and selling calls and puts. As for the value of options, the intrinsic value of calls and puts was explained with examples and a graphical presentation. Also, the time value of calls and puts and the determinants of option value were addressed. Finally, the option valuations were put together and the combined intrinsic values were calculated.

Mathematical Finance, First Edition. M. J. Alhabeeb.
© 2012 John Wiley & Sons, Inc. Published 2012 by John Wiley & Sons, Inc.

The last chapter in this unit covered cost of capital and ratio analysis. The cost of capital is a central issue in investment and in finance in general, and that was why it deserved its room in this unit, especially what the cost of capital meant, first, as a rate of return before and after taxes, and then as a weighted average.

In the ratio analysis section we explained many financial ratios with solved examples to emphasize their direct meanings and how they can serve as indicators for financial performance. This topic specifically and the unit generally concluded with the DuPont model, which wrapped up most of the concepts in financial analysis.

List of Formulas

Stocks

Market capitalization rate (expected rate):

$$\text{MCR} = \text{Er} = r = \frac{D + P_1 - P_0}{P_0}$$

Current price:

$$P_0 = \frac{D_1 + P_1}{1 + r}$$

Next year's price:

$$P_1 = \frac{D_2 + P_2}{1 + r}$$

Collective current price:

$$P_0 = D_1 \sum_{t=1}^{\infty} \frac{1}{(1+r)^t}$$

Current price when dividends grow at a constant rate:

$$P_0 = \frac{D_0(1 + g)}{r - g}$$

Market capitalization rate when dividends grow at a constant rate:

$$r = \frac{D_1}{P_0} + g$$

Two-state dividend growth:

$$P_0 = D_0 \left[\frac{1 - ((1 + g_1)/(1 + r)^n)}{r - g_1} \right] (1 + g_1) + D_{n+1} \left(\frac{1}{r - g_2} \right) \left(\frac{1}{1 + r} \right)^n$$

Cost of common stock—CAMP model:

$$r_c = R_f + \beta(R_m - R_f)$$

Mathematical Finance, First Edition. M. J. Alhabeeb.
© 2012 John Wiley & Sons, Inc. Published 2012 by John Wiley & Sons, Inc.

Value of common stock by P/E multiples:

$$V_c = (\text{EPS}_e)(\text{P/E}_i)$$

Value of preferred stock:

$$P_p = \frac{D_p}{r_p}$$

Cost of preferred stock:

$$r_p = \frac{D_p}{N_p}$$

Bonds

Value of bond:

$$B_0 = I\left[\sum_{t=1}^{n} \frac{1}{(1+i)^t}\right] + M\left[\frac{1}{(1+i)^n}\right]$$

$$B_0 = I(\text{PV/FA}_{i,n}) + M(\text{PVIF}_{i,n})$$

$$B_0 = I(a_{\overline{n}|i}) + M(v^n)$$

Discount price:

$$D_s = M(i-r)a_{\overline{n}|i}$$

Premium price:

$$P_m = M(r-i)a_{\overline{n}|i}$$

Price between interest dates:

$$B_{bd} = B_0 + Y_{bp}$$

$$B_{bd} = B_0[1 + i(bp)]$$

Price quoted:

$$B_q = B_{bd} - I_{ac}$$

Yield rate:

$$\text{YR} = \frac{\text{AII}}{\text{AAI}}$$

$$\text{AII} = Mr - \frac{P_m}{n}$$

$$\text{AII} = Mr + \frac{D_s}{n}$$

$$\text{AAI} = \frac{M + B}{2}$$

Current yield:

$$\text{YR} = \text{Cr}(2 + \text{Cr})$$

$$\text{Cr} = \frac{I - P_m/n}{B_q}$$

Duration:

$$D = \frac{I \sum_{t=1}^{n} \dfrac{t}{(1 + i)^t} + \dfrac{nM}{(1 + r)^n}}{I \sum_{t=1}^{n} \dfrac{t}{(1 + i)^t} + \dfrac{nM}{(1 + i)^n}}$$

Percent change in bond price:

$$\%\Delta B = \Delta i(\text{VL})$$

$$\%\Delta B = \Delta i \left(\frac{D}{1 + i} \right)$$

Mutual funds

Net asset value:

$$\text{NAV} = \frac{1}{S}[(\text{MV} + C) - O]$$

Purchase price:

$$\text{PP} = \text{NAV} \left(\frac{1}{1 - L_F} \right)$$

Selling price:

$$\text{SP} = \text{NAV}(1 - L_B)$$

Expense ratio:

$$\text{ER} = \exp \left[\frac{1}{(A_1 + A_2)/2} \right]$$

Total investment expense:

$$\text{TIE} = \frac{1}{n}(L_F + L_B) + \text{ER}$$

Reward variability ratio:

$$\text{RVR} = \frac{1}{\sigma_j}(R_{jt} - R_{ft})$$

Rate of return:

$$R = \frac{S}{1\text{F}}(\Delta P + D + G)$$

Treynor index:

$$\text{TI} = \frac{1}{\beta_j}(R_{jt} - R_{ft})$$

Specific fund beta:

$$\beta_i = \frac{\text{Cov}(R_i, R_m)}{\sigma_m^2}$$

Market beta:

$$\beta_m = \frac{\sigma_m^2}{\sigma_m^2} = 1$$

Collective beta:

$$\beta_p = \sum_{j=1}^{k} \beta_j w_j \qquad j = 1, 2, \ldots, k$$

Options

Intrinsic value of a buyer's call:

$$\text{IVC}_B = \max[(\text{MP} - \text{SP}), 0]$$

Intrinsic value of a writer's call:

$$\text{IVC}_W = \min[(\text{SP} - \text{MP}), 0]$$

Intrinsic value of a buyer's put:

$$\text{IVP}_B = \max[(\text{SP} - \text{MP}), 0]$$

Intrinsic value of a writer's put:

$$\text{IVP}_w = \min[(\text{MP} - \text{SP}), 0]$$

Time value of a call:

$$TV_c = OP - IVC$$

Time value of a put:

$$TV_p = OP - IVP$$

Delta ratio:

$$D = \frac{\Delta OP}{\Delta MP}$$

Call value:

$$VC = MP[N(d_1)] - (e)^{-rt} \cdot SP[N(d_2)]$$

Put value:

$$VP = [VC + SP(e)^{-rt}] - MP$$

$$d_1 = \frac{\ln(MP/SP) + T(r + \sigma^2/2)}{\sigma\sqrt{T}}$$

$$d_2 = d_1 - \sigma\sqrt{T}$$

Cost of capital

Cost of capital for bonds:

$$CC_b = \frac{I + [(M - NP)/n]}{(NP + M)/2}$$

After-tax cost of capital:

$$CC_a = CC_b(1 - T)$$

Weighted-average cost of capital:

$$CC_{wa} = \sum_{i=1}^{n} w_i k_i$$

Exercises for Unit VI

1. A man invested in 500 shares of a local business stock selling for $15.75 a share. A while later he sold half of his shares at $17.25. Calculate his investment rate of return and his capital gain.

2. An investor pays $65 plus .00675 on her investment of $3,000. She plans to purchase whatever she can get of the stock of Exercise 1 priced at $15.75. Find the net investment and how many shares she is able to purchase. Also calculate the yield on investment.

3. Find the expected rate of return for an investor who purchases 130 shares of stocks at $33.50 per share with dividends expected to be $1.95 per share. Assume that he would sell 80 shares at $35.00.

4. A company is selling its stock at $17.95 per share and expecting it to grow by 5%. It usually distributes 50% of its earnings per share as dividends. Find the value of this stock if the earning per share is $2.15 and if an investor would like to earn a 9% yield.

5. A company decides to sell a new issue of its original stock of $25 per share by lowering the price by 10%. What would be the cost of the new stock if the dividend is $2.15 and the flotation cost is $.50? Assume that the stock grows by 4.5%.

6. An investor is contemplating investing heavily in a stock whose dividend is $27 per share but expected to grow by 18% for the next 3 years and by 12% after that. Find the current price of this stock if this investor requires a rate of return of at least 15%.

7. If the price/earnings ratio of a textile industry is $7\frac{1}{2}$, estimate the value of a share in an apparel firm whose earning per share is $4.15.

8. If the required rate of return is $9\frac{2}{3}$% and the dividend for preferred stock is $6.50, what would be the value of this preferred stock?

9. Find the cost of preferred stock in Exercise 8 if the value of the stock dips to $50 and the flotation cost is $1.75.

10. What is the purchase price of a $1,000 bond that is maturing in 20 years at 12% interest if the required rate of return is 15%?

Mathematical Finance, First Edition. M. J. Alhabeeb.
© 2012 John Wiley & Sons, Inc. Published 2012 by John Wiley & Sons, Inc.

11. An investor who wants to yield of 8% would like to invest in a $1,000 bond maturing in 15 years at a coupon rate of 10.5%. Find the bond's current value.

12. A $2,000 bond is redeemable at a coupon rate of 6.5% in 10 years. Would you purchase it at premium or discount price if you want it to yield 8%?

13. What if another investor requires only a 5% yield. Would he buy the bond of Exercise 12 at a premium or a discount price?

14. A bond has a face value of $2,000 redeemable in 5 years at a coupon rate of 8%. Construct the premium amortization schedule if the bond is to be purchased to yield 6%.

15. Consider a $3,000 bond with a coupon rate of 7% but purchased to yield $8\frac{1}{2}\%$. Construct the discount accumulation schedule for its maturity period of 5 years.

16. A $5,000 bond with a semiannual coupon at $8\frac{1}{2}\%$ is redeemable at par value on November 1, 2015. Find the purchase price on July 15, 2013 to yield $7\frac{1}{2}\%$.

17. A $4,000 bond redeemable at $6\frac{1}{2}\%$ in $7\frac{1}{2}$ years. Find (**a**) the flat price to yield 8%, (**b**) the bond net price, and (**c**) the seller's share of the accrued interest.

18. Find the yield rate for a bond purchased at a quoted price of $2,250 redeemable in 10 years. The face value is $2,000 at a coupon rate of $6\frac{1}{4}\%$.

19. Calculate the duration of a bond with a par value of $2,000 redeemable in 7 years at a coupon rate of $8\frac{1}{4}\%$ when the market yield is 11%.

20. Find the volatility factor (VL) in Exercise 19 and explain what it means for the bond price.

21. Calculate the net asset value of the mutual fund of fire securities with the following information:

Security	No. Shares	Share Price ($)	Liabilities ($)	Cash ($)	Outstanding Shares
A	337,000	10	538,000	444,500	500,000
B	250,000	15			
C	118,500	25.50			
D	120,095	31.95			
E	85,000	23			

22. If the net asset value of a mutual fund is $31.75, the front-end load is 5%, and the back-end load is $5\frac{1}{2}\%$, find the purchase and sale prices.

23. Find the total investment expense for the fund in Exercise 22 if it is held for 5 years given the expense ratio of .075%.

24. If the standard deviation of a fund returns is 13.35 and the return is 6% but the risk-free return in the market is 4.5%, find the reward-to-variability ratio.

25. Suppose that the market beta is 2.05. Find Treynor's index for the fund in Exercise 22 and explain what it means.

26. Find the intrinsic value of a call for both the buyer and writer if the market price of the underlying stock is $37.50 per share, up from $34.00 three months ago.

27. Find the buyer's and writer's intrinsic value of a put whose strike price is $28.50 when the market price of the stock goes up to $32.00.

28. Suppose that the market price of a stock is up to $37.00 and an investor has $550 worth of 100 shares with the strike price of a call at $33.00. Calculate the time value for a buyer's call.

29. If the time value of an option is $4.60 and the cost of 100 shares is $360, what would be the intrinsic value of the option?

30. Suppose that the market price of an underlying stock went from $40.00 to $45.00 and the option price went from $8.00 to $10.00. Find the delta ratio. Explain what it means.

31. If the strike price of a call option is $75.00 with a 6-month maturity and the market price of the underlying stock is $88.00 with a standard deviation of returns of 2.34, what is the value of the call if the risk-free rate is 6%?

32. For a financial operation involving selling bonds at a 10% discount with a flotation cost of 3%, what would be the before- and after-tax costs of capital if the par value of the coupon is $2,000 at a coupon rate of 7% and a redemption period of 15 years?

33. Calculate the weighted-average cost of capital for a firm with the following capital, weight, and cost of capital for the individual sources.

Capital Source	Percent of Capital	Cost of Individual Source ($)
Preferred stocks	17	12.77
Common stocks	60	15.56
Long-term debt	23	10.57

UNIT VII

Mathematics of Return and Risk

1 Measuring Return and Risk

Financial securities such as stocks and bonds, as well as other investment assets, have to be evaluated to determine how good an investment they may offer. Such a process of valuation presents the opportunity to link and assess two of the most significant determinants of the security share price: risk and return, whose assessment is the core of all major financial decisions. **Risk**, in its most fundamental meaning, refers to the chance that an undesirable event will occur. In a financial sense, it is defined as the chance to incur a financial loss. When an asset or an investment opportunity is dubbed as "risky," it would be thought to have a stronger chance of bringing a financial loss. It would refer implicitly to the variability of the returns of that asset. Conversely, the certainty of the return, such as the guaranteed return on a government bond, would refer to a case of no risk. We can say further that the investment risk refers to the probability of having low or negative returns on invested assets, such that the higher the probability of getting a low or negative return on an asset, the riskier that investment would be.

The **return** on an investment asset is defined by the change in value, in addition to any cash distribution, all expressed as a percentage of the asset original value. For example, if 100 shares of stock are purchased at \$15 per share and sold for \$17 per share, the change in value would be \$200 (\$1,700 − \$1,500), and if within the period between purchase and sale, \$75 was received in dividends, then both the change in value, \$200, and the cash dividend, \$75, would be divided by the original value of the investment (\$1,500), to get the return, 18.3%:

$$\text{return} = \frac{\$275}{\$1,500} = 18.3\%$$

1.1. EXPECTED RATE OF RETURN

The **expected rate of return** (k_e) is the sum of products of individual returns (k_i) and their probabilities (Pr_i). It is, therefore, a weighted average of returns.

$$k_e = \sum_{i=1}^{n} k_i \cdot \text{Pr}_i$$

$$k_e = k_1 \cdot \text{Pr}_1 + k_2 \cdot \text{Pr}_2 + \cdots + k_n \cdot \text{Pr}_n$$

Mathematical Finance, First Edition. M. J. Alhabeeb.
© 2012 John Wiley & Sons, Inc. Published 2012 by John Wiley & Sons, Inc.

TABLE 1.1

Asset X	k_i	Pr_i	$k_i \cdot \text{Pr}_i$
k_1	.09	.45	.0405
k_2	.10	.30	.03
k_3	.11	.25	.0275
			$\sum k_i \text{Pr}_i = .098$

Let's calculate the expected return on asset X if there are three probable returns: 9% at 45% probability, 10% at 30% probability, and 11 % at 25% probability (see Table 1.1):

$$k_e = k_1 \cdot \text{Pr}_1 + k_2 \cdot \text{Pr}_2 + k_3 \cdot \text{Pr}_3$$
$$= .0405 + .03 + .0275$$
$$= .098 \text{ or } 9.8\%$$

1.2. MEASURING THE RISK

The first simple and straightforward way to measure the risk of an asset is the range of returns or the **dispersion**, which is the difference between the highest and the lowest returns. If we take asset X above, which has three probable returns, 9%, 10%, and 11%, if we compare it to asset Y, which also has three probable returns, 5%, 10%, and 15%, and if we calculate the ranges of both sets of returns, the X range $= 2(11-9)$ and the Y range $= 10(15-5)$ (see Table 1.2), we can say that asset Y is riskier than asset X because the range of Y returns is larger. The range reflects the variability, which stands for the risk. We can conclude that the greater the range of returns of an asset, the more the variability and the higher the risk.

Building on the variability notion, the second measure of risk can be the probability distribution of returns. The tighter the probability distribution, the more likely that the actual return will be close to the value expected and therefore have a lower risk; and vice versa. The wider the probability distribution, the higher the variability and therefore the higher the risk.

TABLE 1.2

k_i	X (%)	Y (%)
k_1	9	5
k_2	10	10
k_3	11	15
Range	$11 - 9 = 2$	$15 - 5 = 10$

TABLE 1.3

	k_i	k_e	$k_i - k_e$	$(k_i - k_e)^2$	Pr_i	$(k_i - k_e)^2 \cdot Pr_i$
Asset X						
k_1	.09	.098	−.008	.000064	.25	.000016
k_2	.10	.098	.002	.000004	.50	.000002
k_3	.11	.098	.012	.000144	.25	.000036
				$\sum_{i=1}^{3} (k_1 - k_e)^2 \cdot Pr_i$		.000054
Asset Y						
k_1	.05	.01	−.05	.0025	.25	.000625
k_2	.10	.01	0	0	.50	0
k_3	.15	.10	.05	.0025	.25	.000625
				$\sum_{i=1}^{3} (k_1 - k_e)^2 \cdot Pr_i$		.00125

Standard Deviation (σ)

The standard deviation (σ) would be an appropriate tool to measure the dispersion around the expected return; that is, the higher the standard deviation, the wider the dispersion and the greater the risk. Let's assign some probabilities, such as 25%, 50%, and 25%, to the last sets of returns for assets X and Y (see Table 1.3 and Figure 1.1). and let's calculate the standard deviations.

$$\text{standard deviation } \sigma = \sqrt{\sum_{i=1}^{3} (k_i - k_e)^2 Pr_i}$$

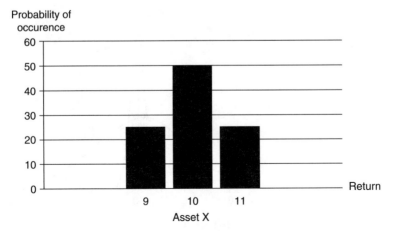

FIGURE 1.1a

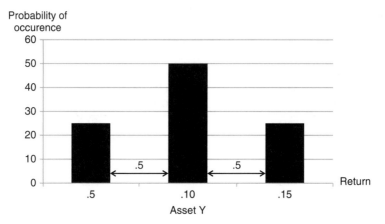

FIGURE 1.1b

$$\sigma_x = \sqrt{.000054} = .0073$$

$$\sigma_y = \sqrt{.00125} = .0353$$

So the standard deviation for the returns on asset Y (σ_y) is larger than the standard deviation of the returns on asset X, which makes asset Y riskier than asset X. In other words, the returns on asset X are closer to their own expected value than are the returns on asset Y to their expected value. Figures 1.1 and 1.2 show this fact visually.

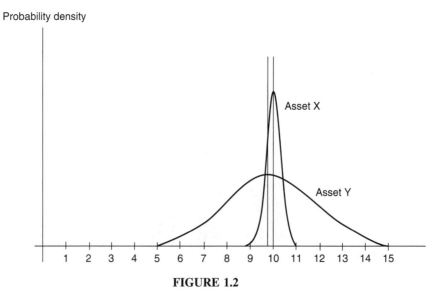

FIGURE 1.2

Furthermore, if we assume that the probability distribution is normal, this would mean that the expected return on asset X (9.8%) would, in fact, be within ± 1 standard deviation 68.26% of the time: that is, between 9.07% (.098 − .0073), and 10.53 (.098 − .0073). It would also be within ± 3 standard deviations .0219 (3 × 0073) 99.74% of the time: that is, between 7.61 (.098 − .0219) and 11.99 (.098 + .0219).

There is another more reliable measure of risk than the standard deviation, especially when the expected returns of the assets are not equal, as in the example above where they were 9.8% and 10%, respectively, for assets X and Y. This additional measure is called the **coefficient of variation** (Coef$_v$) which considers the relative dispersion of data around the value expected. It is obtained by dividing the standard deviation (σ) by the expected return (k_e):

$$\boxed{\text{Coef}_v = \frac{\sigma}{k_e}}$$

So for assets X and Y, we get

$$\text{Coef}_v^x = \frac{\sigma_x}{k_e^x} = \frac{.0073}{.098} = .074$$

$$\text{Coef}_v^y = \frac{\sigma_y}{k_e^y} = \frac{.0353}{.10} = .353$$

The high coefficient of variation for asset Y (.353) confirms that it is more risky than asset X. Generally, the higher the coefficient of variation, the greater the risk associated with the asset. However, it is worthwhile to mention here that when we compare assets with that have an equivalent expected return, the test of coefficient of variation would not differ from the standard deviation test, but it does make a difference when the expected returns are different. In our example here, it just confirmed the standard deviation test, but it is not necessarily the case. Sometimes it would reverse the standard deviation case.

Example 1.2.1   Which of the two assets shown in Table E1.2.1 is riskier? Use both the standard deviation and coefficient of variation tests.

Based on the standard deviation, asset II has a higher standard deviation (5.5) than asset I (4.9) and therefore asset II is riskier. But based on the coefficient of variation test, asset I has a higher coefficient (1.63) than asset II (.46), so asset

TABLE E1.2.1

	Asset I	Asset II
Expected return	3%	12%
Standard deviation	4.9%	5.5%
Coefficient of variation	1.63	.46

I is riskier. Which test is more reliable? The coefficient of variation test is more reliable.

Long-Run Risk

In the long run, asset risk seems to be an increasing function of time. In other words, as time goes by, the variability of returns gets wider and the risk gets greater. That is why experience shows that the longer the life of an investment asset, the higher the risk involved in that asset. Figure 1.3 shows how the dispersion of return distribution of an asset gets wider and how the 1 standard deviation around the expected value gets larger over a 20-year period, assuming that the return expected stays the same.

1.3. RISK AVERSION AND RISK PREMIUM

Risk aversion is a general common behavior that refers to avoiding risky situations. Most investors are risk averse in the sense that, on average, they choose the less risky investment. This tendency comes with the understanding that the returns expected from the less risky investment would not be high. This also implies that if an investor chooses to invest in a higher risk asset, she or he would expect to collect a reward in terms of a high expected return and lower cost of investment. Generally speaking, we can say that a primary implication of the risk aversion dictates that the higher a security risk, the higher its expected return and the lower its price. Let's suppose that investment opportunity A is riskier than investment opportunity B, and let's suppose that both have the same share price of $80 and the same expected return of $8.00 per share. Since investors generally would choose the less risky investment, there would be a high demand for investment B and less demand for investment A. Over a reasonable amount of time, the higher demand on B and lower demand on A would increase the price of B and decrease the price of A. Let's assume that the price of B would rise to $100 and the price of A would go down to $60. Now the rates of return would no longer be 10% (8/80) for each. The rate of return for B would be $8/100 = 8\%$ and the rate for A would be $8/60 = 13.3\%$. The difference in the rates of return for the two investments, brought about by the change in prices due to changes in demand following the risk aversion, is called the **risk premium**. In this case it is $13.3\% - 8\% = 5.3\%$. This is to enforce the notion that riskier securities must attract and reward their investors with a higher expected return than that which is obtained by less risky securities.

1.4. RETURN AND RISK AT THE PORTFOLIO LEVEL

So far, the discussion has been on the return and risk of an individual asset or a single investment opportunity. In reality, it is uncommon to address a single asset in isolation. Financial assets and investment opportunities are often addressed in a

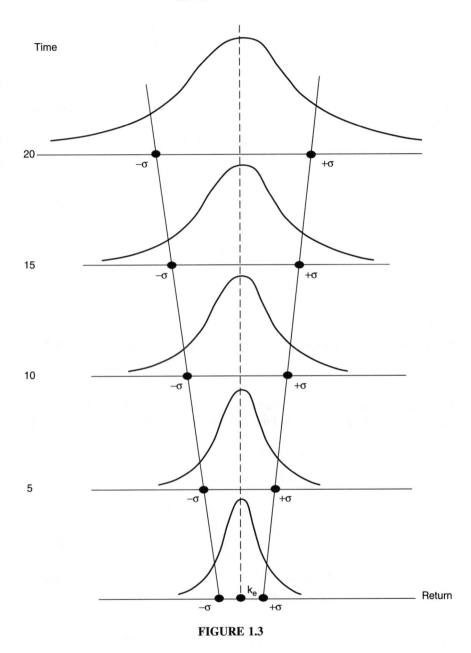

FIGURE 1.3

group and managed in a financial portfolio. Corporate investment, business funds, bank accounts, insurance and pension funds, and individual investments are all held in portfolios that are most likely to be diversified. Therefore, it would be more practical to talk about the return and risk of the entire portfolio as opposed to a single asset. The return or risk of a single component of the portfolio would become important only by its impact on the entire portfolio.

Portfolio Return

Portfolios are most likely to contain a number of individual assets, each in a different proportion. The market value of the entire portfolio (V_p) would be the summation of all values (V_i) of the individual assets.

$$V_p = V_1 + V_2 + \cdots + V_n$$

Each asset value would have its own proportion that would represent its own individual weight (w_i):

$$w_i = \frac{V_i}{V_p}$$

where

$$w_i = w_1, w_2, w_3, \ldots, w_n$$

and all individual weights would make up the entire weight of the portfolio:

$$w_1, w_2, w_3, \ldots, w_n = 100$$

In this same manner, the returns of all these individual assets (V_i) would constitute the return of the portfolio (r_p), and in a way that is commensurate to their weights.

$$r_p = r_1 + r_2 + \cdots + r_n$$
$$r_p = r_1 w_1 + r_2 w_2 + \cdots + r_n w_n$$

$$\boxed{r_p = \sum_{i=1}^{n} r_i w_i}$$

That is, the **portfolio return** is the summation of the weighted individual returns of its component assets, the weights being the proportions of those assets in the entire portfolio.

Example 1.4.1   An investor's portfolio contains two different stocks, two different bonds, and two different mutual funds, with the proportions and returns shown in Table E1.4.1. Calculate the portfolio return.
   The portfolio return is 9.26%.

   There is another aggregate way to calculate the portfolio return based on the change in its entire market value within a certain period of time. The ratio of

TABLE E1.4.1

Asset	r_i (%)	w_i (%)	$r_i w_i$
Stock I	13.5	23	.0311
Stock II	12	18	.0216
Bond I	6.5	15	.00975
Bond II	6	15	.009
MF I	7.5	17	.01275
MF II	7	12	.0084

$$\text{Portfolio return} = \sum_{i=1}^{6} r_i w_i = .0926$$

the change in value to the original value would estimate the rate of return. In other words, the portfolio rate of return, here, is just the percentage change in the value of the entire portfolio between two points in time:

$$r_p = \frac{V_p^2 - V_p^1}{V_p^1}$$

where V_p^1 is the market value of the entire portfolio at the start of the year, and V_p^2 is the market value of the portfolio at the end of the year.

Example 1.4.2 Suppose that the investment portfolio for a small business is estimated at one point by $425,000, and its market value went to $538,620 one year later. That is, $V_p^1 = \$425,000$; $V_p^2 = \$538,620$. What is the portfolio rate of return?

$$r_p = \frac{V_p^2 - V_p^1}{V_p^1}$$

$$r_p = \frac{\$538,620 - \$425,000}{\$425,000}$$

$$r_p = \frac{\$113,620}{\$425,000} = 26.7\%$$

The portfolio expected rate of return (k_p) would be calculated in the same manner as the portfolio rate of return: that is, as the summation of the weighted individual expected returns (k_{ei}):

$$k_p = \sum_{i=1}^{n} k_{ei} w_{ei}$$

Portfolio Risk

Unlike the portfolio return, portfolio risk is not a weighted average of the individual risk of components. **Portfolio risk** can be reduced by adding more and different assets to the portfolio. In other words, the more diversified the portfolio, the less the total portfolio risk. More important is the degree of correlation among the individual assets in a portfolio. The less correlated the assets, the less the portfolio risk. In this sense, diversification would not be as effective in reducing risk unless it involves either negatively correlated assets, or at least the lowest positively correlated assets. The correlation coefficient (Corr) can measure how two variables move against each other. The Corr value ranges from -1, referring to a perfectly negative correlation when variables move in opposite directions, to 1, a perfectly positive correlation where variables move in the same direction. If diversification includes the assets that are negatively correlated, they would move opposite each other and cancel each other out, resulting in risk reduction. However, if diversification brings assets that are strongly positively correlated, risk cannot be diversified away. Let's take a look at the following examples and reflect on how the positively and negatively correlated asset returns affect the portfolio risk.

Example 1.4.3   Let's take a look at two pairs of assets, X and Y and Z and W. Table E1.4.3a shows the returns for assets X and Y where the variances of the two sets of return were calculated individually in columns (7) and (12) as

$$\sigma_x^2 = \sum_{i=1}^{5} (k_i^x - k_e^x)^2 \cdot \mathrm{Pr}_i = .046$$

and

$$\sigma_y^2 = \sum_{i=1}^{5} (k_i^y - k_e^y)^2 \cdot \mathrm{Pr}_i = .067$$

The standard deviations of the two sets of return were calculated as

$$\sigma_x = \sqrt{\sum_{i=1}^{5} (k_i^x - k_e^x)^2 \cdot \mathrm{Pr}_i}$$
$$= \sqrt{.046} = .214$$

$$\sigma_y = \sqrt{\sum_{i=1}^{5} (k_i^y - k_e^y)^2 \cdot \mathrm{Pr}_i}$$
$$= \sqrt{.067} = .259$$

TABLE E1.4.3a Returns for Assets X and Y

| (1) | (2) | Asset X | | | | | Asset Y | | | | | |
| | | (3) | (4) | (5) | (6) | (7) | (8) | (9) | (10) | (11) | (12) | (13) |
	Pr_i	k_i^x	k_e^x $\frac{\sum k_i^x}{5}$	$k_i^x - k_e^x$ $(3)-(4)$	$(k_i^x - k_e^x)^2$ $(5)^2$	$(k_i^x - k_e^x)^2 \cdot Pr_i$ $(6)(2)$	k_i^y	k_e^y $\frac{\sum k_i^y}{5}$	$k_i^y - k_e^y$ $(8)-(9)$	$(k_i^y - k_e^y)^2$ $(10)^2$	$(k_i^y - k_e^y)^2 \cdot Pr_i$ $(11)(2)$	$(k_i^x - k_e^x)$ $(k_i^y - k_e^y) \cdot Pr_i$ $(5)(10)(2)$
k_1	.20	-.12	.155	-.275	.0756	.015	-.15	.174	-.324	.105	.021	.0178
k_2	.15	.405	.155	.25	.0625	.0094	.50	.174	.326	.1063	.016	.0122
k_3	.25	-.07	.155	-.225	.0506	.0126	-.08	.174	-.254	.0645	.016	.0143
k_4	.18	.38	.155	.225	.0506	.0091	.45	.174	.276	.0762	.014	.0112
k_5	.22	.18	.155	.025	.000625	.00014	.15	.174	-.024	.00058	.00013	-.00013
						.046					.067	.0554

The covariance between the two sets of return, $\text{Cov}(x, y)$, is calculated in column (13) as

$$\text{Cov}(x, y) = \sum_{i=1}^{5} (k_i^x - k_e^x)(k_i^y - k_e^y) \cdot \text{Pr}_i$$

$$= .0554$$

and finally, the correlation coefficient (Corr) between the two sets of return is calculated as

$$\text{Corr}_{x,y} = \frac{\text{Cov}(x, y)}{\sigma_x \sigma_y}$$

$$= \frac{.0554}{.214(.259)}$$

$$= 99.9\%$$

A correlation coefficient this close to $+100$ would be considered an indication of perfectly positively correlated assets which exhibit similar dynamics, which makes them move up and down in tandem. This type of matching pattern would not benefit from diversification in risk reduction. Figure E1.4.3a shows such a synchronized movement in the returns of those assets.

The second set of assets, Z and W, is presented in Table E1.4.3b, and the same parameters are calculated in the same manner as that in which they were calculated in Table E1.4.3a. The variances of the assets are at columns (7) and (12):

$$\sigma_z^2 = \sum_{i=1}^{5} (k_i^Z - k_e^Z)^2 \cdot \text{Pr}_i$$

$$\doteq .0339$$

$$\sigma_w^2 = \sum_{i=1}^{5} (k_i^w - k_e^w)^2 \cdot \text{Pr}_i$$

$$= .0129$$

and the standard deviations are

$$\sigma_z = \sqrt{.0339} = .1841$$

$$\sigma_w = \sqrt{.0129} = .1136$$

The covariance is calculated in column (13) as

$$\text{Cov}(z, w) = \sum_{i=1}^{5} (k_i^z - k_e^z)(k_i^w - k_e^w)^2 \cdot \text{Pr}_i$$

$$= -.020353$$

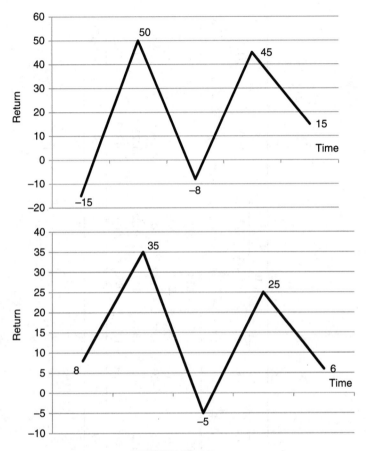

FIGURE E1.4.3a

The correlation coefficient (Corr) is

$$\text{Corr}_{z,w} = \frac{\text{Cov}(z, w)}{\sigma_x \sigma_w}$$

$$= \frac{-.020353}{.1841(.1136)} = 97.3\%$$

A correlation coefficient of -97.3% shows the case opposite to that of the X and Y combination. It indicates that assets Z and W are almost perfectly negatively correlated, which means that the return changes of these assets go up and down opposite to each other. This is the ideal opportunity for these assets to cancel each other out. If one is down, the other is up to compensate—that is the beauty of diversification. The combination of such assets in a portfolio gives the opportunity to have an optimal impact of diversifying the risk away. Figure E1.4.3b shows how the return patterns act opposite each other in a consistently contrasting manner.

TABLE E1.4.3b Returns for Assets Z and W

| (1) | (2) | Asset Z | | | | | Asset W | | | | | (13) |
| | | (3) | (4) | (5) | (6) | (7) | (8) | (9) | (10) | (11) | (12) | $(k_i^z - k_e^z)$ |
	Pr_i	k_i^z	k_e^z	$k_i^z - k_e^z$	$(k_i^z - k_e^z)^2$	$(k_i^z - k_e^z)^2 \cdot \text{Pr}_i$	k_i^w	k_e^w	$k_i^w - k_e^w$	$(k_i^w - k_e^w)^2$	$(k_i^w - k_e^w)^2 \cdot \text{Pr}_i$	$(k_i^w - k_e^w) \cdot \text{Pr}_i$
k_1	.20	.40	.20	.20	.04	.008	.08	.138	−.058	.0034	.00068	−.00232
k_2	.15	.10	.20	−.10	.01	.0015	.35	.138	.212	.045	.00675	−.00318
k_3	.25	.38	.20	.18	.0324	.0081	−.05	.138	−.188	.0077	.001925	−.00846
k_4	.18	−.10	.20	.30	.09	.0162	.25	.138	.112	.0125	.00225	−.00605
k_5	.22	.22	.20	.02	.0004	.000088	.06	.138	−.078	.0061	.001342	−.000343
						.0339					.0129	−.020353

402

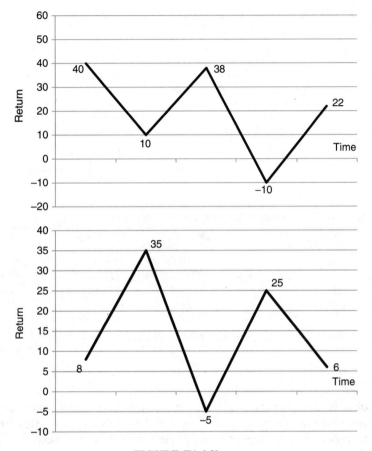

FIGURE E1.4.3b

Combining assets into portfolios would probably reduce the risk even for those assets that are positively correlated. In Tables E1.4.3c and E1.4.3d we combined assets X and Y and obtained an average vector of returns for the combination XY. We also combined assets Z and W and obtained an average vector of returns for the combination ZW. The standard deviation test shows that the combination helps reduce risk even for combining X and Y, which are perfectly positively correlated, as we have seen. The standard deviation of the combined set XY ($\sigma_{xy} = .195$) is still less than either of the assets taken individually, where $\sigma_x = .214$ and $\sigma_y = .259$. This means that the combined assets showed that it is not as risky as either of the individual assets alone. This standard deviation test shows that much better results can be obtained when we combine the negatively correlated assets Z and W. The standard deviation of the combined set ZW ($\sigma_{zw} = .056$), is much less than either of the assets' standard deviation, where $\sigma_z = .1841$ and $\sigma_w = .1136$. It is a further proof that combining assets into portfolios would increase diversification and reduce risk. However,

TABLE E1.4.3c   Combined Assets X and Y with Average Returns

Return	Pr_i	k_i^{xy}	k_e^{xy}	$k_i^{xy} - k_e^{xy}$	$(k_i^{xy} - k_e^{xy})^2$	$(k_i^{xy} - k_e^{xy})^2 \cdot Pr_i$
k_1	.20	$-.135$	.164	.029	.00084	.00017
k_2	.15	.452	.164	.288	.083	.0124
k_3	.25	$-.075$	.164	$-.239$	.057	.0142
k_4	.18	`.415	.164	.251	.063	.0113
k_5	.22	.165	.164	.001	.000001	.00000022

$$\sum_{i=1}^{5} (k_i^{xy} - k_e^{xy})^2 \cdot Pr_i \qquad .038$$

TABLE E1.4.3d   Combined Assets Z and W with Average Returns

Return	Pr_i	k_i^{zw}	k_e^{zw}	$k_i^{zw} - k_e^{zw}$	$(k_i^{zw} - k_e^{zw})^2$	$(k_i^{zw} - k_e^{zw})^2 \cdot Pr_i$
k_1	.20	.24	.169	.071	.00504	.001
k_2	.15	.225	.169	.056	.00314	.00047
k_3	.25	.165	.169	$-.004$	.000016	.000004
k_4	.18	.075	.169	$-.094$	.00884	.00159
k_5	.22	.14	.169	$-.029$	.00084	.00018

$$\sum_{i=1}^{5} (k_i^{zw} - k_e^{zw})^2 \cdot Pr_i \qquad .0032$$

the extent of risk reduction depends primarily on the degree and sign of the correlation between assets. In reality, most of the assets are positively correlated. Studies show that on average, randomly selected assets shows a correlation coefficient around .60. The lower the positive correlation, the better the results of the combination.

In another abstract presentation, Figure 1.4 shows three possible ways to combine two assets in a portfolio: two extreme combinations and one common combination.The assets are A with an expected return of k_A and a risk level of σ_A, and B with an expected higher return of k_B and a higher risk level σ_B.

- The first extreme case of combination occurs at any point along the straight line AB if assets A and B are perfectly positively correlated. This combination cannot benefit much from diversification.
- The second extreme case of combination occurs at any point along BCA where a zero risk can be achieved with a rate of return equal to k_c when the allocation of the two assets can be achieved in reverse proportion to their risk levels. This combination is the show case for the benefit of diversification.
- The third case of combination occurs at any point along curve BA. It is the most likely to occur because assets are often neither negatively nor positively perfectly correlated. The correlation would often be on a moderate level, and

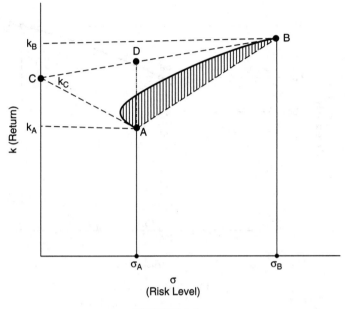

FIGURE 1.4

the combination of assets can enjoy a wide range of returns, from k_A to k_B, for a wide range of risk levels, from σ_A to σ_B. The curve would include all the possible combinations that are better alternatives to any point along straight line AB but lower alternatives to most of the points along BCA, which would offer higher rates of return for the same level of risk, especially along segment BD.

1.5. MARKOWITZ'S TWO-ASSET PORTFOLIO

A great deal of what has been known about risk and return and portfolio construction traces back to a pioneering study by Harry Markowitz published in 1952 (*Journal of Finance*, 7, pp. 77–91). A major issue that was addressed is portfolio diversification of assets and the positive outcome on the portfolio return through the compensatory effect of assets that move in different directions. A well-known fundamental example illustrates the effect of combining two individual assets of different return and risk rates on the portfolio return and risk.

Figure 1.5 shows what happens if an investor decides to invest in two different choices of stocks: stock I (SI), with an expected return of 8% and a low risk (represented by the standard deviation of return) of 15%, and stock II (SII), which offers a higher return of 12% but at a higher risk of 22%. The logical expectation is to calculate the combined return and risk for the mix if we know how much investment an investor is willing to dedicate to each stock. Let's assume that this investor or this portfolio manager is willing to dedicate 55% of investment to

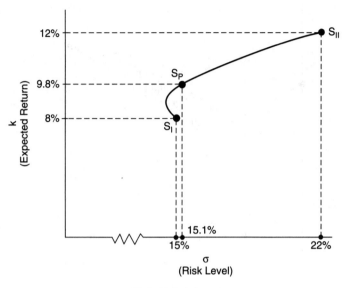

FIGURE 1.5

stock I and 45% to stock II. The portfolio rate of return would be calculated as the weighted average of two returns:

$$k_p = w_1 k_1 + w_2 k_2$$
$$= .55(.08) + .45(.12)$$
$$= 9.8\%$$

As for the portfolio risk, it would be determined by the standard deviation of the combined assets given a correlation between the two assets of .38.

$$\sigma_{I,II} = \sqrt{\sigma_I^2 w_I^2 + \sigma_{II}^2 w_{II}^2 + 2\text{Corr}_{I,II}(w_I\sigma_I)(w_{II}\sigma_{II})}$$
$$= \sqrt{(.15)^2(.55)^2 + (.22)^2(.45)^2 + 2(.38)(.55)(.15)(.45)(.22)}$$
$$= \sqrt{.0228} = 15.1\%$$

So the risk level of the combined stocks in an asset is less than the weighted average risk of the two individual assets, which would have been:

$$(.55)(.15) + (.45)(.22) = 18.2\%$$

Therefore, a combination of assets at S_p would yield a 9.8% rate of return at a reasonable level of risk (15.1%). If we move from this hypothetical example of only two assets in a portfolio to the reality of the investment in the market,

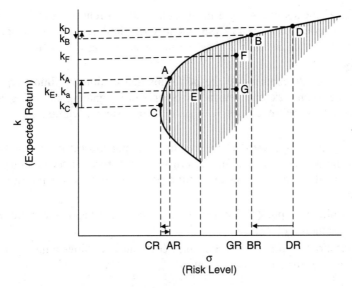

FIGURE 1.6

we would find that there are a large number of assets forming a large number of combinations and producing a large number of portfolios. The broken-egg-shaped area in Figure 1.6 represents all the combinations of assets that are attainable to all investors with their different objectives and different risk and return preferences.

Following are some major observations on Figure 1.6:

- The shaded broken-egg-shaped area is a locus of portfolios with all possible combinations of assets representing a wide range of investor preferences regarding risk and return.
- The solid curve represents the diversified portfolios with the highest returns for any given risk level between CR and DR. Markowitz called this the **efficient portfolio curve**. It is also called the **frontier of risky portfolios**.
- Point D is the portfolio that yields the highest return (k_D) but bears the highest level of risk (DR).
- Point C is the portfolio that yields the lowest return (k_C) but enjoys the lowest level of risk (CR).
- Segment DB contains a collection of portfolios that enjoy a trade-off between risk and return in favor of the risk side. For example, moving from D to B means getting a slightly lower return than k_D but for a greater reduction in the risk level, from DR to BR. Similarly, moving from B to D means gaining a slightly higher return than k_B but carrying more risk from BR to DR.
- Segment AC contains a collection of portfolios that enjoy a trade-off between risk and return in favor of the return. For example, moving from

A to *C* means accepting a greater reduction in return, from k_A to k_C for a lower reduction in risk, from AR to CR. Similarly, moving from *C* to *A* means getting a much higher return for accepting a little more risk, from CR to AR.

- Segment *AB* contains all the portfolios that exhibit an almost equal trade-off between risk and return. In other words, gaining or losing a certain amount of return comes with gaining or losing a compatible amount of risk.

- Inside the shape we can observe that moving toward the northeast means getting portfolios with higher return and higher risk. On the contrary, moving toward the southwest means getting portfolios with lower return and lower risk.

- Portfolio *F* is preferred to portfolio *G* because it yields more return for the same amount of risk.

- Portfolio *E* is preferred to portfolio *G* because it enjoys a much lower level of risk for the same rate of return.

1.6. LENDING AND BORROWING AT A RISK-FREE RATE OF RETURN

Looking at Figure 1.7, let's assume that an investor wants to split his initial investment between asset *A* on the efficient portfolio curve and U.S. Treasury bills, which offer a risk-free rate of return of 5%. Suppose that *A* yields 12% at

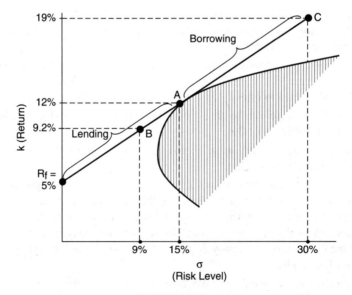

FIGURE 1.7

a risk level of 15%. The investor would like to have 60% of his money invested in asset A and 40% invested in Treasury bills.

The investor in this case is lending 40% of his money to Treasury bills. His rate of return would be

$$.40(.05) + .60(.12) = 9.2\%$$

and his level of risk would be

$$.40(0) + .60(.15) = 9\%$$

He would be at point B. This means that he could be at any point along the line AR_f, depending on the proportions of his investment between asset A and Treasury bills.

Now, let's assume that he borrows at the risk-free rate of 5% an amount of money equal to his own money and invests the total (his own and the borrowed money) in asset A alone. His return would be

$$2(.12) - .05 = 19\%$$

and his risk would be

$$2(.15) + .05(0) = 30\%$$

He would be at point C, which means that he could be at any point along CA depending on how much he borrows and how much risk he tolerates.

1.7. TYPES OF RISK

The risk of the individual assets and of the portfolios that we talked about so far, is the type of **diversifiable risk**, which can be reduced by diversification of assets within portfolios. This type is usually firm-specific risk. It is also called **unsystematic risk**, for it is related to internal conditions and circumstances and varies from firm to firm. It is due to a random set of specifications unique to a specific firm, such as lawsuits in which the firm is involved, the marketing program it conducts, a workers' strike that it faces, or the type and quality of contracts it wins or loses. All these events and circumstances can be mitigated with a certain degree of the firm's diversification of assets. The other type of risk is the **undiversifiable** or **systematic risk**, which is general and market related. It is due to circumstances and conditions that all firms are affected by simultaneously and with no discrimination. The state of the economy is the classic example, highlighted by an impact such as inflation, recession, unemployment, interest rate fluctuation, war, severe weather, or political unrest. Nothing can be done to eliminate or even reduce the risk stemming from such external factors,

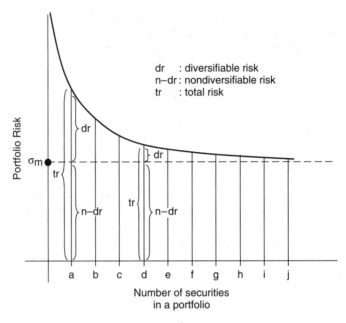

FIGURE 1.8

nor can it be avoided by diversification of assets. It can, however, be assessed by monitoring how a particular asset tends to respond to market circumstances and changes, and can be addressed by an analysis of the capital asset pricing model and measured by the financial beta. Figure 1.8 shows that as a financial portfolio contains more and more individual assets and securities, portfolio risk tends to decline until asymptotically it approaches the systematic nondiversifiable risk. At any number of individual securities in the portfolio, such as *a* or *d*, we can see how the diversifiable and nondiversifiable risks are distributed.

2 The Capital Asset Pricing Model (CAPM)

The capital asset pricing model (CAPM) is a technical tool used to analyze the relationship between a financial asset's expected return and the nondiversifiable market risk. A major component in this model is the financial beta (β), which we hinted to before but will address here in more detail.

2.1. THE FINANCIAL BETA (β)

Beta (β) is a mathematical tool to measure systematic undiversifiable market risk. It is, in this sense, an index of the extent to which a security return moves in response to changes in the overall market. This would make it as a measure of securities volatility, in relation to an average security, represented by the state of the market. Market return is an aggregate measure of the return of all securities traded in a market at a specific time. The beta value can be positive or negative and generally ranges between -2.5 and 2.5. A value of 1.00 denotes the full impact of market risk. Any individual security with a beta of 1.00 indicates that the return pattern of that security moves up and down perfectly with the market return. A value of zero refers to total independence from market impact. A value of more than 1.00, such as 2.00, reveals that a security is twice as volatile as the average security in the market. A negative value says that the asset return pattern moves in the opposite direction from the market. Table 2.1 shows beta coefficients of selected American companies estimated at some point in time. The estimation changes for the same company from time to time.

Mathematically, beta is obtained by dividing the covariance between the individual security return (k_i) and the market return (k_m) by the variance of market return.

$$\beta = \frac{\text{cov}(k_i, k_m)}{\text{Var}(k_m)}$$

In this sense, beta is a concept of correlation to assess how one security return is correlated with the rest in the market. From another perspective, beta measures the percentage change in one security return as it responds to changes in the external market. It can therefore be interpreted as the financial elasticity of the

Mathematical Finance, First Edition. M. J. Alhabeeb.
© 2012 John Wiley & Sons, Inc. Published 2012 by John Wiley & Sons, Inc.

TABLE 2.1   Beta Estimates for Selected American Companies

Company	Beta
AOL	2.46
Dell	2.23
Microsoft	1.82
Texas Instrument	1.75
Intel	1.70
GE	1.16
GM	1.10
Colgate-Palmolive	1.03
Family Dollar Store	.99
K-Mart	.98
ATT	.98
McGraw-Hill	.81
Gillette	.76
MY Times	.71
JC Penney	.52
Johnson & Johnson	.49
Campbell	.41
Exxon	.36

change in a given asset relative to market change. Accordingly, beta becomes the slope of the regression line between the changes in market return and the corresponding response of the asset return. In figure 2.1 the changes in market return are tracked down on the horizontal axis, and the responses to them by a given security are tracked down on the vertical axis. The result would be the regression line $y = \alpha + \beta x + e$, with its slope standing for beta, calculated by dividing the change in the vertical axis over the change in the horizontal axis:

$$\text{beta} = \beta = \frac{\Delta y}{\Delta x}$$

If we track down the change in the market rate when it is increased from 7.3 to 9.3, the linear equation line of $y = -8 + 1.5x$ would allow the return of the asset k_i to increase from 3 to 6. Therefore, we can obtain the slope of the line:

$$\text{slope} = \frac{\Delta y}{\Delta x} = \frac{6 - 3}{9.3 - 7.3} = \frac{3}{2} = 1.5$$

which is the value of β in the equation of the line. This says that the asset rate of return follows the market return, but even more robustly—its volatility is one and a half times as much as that of the volatility of the market return. For example, if the market rate increases by 5%, this asset's rate would increase by 7.5%.

We can also calculate the beta value by the formula method.

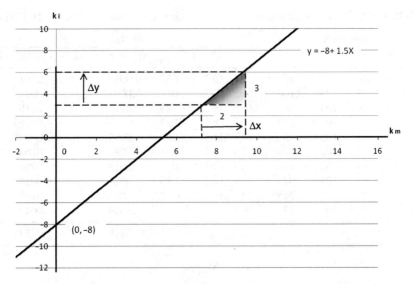

FIGURE 2.1

Example 2.1.1 We can calculate beta for corporation X given 10 periodic rates of return (k_i^x) and market rates for the same period (k_i^m). In Table E2.1.1 we calculate the expected return for both as the averages k_e^x and k_e^m, and we proceed to calculate the covariance between the two sets of rates and the variance of the market. Beta would be calculated by dividing the covariance by the market variance.

$$\text{Cov}(x, m) = \frac{\sum_{i=1}^{10} (k_i^x - k_e^x)(k_i^m - k_e^m)}{N}$$

$$= \frac{.04773}{10} = .0048$$

$$\text{Var}(m) = \frac{\sum_{i=1}^{10} (k_i^m - k_e^m)^2}{N}$$

$$= \frac{.0348}{10} = .0035$$

$$\beta_x = \frac{\text{Cov}(x, m)}{\text{Var}(m)}$$

$$= \frac{.0048}{.0035} = 1.37$$

Note that β_x is just the beta of asset X. The beta for the portfolio would be equal to the weighted average of betas for all individual assets within the portfolio:

$$\beta_p = \sum_{i=1}^{n} \beta_i w_i$$

TABLE E2.1.1   Rates of Return for Corporation x and the Market for 10 Periods

	Corporation x			Market				
Period	k_i^x	k_e^x	$k_i^x - k_e^x$	k_i^m	k_e^m	$k_i^m - k_e^m$	$(k_i^m - k_e^m)^2$	$(k_i^x - k_e^x)(k_i^m - k_e^m)$
1	−.06	.105	−.165	.027	.082	−.055	.003	.0091
2	.27	.105	.165	.095	.082	.013	.00017	.0021
3	.065	.105	−.04	.038	.082	−.044	.0019	.0018
4	.13	.105	.025	.055	.082	−.027	.00073	−.00067
5	.055	.105	−.05	−.017	.082	−.099	.0098	.0049
6	.28	.105	.175	.176	.082	.094	.0088	.0164
7	−.045	.105	−.15	.119	.082	.037	.0014	−.0055
8	.03	.105	−.075	.128	.082	.046	.0021	−.0034
9	.35	.105	.245	.156	.082	.074	.0055	.0181
10	−.025	.105	−.13	.044	.082	−.038	.0014	.0049
							.0348	.04773

where β_p is the portfolio beta, β_i is the beta for any individual asset within the portfolio, and w_i is the proportion of asset i in the entire portfolio, which contains n assets.

2.2. THE CAPM EQUATION

Now that we know what beta is, we can write the central equation of the capital asset pricing model, where beta is an essential factor to calculate the required rate of return on any asset (k_i) given the asset's beta (β_i), the market's required rate of return (k_m), and the riskless rate of return, which is traditionally the rate of return on a U.S. Treasury bond (R_f):

$$k_i = R_f + \beta_i(k_m - R_f)$$

In this model the required rate of return for any asset is obtained by adding the risk-free rate of return to the market risk premium, given that this premium is:

1. The difference between the market required rate of return and the risk-free rate of return: $k_m - R_f$
2. Adjusted to that asset's index of risk by being multiplied by beta: $\beta_i(k_m - R_f)$

Example 2.2.1   What would be the required rate of return for corporation Y, which has a beta of 1.85, given that the return on the market portfolio of assets is 12% and the risk-free rate is 6.5%?

$$k_i = R_f + \beta_i(k_m - R_f)$$
$$= .065 + 1.85(.12 - .065)$$
$$= 16.67\%$$

So the market risk premium is 5.5% (.12 − .065), and it went to a little more than 10% when it was adjusted to the asset's index of risk, a beta of 1.85. When the result of the adjustment was added to the risk-free rate of 6.5%, we got the corporation required rate of 16.67%.

Algebraically, we can obtain any of β_i, R_f, and k_m if the other variables in the equation are available. To find beta:

$$k_i = R_f + \beta_i(k_m - R_f)$$
$$k_i - R_f = \beta_i(k_m - R_f)$$

$$\boxed{\beta_i = \frac{k_i - R_f}{k_m - R_f}}$$

$$\beta_i = \frac{.1667 - .065}{.12 - .065}$$
$$= 1.85$$

To find the free risk of return (R_f):

$$k_i = R_f + \beta_i(k_m - R_f)$$
$$k_i = R_f - \beta_i R_f + \beta_i k_m$$
$$k_i - \beta_i k_m = R_f(1 - \beta_i)$$

$$\boxed{R_f = \frac{k_i - \beta_i k_m}{1 - \beta_i}}$$

$$= \frac{.1667 - 1.85(.12)}{1 - 1.85}$$
$$= \frac{-.0553}{-.85}$$
$$= .065$$

To find the market rate of return (k_m):

$$k_i = R_f + \beta_i(k_m - R_f)$$

$$k_i = R_f + \beta_i k_m - \beta_i R_f$$

$$k_i + R_f(\beta_i - 1) = \beta_i k_m$$

$$\boxed{k_m = \frac{k_i + R_f(\beta_i - 1)}{\beta_i}}$$

$$= \frac{.1667 + .065(1.85 - 1)}{1.85}$$

$$= .12$$

2.3. THE SECURITY MARKET LINE

When we graph the CAPM equation, we get a straight line with a positive slope equal to beta (1.85 in the last example). This line is called the **security market line** (SML). From Example 2.2.1 we have all the points we need to draw the SML. As shown in Figure 2.2, the level of risk, as measured by beta, is on the horizontal axis, and the required rates of return are on the vertical axis. The risk-free rate of 6.5% is associated with zero beta, the market rate of 12% is associated with a beta value of 1.00, and the return k_i of 16.67% is associated with a beta of 1.85. The vertical line BD represents the market risk premium, which is obtained as $BE - DE$, standing for $k_m - R_f$ in the equation. All assets that have betas higher than 1.00 would have a risk premium higher than the market risk premium. Our beta is 1.85, which makes the risk premium much higher than the market risk premium, as depicted by the vertical line CE, which is $CF - EF = 16.67 - 6.5 = 10.17$. Similarly, all assets that have betas below 1.00 will have risk premiums below the market risk premium: For example, firm S, with a beta of. 5. The risk premium for this firm would be represented by the line GH, which is $GI - HI$, and is lower than the market risk premium.

The slope of SML stands for the degree of risk aversion. A steeper SML would reflect a higher degree of risk aversion, and a flatter SML would reflect a lower degree of risk aversion in the economy. Also, the steeper the SML, the higher the risk premium and the higher the required rate of return on the risky assets with higher betas.

SML Shift by Inflation

The risk-free rate of return (R_f) is considered the price of money to a riskless borrower. It is therefore, like any other price, affected by inflation. In fact, it includes a built-in component designed to absorb the impact of inflation. This

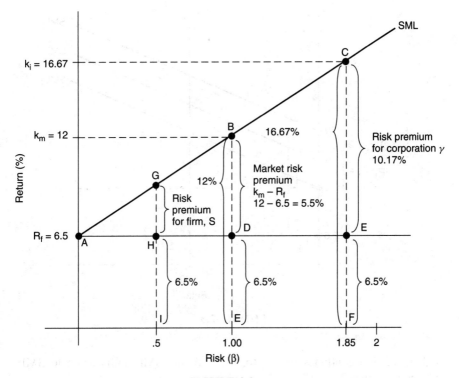

FIGURE 2.2

component, called the **inflation premium** (IP), is to protect the investor's purchasing power from declining as prices rise. The second component is the real core rate (k^0), which is inflation free.

$$R_f = k^0 + IP$$

Let's assume that the risk-free rate of Example 2.2.1, which is 6.5%, is in fact a combination of the real rate k^0, which is 2.5%, and the inflation premium 4%.

Since the SML originates from the R_f point, any increase in inflation would lead to an increase in the IP component and to R_f, and it would cause the point of SML origin to go up, shifting the entire line up. Figure 2.3 shows how the SML shifts to a higher position (SML₂), originating from the 8% rate if the inflation rises by 1.5 points, from 4% to 5.5%.

$$R_f^1 = k^0 + IP_1$$

$$= 2.5 + 4 = 6.5$$

$$R_f^2 = k^0 + IP_2$$

$$= 2.5 + 5.5 = 8.00$$

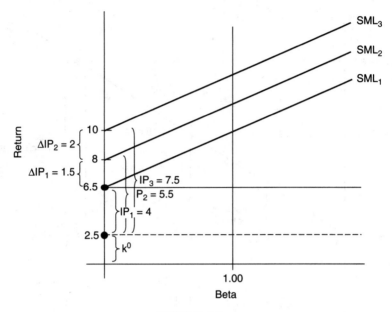

FIGURE 2.3

and if inflation continues to rise to, say, 7.5, the SML shifts again to SML$_3$, originating from $R_f = 10$.

$$R_f^3 = k^0 + IP_3$$
$$= 2.5 + 7.5 = 10.0$$

2.4. SML SWING BY RISK AVERSION

The slope of the security market line (SML) represents how much risk aversion investors usually exhibit. The line would therefore swing up and down to reflect the change in investor's risk aversion. The slope would be equal to zero and the SML would be horizontal at the risk-free level of return (R_f), and there would be no risk premium to the point where even risky assets would sell at the risk-free level of return. As the risk aversion starts to rise and the risk premium starts to grow, the SML would pivot at the R_f point and its other end would start to rise according to how much risk premium there is. From that point, the line would continue to swing up. Figure 2.4 shows that when risk aversion rises, the market risk premium (MRP) would rise—in this example from 4% to 9% (vertical *FG* to *EG*)—and consequently, the market required rate (k_m) would jump from 10% to 15% (vertical *FK* to *EK*), and our risky asset return (k_i) of the 1.75 beta would be up from 13% to 21.75%. This asset risk premium (RAP) would shoot

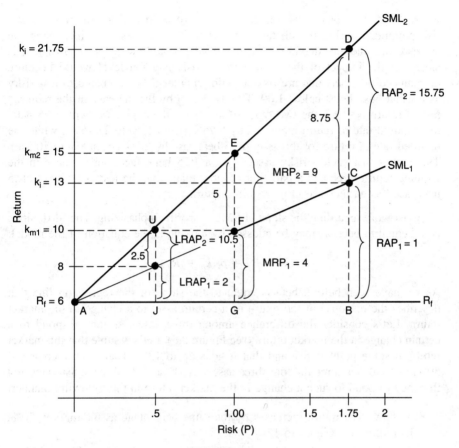

FIGURE 2.4

from 7% to 15.75% (vertical *CB* to *DB*).

$$k_i^1 = R_f + \beta_i(k_m^1 - R_f)$$
$$= .06 + 1.75(.10 - .06)$$
$$= .06 + .07$$
$$= .13$$

where

$$k_i^2 = R_f + \beta_i(k_m^2 - R_f)$$
$$= .06 + 1.75(.15 - .06)$$
$$= .06 + .1575$$
$$= .2175$$

where $1.75(.10 - .06) = 7\%$ and $1.75(.15 - .06) = 15.75\%$ are the risky asset's risk premiums during the change. This change would result in increasing the slope of the line and making it steeper. SML_1 would swing up to SML_2. It is clear that the impact of the change in the risk aversion level would be more pronounced on assets that are riskier, with a higher beta, compared to less risky assets that have betas below 1.00. This is shown by the increase in the required rate of return between the two types of assets for the risky asset with 1.75 beta, the required rate of return increased by 8.75% (from 13% to 21.75%), while the required rate of return for the asset of .5 beta rose by 2.5% (from 8% to 10.5%). This means that such a risky asset, with a 1.75 beta, faced an increase in the required return three and a half times higher than what the less risky asset of .5 beta faced as an increase in its required return.

A last word regarding the slope of SML is worth emphasizing. The SML slope is not equal to beta, as may be guessed by looking at the equation of the model:

$$k_i = R_f + \beta_i(k_m - R_f)$$

As we have seen before, beta is equal to the slope of the regression line that describes the response of the return of a certain asset to a change in the market return. Let's consider the difference among three assets as they respond to a certain change in the market return (see Figure 2.5). Let's assume that the market return at some point is 8% and that it goes up to 12%. That is an increase of 50%. Let's also assume that our three assets, X, Y, and Z, all have 8% rates and that they respond to such a change in the market return in the following manner:

- *Asset X*: The return increases by the same percentage as the market, 50%. Its return of 8% goes to 12%.
- *Asset Y*: The return increases by 100%. Its return goes from 8% to 16%.
- *Asset Z*: The return increases by 25%. Its return goes from 8% to 10%.

Asset X would have a 45° line going through the original point. Its slope is equal to 1 because the change in the asset return is the same as the change in the market return:

$$\text{slope}_x = \frac{\text{rise}}{\text{run}} = \frac{\Delta \text{ asset return}}{\Delta \text{ market return}}$$

$$= \frac{12 - 8}{12 - 8} = 1 = \text{beta}$$

The line equation would be

$$Y = a + b_x$$

$$= 0 + 1(x)$$

$$\boxed{Y = x}$$

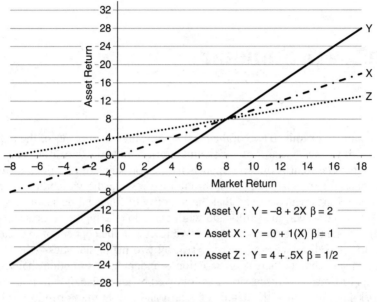

FIGURE 2.5

Asset Y would have a line with a slope of 2, an x-intercept of 4, and a y-intercept of -8:

$$\text{slope}_y = \frac{16 - 8}{12 - 8} = \frac{8}{4} = 2$$

The line equation would be

$$\boxed{Y = 8 + 2x}$$

Asset Z would have a line with a slope of. 5, a y-intercept of 4, and an x-intercept of -8:

$$\text{slope}_z = \frac{10 - 8}{12 - 8} = \frac{2}{4} = \frac{1}{2}$$

The line equation would be

$$\boxed{Y = 4 + .5x}$$

Since all three of these equations are linear equations of regression lines, beta can be read off the equation directly as it corresponds to the b value in the standard format of the linear equation:

$$\boxed{Y = a + bx}$$

Unit VII Summary

In this unit we examined two of the most significant determinants of the price and stability of financial securities—risk and return—and how they affect each other. While financial returns directly express the performance of securities as to gain, risk stands for the chance of loss through expressing the variability of returns, which is a fundamental reason that the majority of financial decision makers are risk averse. However, the ability and willingness to take higher risks is associated with the potential to earn high returns. We reviewed the concept of expected rate of return as well as ways to measure risk by two major statistics—the standard deviation and the coefficient of variation—under the assumption that risk is a function of time. This discussion was followed by an explanation of risk aversion and risk premium, then the details of return and risk at the portfolio level.

Types of risk were discussed to stress that the only relevant risk is nondiversifiable risk, because diversifiable risk can be greatly reduced, and even eliminated, by a sensible diversification of assets. In this context, the capital asset pricing model was explained, and beta as a risk measure was highlighted, theoretically and practically. Also, the concept of the security market line was clarified and illustrated, as it responds to inflation and risk aversion.

we also paid homage to the pioneer contribution in risk and return theory. This was achieved by a brief review of Markowitz's two-asset portfolio explanation and illustrations of portfolio diversification assets and its positive outcome on portfolio return. It was shown that a positive impact was achieved by the compensatory effect of assets that move in different directions from each other.

Mathematical Finance, First Edition. M. J. Alhabeeb.
© 2012 John Wiley & Sons, Inc. Published 2012 by John Wiley & Sons, Inc.

List of Formulas

Expected rate of return:

$$k_e = \sum_{n=1}^{n} k_i \cdot \mathrm{Pr}_i$$

Coefficient of variation:

$$\mathrm{Coef_v} = \frac{\sigma}{k_e}$$

Portfolio return:

$$r_p = \sum_{i=1}^{n} r_i w_i$$

Portfolio expected rate of return:

$$k_p = \sum_{i=1}^{n} k_{ei} w_{ei}$$

Beta:

$$\beta = \frac{\mathrm{Cov}(k_i, k_m)}{\mathrm{Var}(k_m)}$$

$$\beta = \frac{\Delta y}{\Delta x}$$

Covariance between an individual security return and a market return:

$$\mathrm{Cov}(x, m) = \frac{\sum_{i=1}^{n} (k_i^x - k_e^x)(k_i^m - k_e^m)}{N}$$

Market variance:

$$\mathrm{Var}(m) = \frac{\sum_{i=1}^{n} (k_i^m - k_e^m)^2}{N}$$

Mathematical Finance, First Edition. M. J. Alhabeeb.
© 2012 John Wiley & Sons, Inc. Published 2012 by John Wiley & Sons, Inc.

CAPM equation:

$$k_i = R_f + \beta_i(k_m - R_f)$$

Beta by CAPM:

$$\beta_i = \frac{k_i - R_f}{k_m - R_f}$$

Risk-free rate by CAPM:

$$R_f = \frac{k_i - \beta_i k_m}{1 - \beta_i}$$

$$R_f = K^0 + \text{IP}$$

Market rate by CAPM:

$$k_m = \frac{k_i + R_f(\beta_i - 1)}{\beta_i}$$

Exercises for Unit VII

1. Calculate the expected return on an asset that has the following probable returns:

Order	Return (%)	Probability
1	$6\frac{3}{4}$ %	.39
2	$8\frac{1}{4}$ %	.27
3	$9\frac{1}{2}$ %	.19
4	7%	.09
5	4%	.06

2. If you compare the asset in Exercise 1 to the following asset, can you quickly tell which one is riskier?

Order	Return (%)	Probability
1	9%	.29
2	10%	.25
3	$6\frac{1}{2}$%	.22
4	5%	.15
5	4%	.09

3. If these two assets are in the same portfolio, would that be better or worse for the portfolio return? Can you tell by a quick examination, and how?

4. Calculate the standard deviation of the two assets in Exercises 1 and 2 and explain how you can use the standard deviation to tell which asset is riskier.

5. Calculate the coefficient of variation of the two assets in Exercises 1 and 2 and explain which asset is riskier, and why.

Mathematical Finance, First Edition. M. J. Alhabeeb.
© 2012 John Wiley & Sons, Inc. Published 2012 by John Wiley & Sons, Inc.

6. Calculate the portfolio return of the following five-asset portfolio and how they are making up the portfolio capital.

Asset	Return (%)	Asset % of Portfolio
A	14%	.25
B	$13\frac{1}{2}$ %	.20
C	12%	.15
D	$9\frac{1}{4}$ %	.26
E	10%	.14

7. Calculate the portfolio return for a business whose market value went up from \$720,000 in 2010 to \$985,000 in 2011.

8. Let's combine the last three assets in Exercises 1, 2, and 6 into a portfolio using the probabilities of the first asset. Calculate the standard deviation, variance, covariance, and correlation between assets I and II and explain the risk involved in this portfolio regarding these two assets.

	Return (%)		
Asset I	Asset II	Asset III	Probability (%)
$6\frac{3}{4}$	9	14	.39
$8\frac{1}{4}$	10	13	.27
$9\frac{1}{2}$	$6\frac{1}{2}$	12	.19
7	5	$9\frac{1}{4}$	.09
4	4	10	.06

9. Calculate standard deviation, variance, covariance, and correlation between assets II and III and explain portfolio risk in terms of these two assets.

10. Calculate standard deviation, variance, covariance, and correlation between assets I and III and explain what would happen to the portfolio risk if these two assets were present.

11. Let's take the information in Exercise 8 and consider the return of the three assets as the returns of one corporation (X) in three different phases. Let's also add to the table a column on the market return during these phases. Calculate beta for corporation X in phase I.

| Returns of Corporation X (%) | | | Market |
Phase 1	Phase 2	Phase 3	Return (%)
$6\frac{3}{4}$	9	14	$5\frac{1}{2}$
$8\frac{1}{4}$	10	13	$6\frac{1}{4}$
$9\frac{1}{2}$	$6\frac{1}{2}$	12	$6\frac{3}{4}$
7	5	$9\frac{1}{4}$	7
4	4	10	6

12. Calculate beta for corporation X in phases 2 and 3.

13. Suppose that corporation A has a beta of 1.97 at a time when the market return is $8\frac{1}{2}$ and the risk-free rate is $5\frac{1}{4}\%$. Calculate the required rate of return for corporation A.

14. What would beta in Exercise 13 be if the market return goes down to 7% and the risk-free rate drops to 5%?

15. Now let's suppose that beta stays at 1.97 and the required rate of return stays as it was found in Exercise 13, but the market rate goes up to 10%. What would be the risk-free rate?

16. Now we should find the market rate of return if we keep the required rate of return as in Exercise 13 but beta goes down to 1.5 and the risk-free rate goes up to 6%.

UNIT VIII

Mathematics of Insurance

1. **Life Annuities**
2. **Life Insurance**
3. **Property and Casualty Insurance**
   Unit VIII Summary
   List of Formulas
   Exercises for Unit VIII

1 Life Annuities

Life annuities are different from the annuities certain that we discussed earlier. In a major distinction, life annuities are more related to life insurance for being contingent. A **contingent contract** involves a sequence of payments that are dependent on an occurrence of a certain event that cannot be foretold. In this case, it is either death or living up to a certain age. This element of uncertainty requires the use of probability distribution, which in this context is in the form of a mortality table. Like life insurance, life annuities are dealt with by insurance companies and the person to whom the annuity would be payable, is called an annuitant, as opposed to the "insured". Life insurance proceeds are payable to survivors upon the death of the insured.

The typical stated premium of life annuities and life insurance is usually a gross premium, which includes the loading costs in addition to the net premium. **Loading costs** include a company's operating expenses and profit margin. The net premium is the pure cost of the ultimate benefits to the annuitant or the insured, usually broken down into installments unless it is paid in a single payment on the day of purchase. In this case it is called a **net single premium**, while a yearly installment is called a **net annual premium**. Our calculations here focus on the net premium but without the loading component, since the loading would vary from one company to another. This is why we assume that the present value of the net premium would equal the present value of all future benefits.

1.1. MORTALITY TABLE

A **mortality table** contains statistical data on people living and dying categorized primarily by age and sometimes by gender. The table's major use is in calculating predictions as to how long a person will live and when he or she will probably die, to be used in estimating the benefits for life annuities and life insurance. The first mortality table in the United States, called the *American Experience Table of Mortality*, was published in New York in 1868. The table we use in this book, the commutation table (Table 10 in the Appendix), is a sample included for the purpose of calculation. It is constructed based on U.S. Internal Revenue Service (IRS) data. We include age group 0 to 100, although the IRS data goes to age 110. Following are some major definitions of the table items:

Mathematical Finance, First Edition. M. J. Alhabeeb.
© 2012 John Wiley & Sons, Inc. Published 2012 by John Wiley & Sons, Inc.

x: the age of a person in years. Age zero is the base sample of 100,000 infants under 1 year of age.

l_x: the number of people alive at age x.

d_x: the number of people who die between age x and age $x+1$. It can be calculated as the difference between l_x and l_{x+1}.

$$d_x = l_x - l_{x+1}$$

Example 1.1.1   The number of people who would die at age 20 (d_{20}) is 118. It is calculated as the difference between people who are alive at age 20 and those who made it to age 21:

$$d_x = l_x - l_{x+1}$$
$$d_{20} = l_{20} - l_{21}$$
$$= 97,741 - 97,623 = 118$$

l_{x+1} : the number of people alive in the age group a year after any age x. It can be calculated as the difference between the number of people living at age $x(l_x)$ and the number of people dying at age $x(d_x)$.

$$l_{x+1} = l_x - d_x$$

Example 1.1.2 l_{x+1} for age 15 is $l_{15+1} = l_{16}$. It is 98,129, and can be calculated as

$$l_{x+1} = l_x - d_x$$
$$l_{16} = l_{15} - d_{15}$$
$$= 98,196 - 67 = 98,129$$

q_x : the probability that a person will die between age x and $x+1$. It is called the **mortality rate** and calculated by dividing the number of people who will die between the ages of x and $x+1(d_x)$ by the number of living people in that age group (l_x):

$$q_x = \frac{d_x}{l_x}$$

Knowing this probability in advance would enable us to obtain either d_x or l_x in terms of the other.

$$d_x = q_x l_x$$

and

$$I_x = \frac{d_x}{q_x}$$

Example 1.1.3 q_x for age 30 is. 001316. It is calculated as

$$q_{30} = \frac{d_{30}}{l_{30}}$$

$$= \frac{127}{96,477}$$

$$= .001316$$

In some tables, this result would be written as multiplied by 1,000 and the category would be called $(1,000q_x)$, which would be 1.316 (.001316 × 1,000). Consequently, we can apply the second formulas for calculating d_x and l_x:

$$d_{30} = q_{30} \cdot l_{30}$$

$$= .001316 \times 96,477$$

$$= 127$$

and

$$l_{30} = \frac{d_{30}}{q_{30}}$$

$$= \frac{127}{.001316}$$

$$= 96,477$$

Since q_x was the probability of dying, we can also calculate the probability of living one more year at any point x. That is how probable a person age x is to live to age $x+1$. It is denoted by p_x and calculated as

$$p_x = \frac{I_{x+1}}{I_x}$$

Example 1.1.4   The probability that a 30-year-old person will live one more year is nearly 100%:

$$p_{30} = \frac{l_{31}}{l_{30}}$$

$$= \frac{96,350}{96,477}$$

$$= .998684$$

It is the complement of q_{30}, which was. 001316. The complementarity relation between p_x and q_x simply means that there would be one of two possibilities for a person either to live or to die:

$$\boxed{q_x + p_x = 1}$$

$$q_x + p_{30} = .001316 + .998684 = 1$$

This would give us another formula for q_x and p_x:

$$p_x = 1 - q_x$$

and

$$q_x = 1 - p_x$$

The last formula would be a proof of the previous q_x formula:

$$q_x = 1 - p_x$$

$$q_x = 1 - \frac{l_{x+1}}{l_x}$$

$$q_x = \frac{l_x - l_{x+1}}{l_x}$$

Since $l_{x+1} = l_x - d_x$, then

$$q_x = \frac{l_x - l_x + d_x}{l_x}$$

$$q_x = \frac{d_x}{l_x}$$

Now, we can turn to p_x, the probability of a person age x to live to age $x + 1$:

$$p_x = \frac{l_{x+1}}{l_x}$$

We can, by the same logic, say that if n is a number of years, the probability of a person age x to live to that n number of years would be

$$\boxed{np_x = \frac{I_{x+n}}{I_x}}$$

Example 1.1.5   If we want to find how probable it is that a person age 50 will live to age 70, we are talking about n as 20 years, from age 50 to age 70, and therefore the probability would be calculated as

$$_{20}p_{50} = \frac{l_{50+20}}{l_{50}} = \frac{l_{70}}{l_{50}}$$

$$= \frac{68,248}{91,526}$$

$$= 74.6\%$$

Because of the complementarity between p_x and q_x, we can conclude that

$$np_x + nq_x = 1$$

$$nq_x = 1 - np_x$$

$$nq_x = 1 - \frac{l_{x+n}}{l_x}$$

$$\boxed{nq_x = \frac{l_x - l_{x+n}}{l_x}}$$

which is defined as the probability that a person age x will die within n years or at age $x + n$.

Example 1.1.6   What is the probability that a person age 50 will die within the next 20 years?

Here also, n is 20, and therefore the probability of death would be calculated as

$$nq_x = \frac{l_x - l_{x+n}}{l_x}$$

$$_{20}q_{50} = \frac{l_{50} - l_{50+20}}{l_{50}}$$

$$= \frac{l_{50} - l_{70}}{l_{50}}$$

$$= \frac{91,526 - 68,248}{91,526}$$

$$= 25.4\%$$

We can confirm that by

$$np_x + nq_x = 1$$

$$_{20}p_{50} + _{20}q_{50} = 1$$

$$.746 + .254 = 1$$

1.2. COMMUTATION TERMS

The mortality table also includes other terms that are used in the calculation of life annuities and life insurance: D_x, N_x, C_x, and M_x.

D_x : the present value of $1.00 for each person alive at age x. Collectively for each age group it would be calculated by multiplying the present value for $1.00 ($v^x$) or $1/(1+r)^x$ by the number of living people in age group x:

$$\boxed{D_x = l_x \cdot v^x}$$

$$\boxed{D_x = \frac{l_x}{(1+r)^x}}$$

Note that our mortality table is based on a 5% rate of interest.

Example 1.2.1   The D_x value for a person at age 40 in Table 10 is $13,483.83. It is obtained by

$$D_x = l_x \cdot v^x$$

$$D_{40} = l_{40} \cdot v^{40}$$

$$= 94{,}926 \left[\frac{1}{(1+.05)^{40}} \right]$$

$$= 13{,}483.83$$

We can also obtain the v^x value from the v^n table, which happens to be .14204568.

$$D_{40} = 94{,}926(.14204568)$$

$$= 13{,}483.83$$

N_x : the present value of annuity of payments for all persons living at each age group from x to the end of the table, age 100. So it can be viewed as the summation of D's above.

$$N_x = D_x + D_{x+1} + D_{x+2} + \cdots + D_{x+100}$$

$$\boxed{N_x = \sum_{k=x}^{x+100} D_k}$$

Example 1.2.2   Suppose that the annuity for each person in the cohort of 70 years is $500 per year. Then the present value of those annuities back at age zero would be calculated by multiplying the annuity by the N_{70} value of the table:

$$500N_{70} = 500(21,427.28)$$

$$= 10,713,640$$

C_x : the present value for $1.00 of a payment to the beneficiaries of people who die at age x. It is calculated by multiplying the number of people who will die at age x, (d_x) by the present value of $1.00 for the time $x+1$.

$$\boxed{C_x = d_x \cdot v^{x+1}}$$

Example 1.2.3   The C_x value for age 55 is 51.86 in Table 10. It is calculated as

$$C_{55} = d_{55}v^{55+1}$$

$$= d^{55}v^{56}$$

$$= 797\left[\frac{1}{(1+.05)^{56}}\right]$$

$$= 51.86$$

Using the table value of v^{56} from the v^n table would give the same result:

$$C_{55} = 797(9.06507276)$$

$$= 51.86$$

M_x : the summation of all C_x. It represents the present value of a $1.00 payment for all people who eventually die but are still alive at age x. It is like N_x, an accumulation of the values back from age x to age 0.

$$M_x = C_x + C_{x+1} + C_{x+3} + \cdots + C_{x+100}$$

$$\boxed{M_x = \sum_{k=x}^{x+100} C_k}$$

Example 1.2.4   If the payment for each person who dies in the 85-year age group is $1,000, the accumulated present value for all back at age 0 would be

$$1,000(329.76) = 329,760$$

where 329.76 is M_{85} in Table 10.

L.E.: the life expectancy at each age group.

1.3. PURE ENDOWMENT

A **pure endowment** is a single payment received at a certain future time by a person who has to be alive at that time. Since this payment is received at a future time, we should be concerned about its present value. That is why it is equivalent to the discounted value of the payment. If we assume that nE_x is the contribution of \$1.00 for n years by each person age x, the total premiums would be multiplying that contribution by the number of people alive at age x ($l_x \cdot nE_x$), and if we discount it to age x since it will be received at age $x + n$ (see Figure 1.1), we get

$$l_x \cdot nE_x = \frac{l_{x+n}}{(1+r)^n}$$

$$nE_x = \frac{\left[\dfrac{l_{x+n}}{(1+r)^n}\right]}{l_x}$$

$$nE_x = \frac{l_{x+n}\left[\dfrac{l}{(1+1)^n}\right]}{l_x}$$

$$nE_x = \frac{l_{x+n} \cdot v^n}{l_x}$$

This is for a hypothetical \$1.00. If it is for any other actual amount of p, the present value would be

$$\boxed{nE_x = P\left(\frac{l_{x+n}}{l_x}\right) v_n}$$

and for more simplicity, we can use the commutation values:

$$nE_x = \frac{l_{x+n} \cdot v^n \cdot v^x}{l_x \cdot v^x} = \frac{l_{x+n} \cdot v^{x+n}}{l_x \cdot v^x}$$

and that would be

$$nE_x = \frac{D_{x+n}}{D_x}$$

and for payment P

$$\boxed{nE_x = P\left(\frac{D_{x+n}}{D_x}\right)}$$

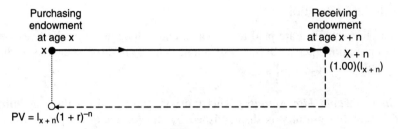

FIGURE 1.1

Example 1.3.1 Glenn is 55 years old. He would like to receive a pure endowment of $50,000 when he retires at 65. How much would he have to pay now if the rate is 5%?

The time between purchasing and receiving the endowment is $n = 10$.

$$nE_x = P\left(\frac{l_{x+n}}{l_x}\right)v^n$$

$$10E_{55} = 50,000\left(\frac{l_{55+10}}{l_{55}}\right)v^{10}$$

$$= 50,000\left(\frac{77,107}{88,348}\right)(0.61391325)$$

$$= 26,790$$

and we can also use the commutation formula:

$$nE_x = P\left(\frac{D_{x+n}}{D_x}\right)$$

$$10E_{55} = 50,000\left(\frac{D_{65}}{D_{55}}\right)$$

$$= 50,000\left(\frac{3,234.37}{6,036.50}\right)$$

$$= 26,790$$

1.4. TYPES OF LIFE ANNUITIES

The life annuities discussed here are called **single-life annuities**, referring to their distinct characteristic: that they are for a single living person and they cease when the person dies. There are two major types, whole life annuities and temporary life annuities.

Whole Life Annuities

Whole life annuities are paid to the annuitant as long as he or she lives and do not stop until the person dies. These annuities come in three categories: ordinary, due, and deferred.

Ordinary Whole Life Annuity   This type of annuity, also called an **immediate whole life annuity** is distinguished by the receipt of its payments at the end of each year of the annuitant's life. We can look at these payments as a series of individual pure endowments, and for this reason, let's consider a as the present value of $1.00 a year for each person in the age group x. The collective endowments would be

$$a_x = 1E_x + 2E_x + 3E_x + \cdots + 100E$$

$$a_x = \frac{D_{x+1}}{D_x} + \frac{D_{x+2}}{D_x} + \frac{D_{x+3}}{D_x} + \cdots + \frac{D_{x+100}}{D_x}$$

$$a_x = \frac{D_{x+1} + D_{x+2} + D_{x+3} + \cdots + D_{x+100}}{D_x}$$

$$a_x = \frac{N_{x+1}}{D_x}$$

and if the payment is P instead of $1.00, then

$$\boxed{a_x = P\left(\frac{N_{x+1}}{D_x}\right)}$$

Example 1.4.1   Nicole is 45 years old. She is thinking of buying a whole life annuity that would pay her $2,000 at the end of each year for the rest of her life. What would the premium be for this annuity?

We immediately knew that this is an ordinary whole life annuity since the payments are to be made at the end of each year.

$$a_x = P\left(\frac{N_{x+1}}{D_x}\right)$$

$$a_{45} = 2,000\left(\frac{N_{46}}{D_{45}}\right)$$

$$= 2,000\left(\frac{154,684.29}{10,417.24}\right)$$

$$= 29,697.75$$

Whole Life Annuity Due   This is the same as a whole life annuity except that the first payment is made at the time of purchase and continues to be made after that on the same time of each year for the rest of the annuitant's life. Since we designated our formulas based on a $1.00 payment, and we stated that this annuity has the first payment made one year earlier, we can conclude that the whole life annuity due ($\ddot{a}_x$) is different from the ordinary whole life annuity (a_x):

$$\ddot{a}_x = 1 + a_x$$

$$\ddot{a}_x = 1 + \frac{N_{x+1}}{D_x}$$

$$\ddot{a}_x = \frac{D_x + N_{x+1}}{D_x}$$

and since $N_{x+1} = D_{x+1} + D_{x+2} + \cdots + D_{x+100}$, then

$$\ddot{a}_x = \frac{D_x + D_{x+1} + D_{x+2} + \cdots + D_{x+100}}{D_x}$$

which made the numerator N_x:

$$\ddot{a}_x = \frac{N_x}{D_x}$$

and for payment P instead of $1.00, we obtain the formula for the whole life annuity due:

$$\boxed{\ddot{a}_x = P\left(\frac{N_x}{D_x}\right)}$$

Example 1.4.2   Suppose that Nicole of Example 1.4.1 wanted to get her first payment right at the time of purchase. Would that affect the amount that she would receive each year?

This change would make the annuity an annuity due, and therefore

$$\ddot{a}_{45} = P\left(\frac{N_x}{D_x}\right)$$

$$= 2,000\left(\frac{165,101.53}{10,417.24}\right)$$

$$= 31,697.75$$

So the payment got bigger! But by how much? It got bigger by

$$31,697.75 - 29,697.75 = 2,000$$

and that is exactly the one payment difference between the ordinary and the due.

Example 1.4.3 A 62-year-old woman who was hit by a car received an insurance settlement which she wanted to put in a whole life annuity so that an annual payment of $12,000 is made to her starting immediately and continuing to be paid at the same time of each year for the rest of her life. How much was her settlement?

$$\ddot{a}_x = P\left(\frac{N_x}{D_x}\right)$$

$$\ddot{a}_{62} = P\left(\frac{N_{62}}{D_{62}}\right)$$

$$= 12,000\left(\frac{46,632.42}{3,950.12}\right)$$

$$= 12,000(11.81)$$

$$= 141,720$$

Deferred Whole Life Annuity This is an annuity whose first payment is deferred to a period (n) that is beyond one year after the purchase date. Since the purchase occurs at age x, the deferment period would be $x + n$ for the annuity due, which starts on the day of purchase. It could also be $x+1 + n$ for the ordinary annuity, which starts one year after the purchase date, at $x+1$ (see Figure 1.2).

The present value of the deferred ordinary whole life annuity would be $n|a_x$, which can be obtained by

$$n|a_x = (n + 1E_x) + (n + 2E_x) + (n + 3E_x) + \cdots + (n + 100E_x)$$

$$n|a_x = \frac{D_{x+1+n} + D_{x+2+n} + D_{x+3+n} + \cdots + D_{x+100+n}}{D_x}$$

$$n|a_x = \frac{N_{x+1+n}}{D_x}$$

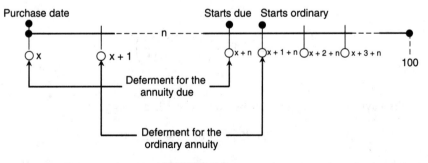

FIGURE 1.2

and for payment P:

$$\boxed{n|a_x = P\left(\frac{N_{x+1+n}}{D_x}\right)}$$

In the same manner, we can obtain the deferred whole life annuity due $(n|\ddot{a}_x)$ by

$$n|\ddot{a}_x = nE_x + (n+1E_x) + (n+2E_x) + \cdots + (n+100E_x)$$

$$n|\ddot{a}_x = \frac{D_{x+n} + D_{x+1+n} + D_{x+2+n} + \cdots + D_{x+100+n}}{D_x}$$

$$n|\ddot{a}_x = \frac{N_{x+n}}{D_x}$$

and for payment P it would be

$$\boxed{n|\ddot{a}_x = P\left(\frac{N_{x+n}}{D_x}\right)}$$

Example 1.4.4 Linda, who is 25, wishes to purchase an ordinary whole life annuity that pays \$2,400 a year, but she wants to start collecting her payments 25 years later. How much would she have to pay for the annuity?

$$n|a_x = P\left(\frac{N_{x+1+n}}{D_x}\right)$$

$$25|a_{25} = P\left(\frac{N_{25+1+25}}{D_{25}}\right)$$

$$= 2,400\left(\frac{N_{51}}{D_{25}}\right)$$

$$= 2,400\left(\frac{110,125.43}{28,676.85}\right)$$

$$= 9,216.53$$

Example 1.4.5 A 27-year-old man purchased a whole life annuity due so he will be able to receive \$10,000 a year when he is 65. How much would his net single premium be?

$$n|\ddot{a}_x = P\left(\frac{N_{x+n}}{D_x}\right)$$

$$38|\ddot{a}_{27} = 10,000\left(\frac{N_{65}}{D_{27}}\right)$$

$$38|\ddot{a}_{27} = 10,000 \left(\frac{35,523.27}{25,942.72} \right)$$

$$= 13,692$$

Example 1.4.6 An inheritance of \$65,000 went to a 12-year-old girl. Her guardians decided to buy her a whole life annuity that starts paying her annually when she turns 18. How much would she receive annually?

The net single premium is \$65,000 and $n = 6$.

$$65,000 = P \left(\frac{N_{18}}{D_{12}} \right)$$

$$= P \left(\frac{782,546.42}{54,742.13} \right)$$

$$= P(14.30)$$

$$P = \frac{65,000}{14.30} = 4,547$$

Temporary Life Annuities

Whereas whole life annuities pay the annuitant for the rest of his or her life, temporary annuities pay only for a certain contracted period of time, given that the annuitant is alive throughout that period. When the person dies, the payments cease. Depending on the date of the first payment, temporary life annuities are also classified into ordinary, due, and deferred.

Ordinary Temporary Life Annuity If the term of annuity is n, the present value of the temporary life annuity $(a_{x:\overline{n}|})$ can be the difference between an ordinary whole life annuity (a_x) and a deferred whole life annuity $(n|a_x)$:

$$a_{x:\overline{n}|} = a_x - n|a_x$$

$$a_{x:\overline{n}|} = \frac{N_{x+1}}{D_x} - \frac{N_{x+1+n}}{D_x}$$

$$a_{x:\overline{n}|} = \frac{N_{x+1} - N_{x+1+n}}{D_x}$$

and for payment P it would be

$$\boxed{a_{x:\overline{n}|} = P \left(\frac{N_{x+1} - N_{x+1+n}}{D_x} \right)}$$

Temporary Life Annuity Due   In an ordinary temporary life annuity, the payments are made at the end of each year, but if the payments are made at the beginning of each year, the annuity would be considered a temporary annuity due ($\ddot{a}_{x:\overline{n}|}$), and it would be calculated as the difference between a whole life annuity due ($\ddot{a}_x$), and a deferred annuity due ($n|\ddot{a}_x$):

$$\ddot{a}_{x:\overline{n}|} = \ddot{a}_x - n|\ddot{a}_x$$

$$\ddot{a}_{x:\overline{n}|} = \frac{N_x}{D_x} - \frac{N_{x+n}}{D_x}$$

$$\ddot{a}_{x:\overline{n}|} = \frac{N_x - N_{x+n}}{D_x}$$

and for payment P it would be

$$\boxed{\ddot{a}_{x:\overline{n}|} = P\left(\frac{N_x - N_{x+n}}{D_x}\right)}$$

Example 1.4.7   A person aged 33 wants to purchase an ordinary life annuity that would pay him \$3,600 a year for 10 years. How much would his net single premium be?

Since this is specified for 10 years, it is an ordinary temporary annuity.

$$a_{x:\overline{n}|} = P\left(\frac{N_{x+1} - N_{x+1+n}}{D_x}\right)$$

$$a_{33:\overline{10}|} = 3,600\left(\frac{N_{33+1} - N_{33+1+10}}{D_{33}}\right)$$

$$= 3,600\left(\frac{N_{34} - N_{44}}{D_{33}}\right)$$

$$= 3,600\left(\frac{323,068.92 - 176,076.33}{19,205.35}\right)$$

$$= 27,553.43$$

Example 1.4.8   Find the net present value of the annuity in Example 1.4.7 if the annuitant requests that the payments be made at the beginning of the year.

This request makes the annuity a temporary annuity due:

$$\ddot{a}_{x:\overline{n}|} = P\left(\frac{N_x - N_{x+n}}{D_x}\right)$$

$$\ddot{a}_{33:\overline{10}|} = 3,600\left(\frac{N_{33} - N_{43}}{D_{33}}\right)$$

$$= 3,600 \left(\frac{342,274.28 - 187,635.20}{19,205.35} \right)$$

$$= 28,986.75$$

Forborne Temporary Life Annuity Due   Sometimes the annuitant of the temporary life annuity due chooses not to cash the payments, allowing them to accumulate for a certain period of time (n) and become a pure endowment. In this case, when the annuitant forbears to draw the annuity, it would be called a **forborne temporary life annuity** (nF_x), which is calculated as

$$n F_x = \frac{N_x - N_{x+n}}{D_{x+n}}$$

and for payment P it would be

$$\boxed{n F_x = P \left(\frac{N_x - N_{x+n}}{D_{x+n}} \right)}$$

Example 1.4.9   At the start of the year, David turns 50 and is to receive $3,000 as his temporary life annuity payment, which continues until he is 60 years old. David decides to leave the money with the insurance company to accumulate for 10 years. How much will he get when he turns 60?

$$n F_x = P \left(\frac{N_x - N_{x+n}}{D_{x+n}} \right)$$

$$10 F_{50} = 3,000 \left(\frac{N_{50} - N_{60}}{D_{60}} \right)$$

$$= 3,000 \left(\frac{118,106.84 - 55,329.23}{4,482.32} \right)$$

$$= 42,016.82$$

Deferred Temporary Life Annuity   This is the last type of temporary life annuity, along with the ordinary and due types. Payments of this annuity will not start until a period of time (k) has elapsed from the day of annuity purchase; k is more than one year. The deferred temporary life annuity is most likely to be an annuity due, and its present value ($k|\ddot{a}_{x:\overline{n}|}$) would be calculated as

$$k|\ddot{a}_{x:\overline{n}|} = \frac{N_{x+k} - N_{x+k+n}}{D_x}$$

where k is the deferment period, x is the annuitant age, and n is the annuity time. For a payment P, the net single premium or the present value of the deferred temporary life annuity would be

$$k|\ddot{a}_{x:\overline{n}|} = P\left(\frac{N_{x+k} - N_{x+k+n}}{D_x}\right)$$

Example 1.4.10   Find the net single premium for a temporary life annuity that runs for 12 years paying an annual payment of \$2,500 to a 15-year-old boy. This annuity comes with a stipulation that the annuitant will not start to receive the annual payments until he turns 18.

So, x is 15, k is 3 $(18 - 15)$, and n is 12.

$$k|\ddot{a}_{x:\overline{n}|} = P\left(\frac{N_{x+k} - N_{x+k+n}}{D_x}\right)$$

$$3|\ddot{a}_{15:\overline{12}|} = 2,500\left(\frac{N_{18} - N_{30}}{D_{15}}\right)$$

$$= 2,500\left(\frac{782,546.43 - 406,021.84}{47,233.95}\right)$$

$$= 19,928.70$$

2 Life Insurance

The fundamental difference between a life annuity and life insurance is that a life annuity pays to an annuitant whom the policy stipulates to be alive, whereas life insurance pays for the survivors of the insured upon her death. So the life insurance policy stipulates the death of the insured before paying any benefits. The insured would pay either a single premium or annual premiums in purchasing the life insurance policy. The insured or the policyholder's age would be determined by her age at the time of purchase: specifically, her age on the nearest birthday to the formal policy date from which the next years start to count. As we did with life annuities, our calculations skip the operating costs or loadings that would normally be added. We consider only the value of the net premium, which would be equal to the present value of the face of the policy. Therefore, in the case of breaking down the net single premium into annual installments, the present value of all the premiums would be equal to the net single premium.

There are three major types of life insurance policies: the whole life policy, the term policy, and the endowment policy.

2.1. WHOLE LIFE INSURANCE POLICY

According to a whole life insurance policy, the insurance company is obligated to pay the face value of the policy to the survivors of the policyholder upon his death, whenever it occurs. The benefits are usually paid at the end of the year in which the insured's death occurs. As we did with the life annuities, we construct our formulas based on a $1.00 present value so that we can multiply it by the face value in question. The net single premium (A_x) for this policy is the sum of the mathematical expectations that the face value would be paid to the policy beneficiaries. The mathematical expectation is the product of the probability that the insured would die (q_x), and the present value of the policy benefits (v^n):

$$A_x = q_x v^1 + q_{x+1} v^2 + q_{x+2} v^3 + \cdots$$

$$A_x = \left(\frac{d_x}{1_x}\right) v^1 + \left(\frac{d_{x+1}}{1_x}\right) v^2 + \left(\frac{d_{x+2}}{1_x}\right) v^3 + \cdots$$

Mathematical Finance, First Edition. M. J. Alhabeeb.
© 2012 John Wiley & Sons, Inc. Published 2012 by John Wiley & Sons, Inc.

If we multiply the numerator and denominator by v^x, we get

$$A_x = \frac{d_x v^{x+1}}{1_x v^x} + \frac{d_{x+1} v^{x+2}}{1_x v^x} + \frac{d_{x+2} v^{x+3}}{1_x v^x} + \cdots$$

$$A_x = \frac{d_x v^{x+1} + d_{x+1} v^{x+2} + d_{x+2} v^{x+3} + \cdots}{l_x v^x}$$

Since $d_x v^{x+1} = C_x$ and $l_x v^x = D_x$, then

$$A_x = \frac{C_x + C_{x+1} + C_{x+2} + \cdots}{D_x}$$

and since $M_x = \sum C_k$, then

$$A_x = \frac{M_x}{D_x}$$

and for a certain face value of the policy (F), the net single premium or, as it is sometimes called, the **cost of insurance** would be

$$\boxed{A_x = F \left(\frac{M_x}{D_x} \right)}$$

Example 2.1.1   What would be the net single premium for a whole life insurance policy for Tim, who is 49 years old and wants his family to receive $200,000 after his death?

$$A_x = F \left(\frac{M_x}{D_x} \right)$$

$$A_{49} = 200,000 \left(\frac{M_{49}}{D_{49}} \right)$$

$$= 200,000 \left(\frac{2,400.44}{8,425.80} \right)$$

$$= 56,978.33$$

2.2. ANNUAL PREMIUM: WHOLE LIFE BASIS

Paying $56,978.33 at once as in the example above would be difficult for most people to afford. That is why companies break this single premium into annual premiums that would be paid by the insured until his death. Having a large number of payments in this case would lead to having a whole life annuity due

$(\ddot{a}_x)$. To calculate the annual premium for a whole life insurance policy (P_x), we set up the following:

$$A_x = P_x \ddot{a}_x$$

$$P_x = \frac{A_x}{\ddot{a}_x} = \frac{M_x/D_x}{N_x/D_x}$$

$$P_x = \frac{M_x}{D_x} \cdot \frac{D_x}{N_x}$$

$$P_x = \frac{M_x}{N_x}$$

and for face value F:

$$\boxed{P_x = F\left(\frac{M_x}{N_x}\right)}$$

Example 2.2.1 Find the annual premium for Tim in Example 2.1.1 since he cannot afford to pay the entire net premium in one payment.

$$P_x = F\left(\frac{M_x}{N_x}\right)$$

$$P_{49} = 200,000\left(\frac{M_{49}}{N_{49}}\right) = \frac{2,400.44}{126,532.64}$$

$$= 200,000\left(\frac{2,400.44}{126,532.64}\right)$$

$$= 3,794.18 \qquad \text{annual premium for a}$$
$$\text{\$200,000 whole life policy}$$

2.3. ANNUAL PREMIUM: m-PAYMENT BASIS

If the insured wants to limit the number of premiums to a certain number instead of continuing to pay until his death, this can also be arranged with the insurance company. If we consider m to be the number of years to which the net premium will be paid, it will form a temporary life annuity due $(\ddot{a}_{x:\overline{m}|})$, and the annual premium (mP_x) would be obtained by

$$A_x = mP_x \ddot{a}_{x:\overline{m}|}$$

$$mP_x = \frac{A_x}{\ddot{a}_{x:\overline{m}|}}$$

$$m\,P_x = \frac{M_x/D_x}{(N_x - N_{x+m})/D_x}$$

$$m\,P_x = \frac{M_x}{D_x} \cdot \frac{D_x}{N_x - N_{x+m}}$$

cancelling D_x out:

$$m\,P_x = \frac{M_x}{N_x - N_{x+m}}$$

and for face value F it would be

$$\boxed{m\,P_x = F\left(\frac{M_x}{N_x - N_{x+m}}\right)}$$

Example 2.3.1 Suppose that Tim of earlier examples wants to pay all his net premium in 10 years. How much would he be paying annually?

$$m\,P_x = F\left(\frac{M_x}{N_x - N_{x+m}}\right)$$

$$10\,P_{49} = 200{,}000\left(\frac{M_{49}}{N_{49} - N_{49+10}}\right)$$

$$= 200{,}000\left(\frac{2{,}400.44}{126{,}532.64 - 60{,}095.41}\right)$$

$$= 7{,}226.19$$

This is the annual premium if the entire cost of insurance as it is paid in only 10 years. It would be called a 10-payment premium.

2.4. DEFERRED WHOLE LIFE POLICY

A deferred whole life insurance policy is based on the notion that the insured will not die until after a certain period of time (n), called the **deferment period**. The net single premium of this policy for \$1.00 is denoted by $n|A_x$ and determined by

$$n|A_x = \frac{d_{x+n} \cdot v^{n+1}}{l_x} + \frac{d_{x+n+1} \cdot v^{n+2}}{l_x} + \frac{d_{x+n+2} \cdot v^{n+3}}{l_x} + \cdots$$

$$n|A_x = \frac{d_{x+n} \cdot v^{n+1} + d_{x+n+1} \cdot v^{n+2} + d_{x+n+2} \cdot v^{n+3} + \cdots}{l_x}$$

Multiplying by v^x, we get

$$n|A_x = \frac{d_{x+n} \cdot v^{x+n+1} + d_{x+n+1} \cdot v^{x+n+2} + d_{x+n+2} \cdot v^{x+n+3} + \cdots}{l_x \cdot v^x}$$

$$n|A_x = \frac{C_{x+n} + C_{x+n+1} + C_{x+n+2} + \cdots}{D_x}$$

$$n|A_x = \frac{M_{x+n}}{D_x}$$

and for face value F it would be

$$\boxed{n|A_x = F\left(\frac{M_{x+n}}{D_x}\right)}$$

Example 2.4.1 Rita is 35. She wants to buy a $30,000 life insurance policy that would be activated only if she dies when she is 40 or older. How much would her net single premium be?

The period of deferment (n) is 5 years.

$$n|A_x = F\left(\frac{M_{x+n}}{D_x}\right)$$

$$5|A_{35} = 30,000\left(\frac{M_{35+5}}{D_{35}}\right)$$

$$= 30,000\left(\frac{2,717.07}{17,369.06}\right)$$

$$= 4,692.95$$

2.5. DEFERRED ANNUAL PREMIUM: WHOLE LIFE BASIS

We can also calculate the cost of this insurance policy in terms of its annual premiums $(n|P_x)$ involving the annuity due formed by these premiums:

$$n|A_x = n|P_x \cdot \ddot{a}_x$$

$$n|P_x = \frac{n|A_x}{\ddot{a}_x}$$

$$n|P_x = \frac{M_{x+n}/D_x}{N_x/D_x}$$

$$n|P_x = \frac{M_{x+n}}{D_x}\left(\frac{D_x}{N_x}\right)$$

cancelling D_x out:

$$n|P_x = \frac{M_{x+n}}{N_x}$$

and for face value F it would be

$$\boxed{n|P_x = F\left(\frac{M_{x+n}}{N_x}\right)}$$

Example 2.5.1 Suppose that Rita wants to break the premium into annual payments. How much would each payment be?

$$n|P_x = F\left(\frac{M_{x+n}}{N_x}\right)$$

$$5|P_{35} = 30,000\left(\frac{M_{35+5}}{N_{35}}\right)$$

$$= 30,000\left(\frac{2,717.07}{304,804.19}\right)$$

$$= 267.42$$

2.6. DEFERRED ANNUAL PREMIUM: *m*-PAYMENT BASIS

The net single premium of this policy can also be broken down into a certain number of annual payments corresponding to *m*-period during which the policyholder must live. Let's suppose that the annual premium based on this *m* years is $mP(n|A_x)$, which can be calculated as

$$n|A_x = mP(n|A_x) \cdot \ddot{a}_{x:\overline{m}|}$$

$$mP(n|A_x) = \frac{n|A_x}{\ddot{a}_{x:\overline{m}|}}$$

$$mP(n|A_x) = \frac{M_{x+n}/D_x}{(N_x - N_{x+m})/D_x}$$

$$mP(n|A_x) = \frac{M_{x+n}}{D_x}\left(\frac{D_x}{N_x - N_{x+m}}\right)$$

cancelling out D_x:

$$mP(n|A_x) = \frac{M_{x+n}}{N_x - N_{x+m}}$$

and for face value F it would be

$$mP(n|A_x) = F\left(\frac{M_{x+n}}{N_x - N_{x+m}}\right)$$

Example 2.6.1   Let's suppose that Rita wants to pay her net premium in 4 years. How much would the annual premium be?

$$mP(n|A_x) = F\left(\frac{M_{x+n}}{N_x - N_{x+m}}\right)$$

$$4P(5|A_{35}) = 30,000\left(\frac{M_{35+5}}{N_{35} - N_{35+4}}\right)$$

$$= 30,000\left(\frac{2,717.07}{304,804.19 - 240,290.14}\right)$$

$$= 1,263.48$$

With this amount annually, Rita would be able to pay all the policy cost in 4 years, provided that she lives through these 4 years.

2.7. TERM LIFE INSURANCE POLICY

A term life insurance policy would pay the face value of the policy to the survivors only when the insured dies within a specified period of time called the **term** of the policy (n). Technically, the net single premium of this policy $(A^1_{x:\overline{n}|})$ is viewed as the difference between the costs of whole life insurance (A_x) and deferred whole life insurance $(n|A_x)$.

$$A^1_{x:\overline{n}|} = A_x - n|A_x$$

$$A^1_{x:\overline{n}|} = \frac{M_x}{D_x} - \frac{M_{x+n}}{D_x}$$

$$A^1_{x:\overline{n}|} = \frac{M_x - M_{x+n}}{D_x}$$

and for face value F it would be

$$A^1_{x:\overline{n}|} = F\left(\frac{M_x - M_{x+n}}{D_x}\right)$$

Example 2.7.1   Alison is 57. She would like to buy a 13-year term life insurance policy of $50,000. How much would it cost her?

$$A^1_{x:\overline{n}|} = F\left(\frac{M_x - M_{x+n}}{D_x}\right)$$

$$A^1_{57:\overline{13}|} = 50{,}000\left(\frac{M_{57} - M_{57+13}}{D_{57}}\right)$$

$$= 50{,}000\left(\frac{2{,}014.22 - 1{,}222.70}{5{,}372.84}\right)$$

$$= 7{,}365.94$$

Term insurance policies are usually paid for by annual payments where the number of payments (k) should be either equal to or less than the term of the policy (n).

$$\boxed{k \le n}$$

The annual premium would be $kP^1_{x:\overline{n}|}$, which is calculated as

$$A^1_{x:\overline{n}|} = kP^1_{x:\overline{n}|} \cdot \ddot{a}_{x:\overline{k}|}$$

$$kP^1_{x:\overline{n}|} = \frac{A^1_{x:\overline{n}|}}{\ddot{a}_{x:\overline{k}|}}$$

$$kP^1_{x:\overline{n}|} = \frac{(M_x - M_{x+n})/D_x}{(N_x - N_{x+k})/D_x}$$

$$kP^1_{x:\overline{n}|} = \frac{M_x - M_{x+n}}{D_x}\left(\frac{D_x}{N_x - N_{x+k}}\right)$$

cancelling out D_x:

$$kP^1_{x:\overline{n}|} = \frac{M_x - M_{x+n}}{N_x - N_{x+k}}$$

and for face value F it would be

$$\boxed{kP^1_{x:\overline{n}|} = F\left(\frac{M_x - M_{x+n}}{N_x - N_{x+k}}\right)}$$

Example 2.7.2   Find the annual premium for Alison's term life insurance if she wants to pay it off in 10 years.

$$kP^1_{x:\overline{n}|} = F\left(\frac{M_x - M_{x+n}}{N_x - N_{x+k}}\right)$$

$$10P^1_{57:\overline{13|}} = 50,000 \left(\frac{M_{57} - M_{70}}{N_{57} - N_{67}} \right)$$

$$= 50,000 \left(\frac{2,014.22 - 1,222.70}{70,531 - 29,271.95} \right)$$

$$= 959.21$$

Note that if the number of payments (k) is equal to the term of the policy (n), the formula can be adjusted accordingly:

$$n P^1_{x:\overline{n|}} = F \left(\frac{M_x - M_{x+n}}{N_x - N_{x+n}} \right) \qquad \text{when } k = n$$

Example 2.7.3 Find the annual premium for Alison in Example 2.7.2 if she wants to pay out her insurance cost in annual premiums until her policy term is over.

This means that she would pay off the cost in 13 payments $(n = k = 13)$.

$$n P^1_{x:\overline{n|}} = F \left(\frac{M_x - M_{x+n}}{N_x - N_{x+n}} \right)$$

$$13 P^1_{57:\overline{13|}} = 50,000 \left(\frac{M_{57} - M_{70}}{N_{57} - N_{70}} \right)$$

$$= 50,000 \left(\frac{2,014.22 - 1,222.70}{70,531 - 21,427.28} \right)$$

$$= 805.97$$

2.8. ENDOWMENT INSURANCE POLICY

The endowment insurance is, in fact, a combination of a term life insurance and a pure endowment where both have the same term (n). This mix is designed to assure the receipt of the policy benefits whether the insured dies or lives throughout the specified term. Therefore, the face value of the policy would be paid at any rate. It would be paid to the insured if she survives until the end of the term specified, and it would be paid to the survivors if she dies within the term specified. Therefore, the net single premium of the endowment insurance policy $(A_{x:\overline{n|}})$ would be the sum of the net single premium for the n-term life insurance $(A^1_{x:\overline{n|}})$, and the net single premium for the n-term pure endowment (nE_x):

$$A_{x:\overline{n|}} = A^1_{x:\overline{n|}} + nE_x$$

$$A^1_{x:\overline{n}|} = \frac{M_x - M_{x+n}}{D_x} + \frac{D_{x+n}}{D_x}$$

$$A^1_{x:\overline{n}|} = \frac{M_x - M_{x+n} + D_{x+n}}{D_x}$$

and for face value F it would be

$$A^1_{x:\overline{n}|} = F\left(\frac{M_x - M_{x+n} + D_{x+n}}{D_x}\right)$$

Example 2.8.1   Dan is 46. He has just purchased a \$65,000 endowment policy with a 15-year term. How much would this policy cost him?

$$A^1_{x:\overline{n}|} = F\left(\frac{M_x - M_{x+n} + D_{x+n}}{D_x}\right)$$

$$A^1_{46:\overline{15}|} = 65,000\left(\frac{M_{46} - M_{46+15} + D_{46+15}}{D_{46}}\right)$$

$$= 65,000\left(\frac{2,518.91 - 1,789.21 + 4,210.49}{9,884.83}\right)$$

$$= 32,485.37$$

2.9. ANNUAL PREMIUM FOR THE ENDOWMENT POLICY

Just like with other policies, the large net single premium can be paid off in annual installments that would form an annuity due and result into adjusting the formula as follows:

1. If the number of annual premiums (k) is equal to the policy term (n), $k = n$, then

$$P_{x:\overline{n}|} = F\left(\frac{M_x - M_{x+n} + D_{x+n}}{N_x - N_{x+n}}\right)$$

2. If the number of annual premiums (k) is less than the policy term (n), $k < n$, then:

$$kP_{x:\overline{n}|} = F\left(\frac{M_x - M_{x+n} + D_{x+n}}{N_x - N_{x+k}}\right)$$

Note that k cannot be larger than n; k is always either equal or less than n [$k \le n$].

Example 2.9.1   Find the annual premiums for the endowment policy in Example 2.8.1 as the cost is paid off along the policy term.

$$P_{x:\overline{n}|} = F\left(\frac{M_x - M_{x+n} + D_{x+n}}{N_x - N_{x+n}}\right)$$

$$P_{46:\overline{15}|} = 65,000\left(\frac{M_{46} - M_{61} + D_{61}}{N_{46} - N_{61}}\right)$$

$$= 65,000\left(\frac{2,518.91 - 1,789.21 + 4,210.49}{154,684.29 - 50,846.91}\right)$$

$$= 3,092.45$$

Example 2.9.2   Suppose that Dan wants to pay off his net single premium in 10 years. How much would his annual premium be?

Now $k = 10$, which is less than the policy term, n (15).

$$kP_{x:\overline{n}|} = F\left(\frac{M_x - M_{x+n} + D_{x+n}}{N_x - N_{x+k}}\right)$$

$$10P_{46:\overline{15}|} = 65,000\left(\frac{M_{46} - M_{61} + D_{61}}{N_{46} - N_{56}}\right)$$

$$= 65,000\left(\frac{2,518.91 - 1,789.21 + 4,210.49}{154,684.29 - 76,228.19}\right)$$

$$= 4,092.89$$

2.10. LESS THAN ANNUAL PREMIUMS

In practice, insurance companies agree to collect their net single premium not only in annual installments but also quarterly, monthly, or other terms. This increases the cost to the policyholder, but people can often afford a small premium despite the expense attached to such a convenience. The additional cost to the premium can be figured out as an added certain percentage (j). So if we denote the less than annual premium by $p(m)$ and the annual premium by P,

$$\boxed{P^{(m)} = \frac{P(1+j)}{m}}$$

Example 2.10.1   Let's suppose that Dan, who we met earlier, cannot afford to pay his annual premium of \$3,092.45 and asks the company to let him pay whatever they deem appropriate every other month, that is, in six payments a year. Given that the company charges $6\frac{1}{2}\%$ extra, determine how much his bimonthly premium would be.

$P = \$3,092.45; j = 6\frac{1}{2}\% = .065.$

$$P^{(m)} = \frac{P(1+j)}{m}$$

$$P^{(6)} = \frac{3,092.45(1+.065)}{6}$$

$$= 548.91$$

2.11. NATURAL PREMIUM VS. THE LEVEL PREMIUM

The natural premium is the net single premium for a one-year term life insurance that pledges to pay the face value of the policy only when the insured dies within one year. We can obtain such a premium by adjusting the regular term life insurance formula ($A^1_{x:n\rceil}$) by making $n = 1$:

$$NA^1_{x:1\rceil} = \frac{M_x - M_{x+1}}{D_x}$$

$$NA^1_{x:1\rceil} = \frac{C_x}{D_x}$$

and for face value F it would be

$$\boxed{NA^1_{x:1\rceil} = F\left(\frac{C_x}{D_x}\right)}$$

Example 2.11.1 What would be the natural premium for a 75-year-old person who buys a \$100,000 term life insurance policy?

$$NA^1_{75:1\rceil} = F\left(\frac{C_{75}}{D_{75}}\right)$$

$$= 100,000\left(\frac{62.78}{1,462.66}\right)$$

$$= 4,292.18$$

Since the natural premium is obtained by dividing C_x by D_x, it tends to increase with age. This is because C_x increases and D_x decreases as people advance in age. Normally, the natural premium for one year is supposed to be sufficient to cover all death claims at the end of that year, but this becomes problematic, as the natural premium rises dramatically with older age. One way to deal with this problem and make life insurance affordable is to rely on the regular annual premium, which we described as a level premium. Level premium policies have a higher premium than that of a natural premium in the early years and a lower premium in the later years. On balance, the early years would create an excess fund that would be enough, with the interest it accumulates, to face the deficit that the later years create. In Table 2.1 and Figure 2.1 we show how a natural premium interacts with a level premium to creat that excess and deficit. We take, for example, a 15-year term insurance policy for $100,000 issued to a 50-year-old. We calculate the level premium for the policy in the usual way:

$$n P_x = F \left(\frac{M_x - M_{x+n}}{N_x - N_{x+n}} \right)$$

$$15 P_{50} = 100,000 \left(\frac{M_{50} - M_{50+15}}{N_{50} - N_{50+15}} \right)$$

$$= 100,000 \left(\frac{2,357.27 - 1,542.78}{118,106.84 - 35,523.27} \right)$$

$$= 986.26$$

which would be fixed for the entire term of 15 years, from age 50 to age 64. We also calculate the natural premium for each year by multiplying the face value of the policy, $100,000, by the ratio of C_x to D_x. We see in Table 2.1 that the level premium is higher than the natural premium between ages 50 and 56, and lower between ages 57 and 64. This discrepancy creates the excess and deficit areas on Figure 2.1. The straight line of the level premium first passes above the

TABLE 2.1   Natural Premiums vs. Level Premiums in a $100,000 15-Year Term Insurance Policy at 5%

	(1)	(2)	(3)	(4)	(5)	(6)
				Level	Natural	Excess/
	Age	C_x	D_x	Premium,	Premium	Deficit
				$15 P_{50}$	$100,000\,[(2) \cdot (3)]$	$[(4) - (5)]$
1	50	44.85	7,981.41	986.26	561.93	424.33
2	51	46.19	7,556.49	986.26	611.26	375.00
3	52	47.53	7,150.47	986.26	664.71	321.55
4	53	49.07	6,762.44	986.26	725.62	260.64
5	54	50.49	6,391.34	986.26	789.97	196.29
6	55	51.86	6,036.50	986.26	859.11	127.15
7	56	53.05	5,697.19	986.26	931.16	55.10
8	57	54.24	5,372.84	986.26	1,009.52	−23.26
9	58	55.48	5,062.75	986.26	1,095.85	−109.59
10	59	56.91	4,766.18	986.26	1,194.04	−207.78
11	60	58.38	4,482.32	986.26	1,302.45	−316.19
12	61	59.87	4,210.49	986.26	1,421.92	−435.66
13	62	61.23	3,950.12	986.26	1,550.08	−563.82
14	63	62.32	3,700.79	986.26	1,683.96	−697.70
15	64	63.00	3,462.24	986.26	1,819.63	−833.37

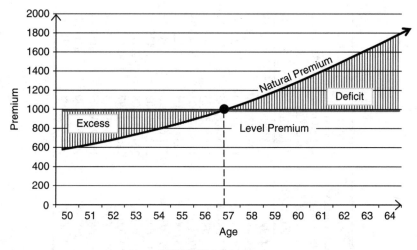

FIGURE 2.1

natural curve up to age 57, then continues below the natural curve until the end of the 15-year term.

2.12. RESERVE AND TERMINAL RESERVE FUNDS

The **insurance reserve fund** is the difference between the accumulated value of insurance premiums and the accumulated cost of the insurance. Specifically, the positive excess of the premium in the early years that we saw in Table 2.1 and Figure 2.1 is allowed to accumulate and gain interest so that it can be used to pay the due death claims and other obligations, especially when the level premium accumulations lag behind the liabilities in the later years. Since the amount of reserve, the interest it earns, the death claims, and the obligations are all calculated in each year of the policy term, the remaining fund forms what is called the **terminal reserve**. Therefore, the terminal reserve is what is left in the reserve fund at the end of any policy years after paying off all death claims and other obligations, and after dividing the result by the number of survivors at that time. In this sense, the terminal reserve would represent the maximum amount that a policyholder would expect to get should the discontinuation of the policy be chosen at year's end. It can also serve as the guide to the maximum amount of loan that a policyholder can obtain from his insurance company if he wants to put his policy up as a security asset for such a loan.

There are two major methods of calculating the terminal reserve: the retrospective and the prospective.

The Retrospective Method

Table 2.2 shows in detail how the terminal reserve is obtained in each year of a 15-year term policy of $100,000 issued to a 50-year-old person. This method

TABLE 2.2 Terminal Reserve by the Retrospective Method

(1)	(2)	(3)	(4)	(5)	(6)	(7)	(8)	(9)	(10)
Policy Year	Age	l_x	d_x	Total Premium at the Start of the Year [(2) × 986.25]	Reserve at the Start of the Year [(5) + (9) ↑]	Interest Earned on Reserve [(6) × .05]	Death Claims at the End of the Year (d_x × 100,000)	Reserve at End of the Year [(6) + (7) = (8)]	Terminal Reserve [(9) ÷ (3)]
1	50	91,526	540	90,268,432	90,268,432	4,513,422	54,000,000	40,781,854	445.58
2	51	90,986	584	89,735,852	130,268,432	6,525,885	58,400,000	78,643,591	864.35
3	52	90,402	631	89,159,876	167,803,467	8,390,173	63,100,000	113,093,640	1,251.00
4	53	89,771	684	88,537,546	201,631,186	10,081,559	68,400,000	143,312,745	1,596.43
5	54	89,087	739	87,862,944	231,175,689	11,558,784	73,900,000	168,843,473	1,895.16
6	55	88,348	797	87,134,098	255,968,573	12,798,428	79,700,000	189,067,002	2,140.03
7	56	87,551	856	86,348,049	275,415,051	13,770,752	85,600,000	203,585,804	2,325.34
8	57	86,695	919	85,503,810	289,089,614	14,454,481	91,900,000	211,644,095	2,441.25
9	58	85,776	987	84,597,437	296,241,532	14,812,077	98,700,000	212,353,609	2,475.68
10	59	84,789	1,063	83,623,999	295,977,608	14,798,880	106,300,000	204,476,488	2,411.59
11	60	83,726	1,145	82,575,604	287,052,092	14,352,605	114,500,000	186,904,697	2,232.34
12	61	82,581	1,233	81,446,337	268,351,034	13,417,552	123,300,000	158,468,585	1,918.98
13	62	81,348	1,324	80,230,278	238,698,863	11,934,943	132,400,000	118,233,806	1,453.43
14	63	80,024	1,415	78,824,470	197,158,276	9,857,914	141,500,000	65,516,190	820.01
15	64	78,609	1,502	77,528,912	143,045,102	8,582,706	150,200,000	1,427,808	18.16

of obtaining the terminal reserve is called the **retrospective method** since its calculations are made at the end of each year of the policy term by looking backward to the beginning of that year.

As an example of the process to calculate the terminal reserve, let's follow how the terminal reserve of the eighth year of the policy, $2,441.25, was calculated:

1. The total premium at the beginning of the eighth year [column (5)] is calculated by multiplying the level premium of $986.26 by the number of survivors at the beginning of that year (1_x; the level premium was calculated in Section 2.11).

$$986.26 \times 86,695 = 85,503,810$$

2. The figure obtained in step 1 (85,503,810) is added to the figure for the reserve at the end of previous year (the seventh) in column (9), which is 203,585,804:

$$85,503,810 + 203,585,804 = 289,089,614$$

This would be the entry in column (6).

3. The figure in step 2 would earn 5% interest, which comes to $14,454,481. It is recorded in column (7).

$$289,089,614 \times .05 = 14,454,481$$

4. Death claims at the end of the year are obtained by multiplying the face value of the policy, $100,000, by the number of dying (d_x) in column (4).

$$919 \times 100,000 = 91,900,000$$

This is recorded in column (8).

5. The reserve fund at the end of the eighth year would be obtained by adding the reserve value at the beginning of the year [column (6)], to the interest earned [column (7)], and then subtracting the death claims [column (8)].

$$(289,089,614 + 14,454,481) - 91,900,000 = 211,644,095$$

This is now the amount of reserve at the end of the eighth year, recorded in column (9).

6. To get the terminal reserve per survivor, the amount of reserve at the end of the year found in step 5 is divided by the number of people living (1_x) in column (3).

$$211,644,095 \div 86,695 = 2,441.25$$

We can also use the following formula to get the terminal reserve for any year (t) without the need to construct a table:

$$V = \frac{P(N_x - N_{x+t}) - (M_x - M_{x+t})}{D_{x+t}}$$

where V is the terminal reserve per survivor at the end of the tth year of the policy term, P is the annual premium per $1.00, t is any year of the policy term, and the rest are the usual commutation values from Table 10 in the Appendix.

Example 2.12.1   Let's try to check this formula in finding the eighth-year terminal reserve per survivor of Table 2.2.

$x = 50, t = 8, x + t = 58$, and the premium per dollar would be obtained by dividing the annual premium of 986.26 by the face value of the policy, $100,000.

$$P = \frac{986}{100,000} = .00986$$

$$V = \frac{P(N_x - N_{x+t}) - (M_x - M_{x+t})}{D_{x+t}}$$

$$V = \frac{P(N_{50} - N_{58}) - (M_{50} - M_{58})}{D_{58}}$$

$$= \frac{.00986(118,106.84 - 65,158.16) - (2,357.27 - 1,959.87)}{5,062.95}$$

$$= .0246$$

which is the terminal reserve for $1.00. So for the face value of $100,000 it would be

$$.0246 \times 100,000 = 2,460$$

which is very close to what we obtained in Table 2.2 (2,441). The difference is due to rounding off.

The Prospective Method

Instead of looking backward as we do in the retrospective method, the **prospective method** would look forward to the future benefits and future liabilities of the policy. In this case, the reserve would be the excess of the present value of the future benefits over the present value of the future premiums. This can be translated as the difference between A_{x+t}, which is the value of insurance at age $x + t$, and $_tA_x$, which is the value of the annuity due of the annual premiums deferred t years.

$$_tV_x = A_{x+t} -_t A_x$$

$$_tV_x = F\left(\frac{M_{x+t} - M_{x+n}}{D_{x+t}}\right) - P_x\left(\frac{N_{x+t} - N_{x+n}}{D_{x+t}}\right)$$

$$\boxed{_tV_x = \frac{F(M_{x+t} - M_{x+n}) - P_x(N_{x+t} - N_{x+n})}{D_{x+t}}}$$

Example 2.12.2 Find the eighth terminal reserve of the earlier examples using the prospective method.

$$F = 100,000; \, x = 50; \, t = 8; \, x + t = 58; \, n = 15; \, x + n = 65.$$

$$_tV_x = \frac{F(M_{x+t} - M_{x+n}) - P_x(N_{x+t} - N_{x+n})}{D_{x+t}}$$

$$_8V_{50} = \frac{100,000\,[M_{58} - M_{65}] - 986.26\,[N_{58} - N_{65}]}{D_{58}}$$

$$= \frac{100,000(1,959.98 - 1,542.78) - 986.26(65,158.16 - 35,523.27)}{5,062.75}$$

$$= 2,467$$

which is close to the value we arrived at earlier except for the slight discrepancies caused by rounding off.

2.13. BENEFITS OF THE TERMINAL RESERVE

Since the terminal reserve is the value of the insurance policy in any year of the insurance term, it can be serving many financial needs for a policyholder. It can, for example, be used to determine the cash that a policyholder may receive in case of terminating payments or surrendering the policy. It can also be used as a guide to determine the maximum amount of loan that a policyholder may get from her insurance company to take advantage of the interest rate, which is usually much lower than the best rate obtained from a commercial bank. The policy and its reserve would be considered a security asset for the loan. The reserve can also be received and used as a single premium to extend a current policy or get a reduced amount of paid-up insurance.

2.14. HOW MUCH LIFE INSURANCE SHOULD YOU BUY?

Life insurance is a protection for survivors who are named as beneficiaries in a policy. It is to provide them with the ability to continue living their lives normally after losing a loved one. There are two approaches to estimating the dollar amount to buy in life insurance.

The Sentimental Approach

The sentimental approach is highly subjective. Basically, it involves the insurance buyer choosing the amount of insurance that reflects the emotional value he places on his survivors—much more than to consider their real needs. In such a case, people may choose $1 million or more just to make a statement that this is the way they want to help their families after the policyholders are dead.

The Rational Approach

In the rational approach, insurance buyers rely on certain criterion as a basis for their consideration to the amount of insurance they buy. They may consider either a very general measure such as the insured's income, or go over more detailed calculations of the needs that survivors may face. Two common methods are used in this rational approach.

The Multiple-Earnings Method   This method is simple and straightforward. It depends on the insured's income as a basis of estimation. The person who wants to buy insurance can multiply her income by a certain multiple such as 5 or 10 or any other number so that the final amount would yield the desired life insurance proceeds to the beneficiaries. Table 2.3 shows general estimates of an average premium for each $100,000 in purchased insurance, arranged by several term policies for men and women. For example, if a 50-year-old man wants to buy a 25-year term life insurance equal to five times his annual income of $75,000, the amount of insurance would be:

$$75,000 \times 5 = 375.000$$

and his premium would be

$$375,000 \div 100,000 = 3.75$$

Looking at Table 2.3 across age 50 and term 25 in the men's section, we find the annual premium of $762 per $100,000. Since we have 3.75, the premium would be

$$3.75 \times 762 = \$2,857.50 \text{ annually}$$

The Needs Method   This method has more objective elements to consider in deciding how much life insurance to buy. It considers the most common factors that would affect the way the survivors will live after the insured's death. It also takes into consideration the time value of money. In addition to the income required for a certain number of years after the insured's death, it considers government benefits, the final expenses related to the insured's death, the re-adjustment expenses for the spouse and children, children's education, family debt (if any), and whatever the insured can specify as an important financial

TABLE 2.3   Average Annual Premium per $100,000 Term Life Insurance

Gender	Age	Term of Insurance (years)					
		5	10	15	20	25	30
Male	20	144	133	154	166	216	187
	30	152	134	157	172	229	250
	40	195	170	208	235	356	375
	50	253	310	409	467	762	825
	55	510	455	610	750	1,174	1,306
Female	20	125	116	154	166	216	187
	30	135	124	142	148	189	210
	40	176	150	186	204	279	301
	50	272	240	317	358	559	604
	55	343	320	429	514	848	984

element, into the calculation of the amount of insurance needed. The following equation sums up most of these elements; LA is the life insurance amount needed.

$$LA = PV(.75Y_d) + OT - [PV(G) + CI]$$

$PV(.75Y_d)$: the present value of a certain stream of future income for the survivors. This income is determined by the notion that the survivors would need about 75% of the insured's disposable income (Y_d) in his working years. This assumes that the 25% deducted used to be his share of the family income.

OT: a vector of needed expenses including but not limited to F, R, D, and E, described below.

F : the final expenses related to the funeral and burial of the deceased.

R : the readjustment expenses for spouses and children related to taking time off, training and getting a job, therapy or psychological treatment and the like.

D : the debt expenses to pay off whatever obligations the insured left.

E : the educational expenses related most likely to the college expenses for the insured's children.

PVG : the present value of the future stream of government benefits, such as Social Security survivor's benefits; this item is to be deducted.

CI: the current insurance policies that exist at the time of the insured's death, such as the employer's insurance and other private insurance (if any); this item is also to be deducted.

The LA equation can be rewritten as

$$LA = \frac{.75Y_d[1 - (1+r)^{-n}]}{r} + OT + \left[\frac{G[1 - (1+r)^{-k}]}{r} + CI\right]$$

Y_d : the insured's disposable income.

r : the interest rate at which the discount is calculated. It is often considered the after-tax, after-inflation investment rate.

n : the time estimated for the needs of the survivors. It is determined by the insured, such as the time until the youngest child finishes college, or when the widow turns 65, and the like.

OT: the sum of whatever expenses are outstanding such as $F + R + D + E$.

G : the annual government benefits.

K : the time until government benefits end, which is determined by the Social Security Administration. Unlike n, it is not determined subjectively by the insured.

CI: the existing insurance.

If we use the table values for the present value factor, the LA formula would be

$$LA = (.75Y_d) \cdot a_{\overline{n}|r} + OT + [G \cdot a_{\overline{k}|r} + CI$$

Example 2.14.1   Bryan and Heather are a married couple in their early thirties. They have two young children. Bryan would like his family to live comfortably for at least 25 years after he dies, and he is thinking of buying a life insurance that would make his goal possible. Here are the data elements:

Gross income : $60,000

Taxes and payroll deductions : $18,000

Social Security benefits : $2,400 a month for 14 years

Education fund for children : $80,000

Debt to be paid off : $12,000

Readjustment expenses for Heather and children : $20,000

Final expenses : $10,000

Given that the couple has existing insurance of $85,000, how much life insurance would they need if the interest rate is 5%?

$$Y_d = Y_g - (T + D)$$

$$= 60,000 - 18,000 = 42,000$$

$$.75Y_d = .75(42,000) = 31,500$$

$$n = 25 \qquad r = .05 \qquad k = 14$$

$$G = 2,400 \times 12 = 28,800$$

$$OT = E + D + R + F$$

$$OT = 80,000 + 12,000 + 20,000 + 10,000 = 122,000$$

$$LA = \frac{.75Y_d[1 - (1+r)^{-n}]}{r} + OT - \left[\frac{G[1 - (1+r)^{-k}]}{r} + CI\right]$$

$$= \frac{31,500[1 - (1 + .05)^{-25}]}{.05} + 122,000$$

$$- \left[\frac{28,800[1 - (1 + .05)^{-14}]}{.05} + 85,000\right]$$

$$= 443,959.25 + 122,000 - (285,079.7 + 85,000)$$

$$= 195,879$$

If we use Appendix Table 10 values for the present value:

$$LA = (.75Y_d) \cdot a_{\overline{n}|r} + OT - (G \cdot a_{\overline{k}|r} + CI)$$

$$= 31,500(a_{\overline{25}|5}) + 122,000 - [28,800(a_{\overline{14}|5}) + 85,000]$$

$$= 31,500(14.0939) + 122,000 - [28,800(9.8986) + 85,000]$$

$$= 195,878.17$$

3 Property and Casualty Insurance

Most property and casualty insurance policies cover both property and liability risks. They reimburse the policyholder for particular financial losses that incur due to any damage, destruction, or loss of use of property owned or controlled by the policyholder. Fire, auto accidents, vandalism, and storms are examples of **casualty damage**. Property and casualty insurance also reimburses property damage and bodily injuries maintained by others but for which the policyholder is responsible. We call this **liability insurance**.

When it comes to reimbursement, two major principles guide the reasoning of the insurance concept.

1. The first is that the financial losses have to be a result of pure vs. speculative risk. Pure risk occurs when there is only the potential for loss and no potential for gain, whereas speculative risk involves a potential for both gain and loss, such as what happens in gambling, which cannot be insured for reimbursing the gambler for his losses.

2. The second principle is that the insurance would not pay more than the actual financial loss, which is called the **principle of indemnity**.

The major concern here is on the matter of how to assess the actual losses. Insurance companies have been instituting the rule that they pay what is called the **actual cash value** as opposed to both the original cost of the property to the policyholder and the replacement cost. The actual cash value (ACV) considers depreciation of the property due to wear and tear and market conditions, while the replacement value would consider the original cost and the appreciation value of the property. The actual cash value can be calculated easily by taking the depreciation effect away from the original cost (OC) of the property. The depreciation effect would consider the current age of property (CA) and its life expectancy (LE).

$$ACV = OC - \left(CA \cdot \frac{OC}{LE} \right)$$

$$\boxed{ACV = OC \left(1 - \frac{CA}{LE} \right)}$$

Mathematical Finance, First Edition. M. J. Alhabeeb.
© 2012 John Wiley & Sons, Inc. Published 2012 by John Wiley & Sons, Inc.

Example 1 A sofa was purchased 4 years ago for $1,500. Determine how much the reimbursement to the policyholder would be, if the insurance adjuster estimates the life expectancy of this sofa by 9 years.

$$OC = 1,500; \quad CA = 4; \quad LE = 9.$$

$$ACV = OC\left(1 - \frac{CA}{LE}\right)$$

$$= 1,500\left(1 - \frac{4}{9}\right)$$

$$= 833.33$$

Example 2 The policyholder in Example 1 was expecting to have his sofa replaced with a new sofa of the type that he lost. Given that the inflation rate has been averaging 1.75% annually, what replacement value would the policyholder hope for?

The replacement value (RV) includes the original cost plus the changes in the original cost (ΔOC) since the sofa was purchased.

$$\boxed{RV = OC + \Delta OC}$$

The change in the original cost here is the rise in prices due to inflation, which would be calculated at 1.75% each year since the sofa was purchased 4 years ago.

$$\boxed{\Delta OC + OC \cdot f \cdot t}$$

where OC is the original cost, f is the annual inflation rate, and t is the time.

$$\Delta OC = 1,500(.0175)(4)$$

$$= 105$$

$$RV = OC + \Delta OC$$

$$= 1,500 + 105$$

$$= 1,605$$

If we plug ΔOC in to the RV formula, we get

$$RV = OC + \Delta OC$$

$$RV = OC + OC \cdot f \cdot t$$

$$\boxed{RV = OC(1 + f \cdot t)}$$

which is more convenient than the earlier formula, and we get exactly the same result:

$$RV = 1,500[1 + (.0175)(4)]$$
$$= 1,605$$

Example 3   What would be the replacement value of a television set that was purchased 7 years ago for $1,350? Inflation has averaged $2\frac{1}{4}$ % annually.

$$RV = OC(1 + ft)$$
$$= 1,350[1 + .0225(7)]$$
$$= 1,562.62$$

Again, the insurance companies would not consider the replacement value in their reimbursement but would probably rely on considering the actual cash value.

3.1. DEDUCTIBLES AND CO-INSURANCE

A **deductible** is a policyholder's initial share of the insurance compensation and the **co-insurance** is the final policyholder's share of the entire compensation. Whereas the deductible is stated as a lump sum, co-insurance is a percentage share. For example, if a policy states a coverage of 85 : 15, it means that the entire reimbursement is split between the insurer and the insured, where the insurer carries 85% and the insured carries 15%. Both the deductible and co-insurance are also strategies to reduce the final cost to the insurer while promoting and pushing for more responsibility from the insured. People would think twice before filing a claim if they are going to be responsible for part of the cost. Some insurance companies make the deductible optional for a policyholder.

The following is specific to dwelling property insurance, where companies require that full coverage of any losses to a dwelling requires that the property has been insured for at least 80% of its replacement value. The specification of 80% is under the assumption that the land is worth 20% of the property and would not suffer any damage even if the property burns to the ground. This is why the 80% would be considered equal to full insurance for the structure and the contents of the property. Therefore, the reimbursement for the dwelling losses (R_d) would be calculated as:

$$R_d = \frac{I}{.80\ RV}(L - D)$$

where I is the insurance purchased, L is the total losses on the claim, D is the deductible, and RV is the replacement value of the property.

Example 3.1.1   A fire causes \$60,000 in damage to a house whose replacement value is \$135,000 but is insured for \$108,000 with a \$1,000 deductible. Determine the insurance reimbursement.

$I = 108{,}000$; $RV = 135{,}000$; $L = 60{,}000$; $D = 1{,}000$.

$$
\begin{aligned}
R_d &= \frac{I}{.80\ RV}(L - D) \\
&= \frac{108{,}000}{.80(135{,}000)}(60{,}000 - 1{,}000) \\
&= 59{,}000
\end{aligned}
$$

The \$59,000 is a full insurance coverage in consideration of the deductible. The reason that the insurance fully compensates the loss because the policyholder has purchased full insurance, as 108,000 is exactly 80% of the replacement value.

Let's consider the following example, where the policyholder buys an insurance policy for less than the replacement value.

Example 3.1.2   Suppose that the policyholder in Example 3.1.1 has purchased only \$96,000 worth of insurance. How much would the reimbursement be for losses of \$60,000?

$$
\begin{aligned}
R_d &= \frac{I}{.80\ RV}(L - D) \\
&= \frac{96{,}000}{.80(135{,}000)}(60{,}000 - 1{,}000) \\
&= 52{,}444
\end{aligned}
$$

\$52,444 would be the reimbursement, commensurate with the amount of coverage the policyholder chose to buy, which is less than the insurance required for full coverage. If we take a closer look, we can quickly discover that because the policyholder purchased only 88.8% of what is required ($96{,}000/108{,}000 = 88.8\%$), the insurance made his imbursement as 88.8% of full coverage ($52{,}444/59{,}000 = 88.8\%$)

3.2. HEALTH CARE INSURANCE

Health care insurance comes in the form of many types of health plans, such as a health maintenance organization (HMO), a preferred provider organization (PPO), a managed care plan (MCP), government health care programs such as Medicare and Medicaid, and private health insurance. Reimbursement to the policyholder is affected first and foremost by the type of plan and the variety of coverage available. In calculating the reimbursement, two elements play a big role: the deductible and the co-insurance. The deductible in the health care insurance domain usually comes in two types: the deductible per event, such as hospitalization or operation, and the annual deductible. The co-insurance is just the percentage share of the insured and can differ from plan to plan. Generally, health insurance reimbursement (R_h) is determined by

$$R_h = (I - CP)(L - D)$$

where CP is the policyholder's co-insurance percentage, L is the total financial loss (entire medical bill), and D is the deductible.

Example 3.2.1   A man had to be hospitalized 6 days for treatment after an accident. The hospital charged $500 per day, doctors' fees were $3,620, lab fees were $987, and ambulance service cost $340. He has a 80 : 20 insurance plan with a $500 deductible per event. Determine how much he would expect his insurance to pay and how much would fall on him to pay.

CP = 20%; D = 500.
First, we have to add up all the medical charges to get L:

$$L = (6 \times 500) + 3,620 + 897 + 340 = 7,857$$
$$R_h = (1 - CP)(L - D)$$
$$= (1 - .20)(7,857 - 500)$$
$$= 5,885.60$$

This is the insurance share, which is 80%—not of the entire bill of $7,857 but of the bill after the deductible, which is $7,357. This would leave the policyholder with

$$7,857 - 5,885.60 = 1,971.50$$

The policyholder's share of $1,971.40 is split between $500 deductible and $1,471.40 co-insurance. It is important to notice here that the insured pays in reality more than the stated 20% of the bill. In this case, he paid $1,971.40 out of $7,857, which is a little more than 25%, and the insurance company paid $5,885.60 out of $7,857, which is a little less than 75% of the bill rather than the stated share of 80%. The reason for this discrepancy, of course, is the deductibles! The 20% share of the policyholder is the $1471.40 out of the total bill after the deductibles [1471.40/(7857−500)].

The policyholder's copayment may come in different categories and different designations. For example, the copayment for a family practitioner's visit is different from the copayment for a specialist's visit. Also, there are different copayments for different medications and different services, such as lab fees and the like. In an aggregate sense, the insurance reimbursement would be to carry

TABLE E3.2.2

Category	n	k	nk	CP	nCP	$k -$ CP	$n(k -$ CP)
Family practitioner	8	75	600	15	120	60	480
Specialist	2	105	210	35	70	70	140
X-ray	4	59	236	20	80	39	156
Lab report	3	70	210	30	90	40	120
Prescription	12	60	720	15	180	45	540
Dietitian	2	55	110	10	20	45	90
			2,086		560	299	1,526

the insurer's share after the policyholder pays his share of the stated co-insurance of a variety of services, plus the deductibles.

Let's consider the following example, which illustrates how to calculate the insurance and the policyholder's shares of a variety of medical services, each of which has its own policy-allowed co-insurance.

Example 3.2.2   Suppose that a patient incurred the following medical expenses in eight visits to his family practitioner at $75 each: two visits to specialists at $105 each, four x-rays at $59 each, three lab reports at $70 each, 12 prescriptions at an average of $60, and two visits to a dietitian at $55 each. The insurance policy specifies the copayments for each category as the following: $15 for a family practitioner; $35 for a specialist, $20 for an x-ray, $30 for a lab service, $15 for a prescription, and $10 for a professional service. Calculate the insurance reimbursement, the policyholder share, and the percentages of both.

The best way to calculate this was to organize the information in Table E3.2.2, where:

n: the number of unit for each category.

k: the cost per unit.

nk: the cost per category.

CP: the co-insurance per category as specified in the policy.

nCP: the co-insurance cost per category.

$k -$ CP: the difference between the cost and the co-insurance per unit.

$n(k -$ CP): the difference between the cost and the co-insurance per category.

This is, in fact, what the insurance company pays per category because they carry the cost after the policyholder's share has been paid. In total, the cost of all categories ($\sum nk$) is $2,086 and the total co-insurance cost ($\sum n \cdot$ CP) is $560, which is the policyholder's share, and the difference would be the insurance share or the reimbursement (R_n):

$$R_n = \sum (nk) - \sum (n \cdot \text{CP})$$
$$= 2{,}086 - 560$$
$$= 1{,}526$$

The reimbursement equation above can be rewritten as

$$R_n = \sum nk - \sum n\text{CP}$$

$$\boxed{R_n = \sum n(k - \text{CP})}$$

which is the total of the last column $n(k - \text{CP})$, \$1,526, to confirm the result. As for the percentages, the insurance company would pay 73% $(1{,}526 \div 2{,}086)$, and the policyholder would pay the rest, 27% $(560 \div 2{,}086)$.

3.3. POLICY LIMIT

Most insurance policies specify a certain limit to what can be reimbursed. The limit is either imposed per event or per year or both. Let's suppose that the insurance in Example 3.2.2 has specified a policy limit of \$1,350 per event (case of illness or injury and the like). Under this policy, the insurance would not pay its share of \$1,526, as calculated. It would pay up to the maximum limit, which is, as specified, only \$1,350. The policyholder has to carry the difference of \$176 (\$1,526 − \$1,350).

The annual limit can work in the same manner. Suppose that there is an annual policy limit of \$12,000 and that throughout the year the insurance has paid \$11,000 for many other cases for this patient. Now that this patient has a new case of treatment in which the insurer's share comes to \$1,526. The insurance would stick to what is left from their coverage limit of \$12,000 which is only \$1,000. They would only pay \$1,000 out of their calculated share of \$1,526. The policyholder has to carry the rest, \$526. In these two examples of imposing the policy limit, the policyholder would end up paying a higher percentage of co-insurance than is stated in the policy.

Unit VIII Summary

This final unit dealt with the mathematics of insurance. We started with life annuities, which are distinct from the earlier annuities certain by being contingent and related to life insurance. We discussed a mortality table, which is a central concept for understanding and calculating all life annuities and insurance. This was followed by an explanation of the commutation terms, which are other important complements to the mortality table, conceptually and technically. Pure endowment as a single payment preceded the discussion of all the common types of life annuities. We discussed and calculated the whole life annuities: ordinary, due, and deferred. The temporary life annuities have the same subdivision: ordinary, due, and deferred.

Life insurance was the subject of the second chapter in the unit. Three major types of life insurance were discussed: the whole life policy, the term policy, and the endowment policy. We calculated the annual premium for regular payments and for m payments, which had the same pattern of deferred premium. The term life policy was discussed next, followed by the endowment policy. A nonannual premium was also discussed and the natural vs. level premium comparison was made. The concepts of reserve and terminal reserve came next in the discussion and their calculation included two methods: retrospective and prospective.

The last topic in the life insurance chapter was how to estimate what is needed to purchase life insurance. Here, four approaches were explained: the sentimental, rational, multiple earnings, and needs approaches.

The last types of insurance discussed in this unit were property and casualty and health care insurance. Two important technical terms were discussed and calculated: actual cash value and replacement value, as well as deductibles, co-insurance, and policy limits.

Mathematical Finance, First Edition. M. J. Alhabeeb.
© 2012 John Wiley & Sons, Inc. Published 2012 by John Wiley & Sons, Inc.

List of Formulas

Mortality table:

$$d_x = l_x - l_{x+1}$$

$$q_x = \frac{d_x}{l_x}$$

$$p_x = \frac{l_{x+1}}{l_x}$$

$$q_x + p_x = 1$$

$$np_x = \frac{l_{x+n}}{l_x}$$

$$nq_x = \frac{1_x - l_{x+n}}{l_x}$$

Commutation terms:

$$D_x = l_x \cdot v^x = \frac{l_x}{(1+r)^x}$$

$$N_x = \sum_{k=x}^{x+100} D_k$$

$$C_x = d_x \cdot v^{x+1}$$

$$M_x = \sum_{k=x}^{x+100} C_k$$

Pure endowment:

$$nE_x = p\left(\frac{l_{x+n}}{l_x}\right)v^n$$

$$nE_x = P\left(\frac{D_{x+n}}{D_x}\right)$$

Mathematical Finance, First Edition. M. J. Alhabeeb.
© 2012 John Wiley & Sons, Inc. Published 2012 by John Wiley & Sons, Inc.

Ordinary whole life annuity:

$$a_x = P\left(\frac{N_{x+1}}{D_x}\right)$$

Whole life annuity due:

$$\ddot{a}_x = P\left(\frac{N_x}{D_x}\right)$$

Deferred whole life annuity:

$$n|a_x = P\left(\frac{N_{x+1+n}}{D_x}\right)$$

$$n|\ddot{a}_x = P\left(\frac{N_{x+n}}{D_x}\right)$$

Temporary life annuity:

$$a_{x:\overline{n}|} = P\left(\frac{N_{x+1} - N_{x+1+n}}{D_x}\right)$$

Temporary life annuity due:

$$\ddot{a}_{x:\overline{n}|} = P\left(\frac{N_x - N_{x+n}}{D_x}\right)$$

Forborne temporary life insurance:

$$nF_x = P\left(\frac{N_x - N_{x+n}}{D_{x+n}}\right)$$

Deferred temporary life annuity:

$$k|\ddot{a}_{x:\overline{n}|} = P\left(\frac{N_{x+k} - N_{x+k+n}}{D_x}\right)$$

Cost of insurance (net single premium):

$$A_x = F\left(\frac{M_x}{D_x}\right)$$

Premium—whole life:

$$P_x = F\left(\frac{M_x}{N_x}\right)$$

Premium—m payments:

$$m P_x = F \left(\frac{M_x}{N_x - N_{x+m}} \right)$$

Deferred whole life policy:

$$n|A_x = F \left(\frac{M_{x+n}}{D_x} \right)$$

Deferred premium—whole life:

$$n|P_x = F \left(\frac{M_{x+n}}{N_x} \right)$$

Deferred premium—m payments:

$$m P(n|A_x) = F \left(\frac{M_{x+n}}{N_x - N_{x+m}} \right)$$

Term life insurance:

$$A^1_{x:\overline{n}|} = F \left(\frac{M_x - M_{x+n}}{D_x} \right)$$

Premium—term life:

$$k P^1_{x:\overline{n}|} = F \left(\frac{M_x - M_{x+n}}{N_x - N_{x+k}} \right)$$

$$n P^1_{x:\overline{n}|} = F \left(\frac{M_x - M_{x+n}}{N_x - N_{x+n}} \right)$$

Endowment insurance:

$$A^1_{x:\overline{n}|} = F \left(\frac{M_x - M_{x+n} + D_{x+n}}{D_x} \right)$$

Premium—endowment:

$$P^1_{x:\overline{n}|} = F \left(\frac{M_x - M_{x+n} + D_{x+n}}{N_x - N_{x+n}} \right)$$

$$k P_{x:\overline{n}|} = F \left(\frac{M_x - M_{x+n} + D_{x+n}}{N_x - N_{x+k}} \right)$$

Nonannual premium:

$$p^{(m)} = \frac{p(1+j)}{m}$$

Natural premium:

$$NA^1_{x:\overline{1}|} = F\left(\frac{C_x}{D_x}\right)$$

Terminal reserve—retrospective:

$$V = \frac{P(N_x - N_{x+t}) - (M_x - M_{x+t})}{D_{x+t}}$$

Terminal reserve—prospective:

$$tV_x = \frac{F(M_{x+t} - M_{x+n}) - P_x(N_{x+t} - N_{x+n})}{D_{x+t}}$$

Needs method:

$$LA = PV(.75Y_d) + OT - [PV(G) + CI]$$

$$LA = \frac{.75Y_d[1 - (1+r)^{-n}]}{r} + OT + \left[\frac{G[1 - (1+r)^{-k}]}{r} + CI\right]$$

$$LA = (75Y_d) \cdot a_{\overline{n}|r} + OT + (G \cdot a_{\overline{n}|r} + CI)$$

Actual cash value:

$$ACV = OC\left(1 - \frac{CA}{LE}\right)$$

Replacement value:

$$RV = OC + \Delta OC$$

$$\Delta OC = OC \cdot f \cdot t$$

$$RV = OC(1 + f \cdot t)$$

Reimbursement—dwelling:

$$R_d = \frac{I}{.80RV}$$

$$R_d = \frac{I}{.80RV}(L - D)$$

Reimbursement—health care:

$$R_h = (1 - CP)(L - D)$$

$$R_n = \sum n(k - CP)$$

Exercises for Unit VIII

1. How probable is it for a person age 35 to live to age 80?

2. What is the probability of a man age 20 dying before his fortieth birthday?

3. What is the present value of the annuities for the cohort of 65 if each person in the cohort has an annuity of $430 per year?

4. What is the present value at age zero for all people who die at age 72 if there is a per person payment of $500?

5. Robert would like his wife to receive a pure endowment of $100,000 when she retires at 55, 15 years from now. How much must he deposit annually if the money is worth 5%?

6. If a person wants to purchase a whole life annuity so that he can be paid $3,600 at the end of each year for the rest of his life, how much of a premium would he have to pay if he is 53 now?

7. Suppose that the person in Exercise 6 increased to $4,000 the amount he wanted to be paid and made it payable at the beginning of the year. How much would his premium be?

8. A person age 37 wishes to set up an ordinary annuity so that she will start to get paid $5,000 a year when she is 47 and throughout the rest of her life. What size premium would she have to pay?

9. Chuck, who is 57, wants an annuity that would pay him $5,000 a year for 15 years. How much would he have to pay for it in a single payment?

10. A temporary life annuity is supposed to pay an annual payment of $4,200 to a young man for a period of 10 years but would not start until 5 years from now. What is the single premium for this annuity?

11. A man wants to buy a whole life insurance policy that pays $150,000 to his wife if he dies. He is 52 now. How much will his single premium be?

12. Calculate the annual premium for a whole life insurance of $177,000 for a 47-year-old woman.

Mathematical Finance, First Edition. M. J. Alhabeeb.
© 2012 John Wiley & Sons, Inc. Published 2012 by John Wiley & Sons, Inc.

13. A 51-year-old man is purchasing a whole life insurance policy of $235,000 but wants to break his single premium down into nine annual payments. How much will each payment be?

14. Sandra, who is 50, would like her $75,000 insurance policy to start paying if she dies 10 years or more later. How much will her single premium be?

15. After finding Sandra's single premium in Exercise 14, calculate the annual premium if Sandra changes her mind about paying a single amount.

16. If Sandra changes her mind toward having the premium paid in only six annual payments, how much will each annual premium be?

17. Blake is 63. He is buying a 10-year term life insurance of $85,000. How much will his premium be?

18. Suppose that Blake wants to pay his premium annually during the policy term. How much will the annual premium be?

19. If Blake decides to break the payment into five annual payments, how much will the annual premium be?

20. Find the cost of a $95,000 endowment policy that has a 10-year term for a 57-year-old woman.

21. What would be the annual premium for a $150,000 endowment policy for 7 years for a 68-year-old man?

22. If an annual premium of a policy is $5,500 and the insurance company charges 8% to allow the insured to pay every 3 months, how much will each payment be?

23. Find the natural premium for an 81-year-old woman who purchases a $50,000 insurance policy.

24. Use the prospective method to find the sixth terminal reserve for an 8-year endowment policy of $2,000 issued to a 30-year-old woman.

25. Take the values from Table 2.2 and use the formula to calculate the terminal reserve for the 11th year of the policy, which should match or be very close to $2,232.34.

26. Use the needs approach method to calculate how much life insurance is needed for Jackie, who receives $50,000 gross income, and pays $8,000 in taxes and deductions. She wants the life insurance policy to cover $50,000 in education expenses for her daughter, $20,000 to pay off the remaining mortgage balance, $10,000 for readjustment expenses for her husband and daughter, and $7,000 in final expenses. The Social Security benefit will be $2,500 a month for 12 years, and she has existing insurance of $25,000. Consider an interest rate of 6%.

27. How much would the insurance reimbursement be for a television system damaged by fire if it was purchased 2 years ago for $6,000 and has a life expectancy of 5 years?

28. What is the replacement value of the television system in Exercise 27 if the inflation rate has been steady at 4%?

29. A storm causes $25,000 in damages to a house that is insured for $180,000; the replacement value of the house is estimated at $250,000. The policy requires a $1,500 deductible.

30. Calculate the health insurance reimbursement for a man who had a car accident and had to spend 10 days in the hospital at a daily cost of $750, with additional physician lab, and ambulance service costs of $2,315. His insurance copayment is $85.15, and his deductible is $560.

References

Aczel, A. (1989). *Complete Business Statistics*. Richard D. Irwin, Homewood, IL.

Bell, C., and L. Adams (1949). *Mathematics of Finance*. Henry Holt, New York.

Bliss, E. (1989). *College Mathematics for Business*. Prentice Hall, Englewood Cliffs, NJ.

Brealey, R., and S. Myers (2003). *Principles of Corporate Finance*. McGraw-Hill, New York.

Brigham, E., and J. Houston (2003). *Fundamentals of Financial Management*. South-Western, Cincinnati, OH.

Cissell, R., H. Cissel, and D. Flaspohler (1990). *Mathematics of Finance*, 8th ed. Houghton Mifflin, Boston.

Copeland, T., and J. Weston (2004). *Financial Theory and Corporate Policy*, 4th ed. Addison-Wesley, Reading, MA.

Cox, D., and M. Cox (2006). *The Mathematics of Banking and Finance*. Wiley, Hoboken, NJ.

Dean, B., M. Sasieni, and S. Gupta (1978). *Mathematics for Modern Management*. R. E. Krieger, Melbourne, FL.

Dowling, E. (1980). *Mathematics for Economists*. Schaum's Outline Series. McGraw-Hill, New York.

Federer Vaaler, L., and J. Daniel (2007). *Mathematical Interest Theory*, 2nd ed. American Mathematical Society, Providence, RI.

Garman, E. T., J. J. Xiao, and B. Branson (2000). *The Mathematics of Personal Financial Planning*. Dame Publications, Cincinnati, OH.

Gitman, L. (2007). *Principles of Managerial Finance*. Addison-Wesley, Reading, MA.

Goodman, V. (2009). *The Mathematics of Finance*. American Mathematical Society, Providence, RI.

Guthrie, G. L., and L. Lemon (2004). *Mathematics of Interest Rates and Finance*. Pearson, Upper Saddle River, NJ.

Hanke, J., and A. Reitsch (1994). *Understanding Business Statistics*. Richard D. Irwin, Homewood, IL.

Johnson, R. (1986). *The Mathematics of Finance: Applied Present Value Concepts*, 2nd ed. Kendall/Hunt, Dubugue, IA.

Joshi, M. (2008). *The Concepts and Practice of Mathematical Finance*. Cambridge University Press, New York.

Mathematical Finance, First Edition. M. J. Alhabeeb.

© 2012 John Wiley & Sons, Inc. Published 2012 by John Wiley & Sons, Inc.

Kellison, S. (1991). *Theory of Interest*. Richard D. Irwin, Homewood, IL.

Kornegay, C. (1999). *Mathematical Dictionary*. Sage Publications, Thousand Oaks, CA.

Mavron, V., and T. Phillips (2000). *Elements of Mathematics for Economics and Finance*. Springer-Verlag, New York.

Mendenhall, W., J. Reinmath, R. Beaver, and D. Duham (1982). *Statistics for Management and Economics*. Duxbury Press, Belmont, CA.

Muksian, R. (2003). *Mathematics of Interest Rates, Insurance, Social Security, and Pensions*. Prentice Hall, Saddle River, NJ.

Parmenter, M. (1999). *Theory of Interest and Life Contingencies*. ACTEX Publications, Winsted, CT.

Reilly, F. K. (1989). *Investment Analysis and Portfolio Management*, 3rd ed. Dryden Press, Hinsdale, IL.

Roman, S. (2004). *Introduction to the Mathematics of Finance*. Springer-Verlag, New York.

Scalzo, F. (1979). *Mathematics for Business and Economics*. Petrocelli Books, New York.

Shao, S., and L. Shao (1998). *Mathematics for Management and Finance*. South-Western, Cincinnati, OH.

Thomself, M. (1989). *The Mathematics of Investing*. Wiley, New York.

Van Matre, J., and G. Gilbreath (1980). *Statistics for Business and Economics*. Richard D. Irwin, Homewood, IL.

Vogt, W. (1999). *Dictionary of Statistics and Methodology*. Sage Publications, Thousand Oaks, CA.

Williams, R. (2006). *Introduction to the Mathematics of Finance*. American Mathematical Society, Providence, RI.

Zima, P., and R. Brown (2001). *Mathematics of Finance*. McGraw-Hill, New York.

Appendix

TABLE 1 Common Aliquot $\left(\frac{1}{100}\%\right)$ and Equivalent Values

Denominator of Common Fraction	Numerator of Common Fraction												
	1	2	3	4	5	6	7	8	9	10	11	12	13
	Unit: $\left(\frac{1}{100}\right)$ or %												
2	50												
3	$33\frac{1}{3}$	$66\frac{2}{3}$											
4	25		75										
5	20	40	60	80									
6	$16\frac{2}{3}$				$83\frac{1}{3}$								
7	$14\frac{2}{7}$	$28\frac{4}{7}$	$42\frac{6}{7}$	$57\frac{1}{7}$	$71\frac{3}{7}$	$85\frac{5}{7}$							
8	$12\frac{1}{2}$		$37\frac{1}{2}$		$62\frac{1}{2}$		$87\frac{1}{2}$						
9	$11\frac{1}{9}$	$22\frac{2}{9}$		$44\frac{4}{9}$	$55\frac{5}{9}$		$77\frac{7}{9}$	$88\frac{8}{9}$					
10	10		30				70		90				
11	$9\frac{1}{11}$	$18\frac{2}{11}$	$27\frac{3}{11}$	$36\frac{4}{11}$	$45\frac{5}{11}$	$54\frac{6}{11}$	$63\frac{7}{11}$	$72\frac{8}{11}$	$81\frac{9}{11}$	$90\frac{10}{11}$			
12	$8\frac{1}{3}$				$43\frac{2}{3}$		$58\frac{1}{3}$				$91\frac{2}{3}$		
13	$7\frac{9}{13}$	$15\frac{5}{13}$	$23\frac{1}{13}$	$30\frac{10}{13}$	$38\frac{6}{13}$	$46\frac{2}{13}$	$53\frac{11}{13}$	$61\frac{7}{13}$	$69\frac{3}{13}$	$76\frac{12}{13}$	$84\frac{8}{13}$	$92\frac{4}{13}$	
14	$7\frac{1}{7}$		$21\frac{3}{7}$		$35\frac{5}{7}$				$64\frac{2}{7}$		$78\frac{4}{7}$		$92\frac{6}{7}$
15	$6\frac{2}{3}$	$13\frac{1}{3}$		$26\frac{2}{3}$			$46\frac{2}{3}$	$53\frac{1}{3}$			$73\frac{1}{3}$		$86\frac{2}{3}$
16	$6\frac{1}{4}$		$18\frac{3}{4}$		$31\frac{1}{4}$		$43\frac{3}{4}$		$56\frac{1}{4}$		$68\frac{3}{4}$		$81\frac{1}{4}$
20	5		15				35		45		55		65

Source: S. Shao and L. Shao (1998). *Mathematics for Management and Finance.* South-Western, Cincinnati, OH.

Mathematical Finance, First Edition. M. J. Alhabeeb.
© 2012 John Wiley & Sons, Inc. Published 2012 by John Wiley & Sons, Inc.

TABLE 2 **Four-Place Common Logarithms**

N	0	1	2	3	4	5	6	7	8	9
10	0000	0043	0086	0128	0170	0212	0253	0924	0334	0374
11	0414	0453	0492	0531	0569	0607	0645	0682	0719	0755
12	0792	0828	0864	0899	0934	0969	1004	1038	1072	1106
13	1139	1173	1206	1239	1271	1303	1335	1367	1399	1430
14	1461	1492	1523	1553	1584	1614	1644	1673	1703	1732
15	1761	1790	1818	1847	1875	1903	1931	1959	1987	2014
16	2041	2068	2095	2122	2148	2175	2201	2227	2253	2279
17	2304	2330	2355	2380	2405	2430	2455	2480	2504	2529
18	2553	2577	2601	2625	2648	2672	2695	2718	2742	2765
19	2788	2810	2833	2856	2878	2900	2923	2945	2967	2989
20	3010	3032	3054	3075	3076	3118	3139	3160	3181	3201
21	3222	3243	3263	3284	3304	3324	3345	3365	3385	3404
22	3424	3444	3464	3483	3502	3522	3541	3560	3579	3598
23	3617	3636	3655	3674	3692	3711	3729	3747	3766	3784
24	3802	3820	3838	3856	3874	3892	3909	3927	3945	3962
25	3979	3997	4014	4031	4048	4065	4082	4099	4116	4133
26	4150	4166	4183	4200	4216	4232	4249	4265	4281	4298
27	4314	4330	4346	4362	4378	4393	4409	4425	4440	4456
28	4472	4487	4502	4518	4533	4548	4564	4579	4594	4609
29	4624	4639	4654	4669	4683	4698	4713	4728	4742	4757
30	4771	4786	4800	4814	4829	4843	4857	4871	4886	4900
31	4914	4928	4942	4955	4969	4983	4997	5011	5024	5038
32	5051	5065	5079	5092	5105	5119	5132	5145	5159	5172
33	5185	5198	5211	5224	5237	5250	5263	5276	5289	5302
34	5315	5328	5340	5353	5366	5378	5391	5403	5416	5428
35	5441	5453	5465	5478	5490	5502	5514	5527	5539	5551
36	5563	5575	5587	5599	5611	5623	5635	5647	5658	5670
37	5682	5694	5705	5717	5729	5740	5752	5763	5775	5786
38	5798	5809	5821	5832	5843	5855	5866	5877	5888	5899
39	5911	5922	5933	5944	5955	5966	5977	5988	5999	6010
40	6021	6031	6042	6053	6064	6075	6085	6096	6107	6117
41	6128	6138	6149	6160	6170	6180	6191	6201	6212	6222
42	6232	6243	6253	6263	6374	6284	6294	6304	6314	6325
43	6335	6345	6355	6365	6375	6385	6395	6405	6415	6425
44	6435	6444	6454	6464	6474	6484	9493	6503	6513	6522
45	6532	6542	6551	6461	6571	6580	6590	6599	6609	6618
46	6628	6637	6646	6656	6665	6675	6684	6693	6702	6712
47	6721	6730	6739	6749	6758	6767	6776	6785	6794	6803
48	6812	6821	6830	6839	6848	6857	6866	6875	6884	6893
49	6902	6911	6920	6928	6937	6946	6955	6964	6972	6981
50	6990	6998	7007	7016	7024	7033	7042	7050	7059	7067
51	7076	7084	7093	7101	7110	7118	7126	7135	7143	7152
52	7160	7168	7177	7185	7193	7202	7210	7218	7226	7235
53	7243	7251	7259	7267	7275	7284	7292	7300	7308	7316
54	7324	7332	7340	7348	7356	7364	7372	7380	7388	7396
55	7404	7412	7419	7427	7435	7443	7451	7459	7466	7474

TABLE 2 (*Continued*)

N	0	1	2	3	4	5	6	7	8	9
56	7482	7490	7497	7505	7513	7520	7528	7536	7543	7551
57	7559	7566	7574	7582	7589	7597	7604	7612	7619	7627
58	7634	7642	7649	7657	7664	7672	7679	7686	7694	7701
59	7709	7716	7723	7731	7738	7745	7752	7760	7767	7774
60	7782	7789	7796	7803	7810	7818	7825	7832	7839	7846
61	7853	7860	7868	7875	7882	7889	7896	7903	7910	7917
62	7924	7931	7938	7945	7952	7959	7966	7973	7980	7987
63	7993	8000	8007	8014	8021	8028	8035	8041	8048	8055
64	8062	8069	8075	8082	8089	8096	8102	8109	8116	8122
65	8129	8136	8142	8149	8156	8162	8169	8176	8182	8189
66	8195	8202	8209	8215	8222	8228	8235	8241	8248	8254
67	8261	8267	8274	8280	8287	8293	8299	8306	8312	8319
68	8325	8331	8338	8344	8351	8357	8363	8370	8376	8382
69	8388	8395	8401	8407	8414	8420	8426	8432	8439	8445
70	8451	8457	8463	8470	8476	8482	8488	8494	8500	8506
71	8513	8519	8525	8531	8537	8543	8549	8555	8561	8567
72	8573	8579	8585	8591	8597	8603	8609	8615	8621	8627
73	8633	8639	8645	8651	8657	8663	8669	8675	8681	8686
74	8692	8698	8704	8710	8716	8722	8727	8733	8739	8745
75	8751	8756	8762	8768	8774	8779	8785	8791	8797	8802
76	8808	8814	8820	8825	8831	8837	8842	8848	8854	8859
77	8865	8871	8878	8882	8887	8893	8899	8904	8910	8915
78	8921	8927	8932	8938	8943	8949	8954	8960	8965	8971
79	8976	8982	8987	8993	8998	9004	9009	9015	9020	9025
80	9031	9036	9042	9047	9053	9058	9063	9069	9074	9079
81	9085	9090	9096	9101	9106	9112	9117	9122	9128	9133
82	9138	9143	9149	9154	9159	9165	9170	9175	9180	9186
83	9191	9196	9201	9206	9212	9217	9222	9227	9232	9238
84	9243	9248	9253	9258	9263	9269	9274	9279	9284	9289
85	9294	9299	9304	9309	9315	9320	9325	9330	9335	9340
86	9345	9350	9355	9360	9365	9370	9375	9380	9385	9390
87	9395	9400	9405	9410	9415	9420	9425	9430	9435	9440
88	9445	9450	9455	9460	9465	9469	9474	9479	9484	9489
89	9494	9499	9504	9509	9513	9518	9523	9528	9533	9538
90	9542	9547	9552	9557	9562	9566	9571	9576	9581	9586
91	9590	9595	9600	9605	9609	9614	9619	9624	9628	9633
92	9638	9643	9647	9652	9657	9661	9666	9671	9675	9680
93	9685	9689	9694	9699	9703	9708	9713	9717	9722	9727
94	9731	9736	9741	9745	9750	9754	9759	9763	9768	9773
95	9777	9782	9786	9791	9795	9800	9805	9809	9814	9818
96	9823	9827	9832	9836	9841	9845	9850	9854	9859	9863
97	9868	9872	9877	9881	9886	9890	9894	9899	9903	9908
98	9912	9917	9921	9926	9930	9934	9939	9943	9948	9952
99	9956	9961	9965	9969	9974	9978	9983	9987	9991	9996

Source: E. T. Dowling (1980). *Mathematics for Economists*. Schaum's Outline Series. McGraw-Hill, New York.

TABLE 3   Selected Values of e^x and e^{-x}

x	e^x	e^{-x}	x	e^x	e^{-x}
0.0	1.000	1.000	5.0	148.4	0.0067
0.1	1.105	0.905	5.1	164.0	0.0061
0.2	1.221	0.819	5.2	181.3	0.0055
0.3	1.350	0.741	5.3	200.3	0.0050
0.4	1.492	0.670	5.4	221.4	0.0045
0.5	1.649	0.607	5.5	244.7	0.0041
0.6	1.822	0.549	5.6	270.4	0.0037
0.7	2.014	0.497	5.7	298.9	0.0033
0.8	2.226	0.449	5.8	330.3	0.0030
0.9	2.460	0.407	5.9	365.0	0.0027
1.0	2.718	0.368	6.0	403.4	0.0025
1.1	3.004	0.333	6.1	445.9	0.0022
1.2	3.320	0.301	6.2	492.8	0.0020
1.3	3.669	0.273	6.3	544.6	0.0018
1.4	4.055	0.247	6.4	601.8	0.0017
1.5	4.482	0.223	6.5	665.1	0.0015
1.6	4.953	0.202	6.6	736.1	0.0014
1.7	5.474	.0183	6.7	812.4	0.0012
1.8	6.050	.0165	6.8	897.8	0.0011
1.9	6.686	.0150	6.9	992.3	0.0010
2.0	7.389	0.135	7.0	1,096.6	0.0009
2.1	8.166	0.122	7.1	1,212.0	0.0008
2.2	9.025	0.111	7.2	1,339.4	0.0007
2.3	9.974	0.100	7.3	1,480.3	0.0007
2.4	11.023	0.091	7.4	1,636.0	0.0006
2.5	12.18	0.082	7.5	1,808.0	0.00055
2.6	13.46	0.074	7.6	1,998.2	0.00050
2.7	14.88	0.067	7.7	2,208.3	0.00045
2.8	16.44	0.061	7.8	2,440.6	0.00041
2.9	18.17	0.055	7.9	2,697.3	0.00037
3.0	20.09	0.050	8.0	2,981.0	0.00034
3.1	22.20	0.045	8.1	3,294.5	0.00030
3.2	24.53	0.041	8.2	3,641.0	0.00027
3.3	27.11	0.037	8.3	4,023.9	0.00025
3.4	29.96	0.033	8.4	4,447.1	0.00022
3.5	33.12	0.030	8.5	4,914.8	0.00020
3.6	36.60	0.027	8.6	5,431.7	0.00018
3.7	40.45	0.025	8.7	6,002.9	0.00017
3.8	44.70	0.022	8.8	6,634.2	0.00015
3.9	49.40	0.020	8.9	7,3332.0	0.00014
4.0	54.60	0.018	9.0	8,103.1	0.00012
4.1	60.34	0.017	9.1	8,955.3	0.00011
4.2	66.69	0.015	9.2	9,897.1	0.00010
4.3	73.70	0.014	9.3	10,938.0	0.00009
4.4	81.45	0.012	9.4	12,088.0	0.00008
4.5	90.02	0.011	9.5	13,360.0	0.00007
4.6	99.48	0.010	9.6	14,765.0	0.00007
4.7	109.95	0.009	9.7	16,318.0	0.00006
4.8	121.51	0.008	9.8	18,034.0	0.00006
4.9	134.29	0.007	9.9	19,930.0	0.00005

TABLE 4   Serial Table of the Number of Each Day of the Year

Days	Jan.	Feb.	Mar.	Apr.	May	June	July	Aug.	Sept.	Oct.	Nov.	Dec.	Days
1	1	32	60	91	121	152	182	213	244	274	305	335	1
2	2	33	61	92	122	153	183	214	245	275	306	336	2
3	3	34	62	93	123	154	184	215	246	276	307	337	3
4	4	35	63	94	124	155	185	216	247	277	308	338	4
5	5	36	64	95	125	156	186	217	248	278	309	339	5
6	6	37	65	96	126	157	187	218	249	279	310	340	6
7	7	38	66	97	127	158	188	219	250	280	311	341	7
8	8	39	67	98	128	159	189	220	251	281	312	342	8
9	9	40	68	99	129	160	190	221	252	282	313	343	9
10	10	41	69	100	130	161	191	222	253	283	314	344	10
11	11	42	70	101	131	162	192	223	254	284	315	345	11
12	12	43	71	102	132	163	193	224	255	285	316	346	12
13	13	44	72	103	133	164	194	225	256	286	317	347	13
14	14	45	73	104	134	165	195	226	257	287	318	348	14
15	15	46	74	105	135	166	196	227	258	288	319	349	15
16	16	47	75	106	136	167	197	228	259	289	320	350	16
17	17	48	76	107	137	168	198	229	260	290	321	351	17
18	18	49	77	108	138	169	199	230	261	291	322	352	18
19	19	50	78	109	139	170	200	231	262	292	323	353	19
20	20	51	79	110	140	171	201	232	263	293	324	354	20
21	21	52	80	111	141	172	202	233	264	294	325	355	21
22	22	53	81	112	142	173	203	234	265	295	326	356	22
23	23	54	82	113	143	174	204	235	266	296	327	357	23
24	24	55	83	114	144	175	205	236	267	297	328	358	24
25	25	56	84	115	145	176	206	237	268	298	329	359	25
26	26	57	85	116	146	177	207	238	269	299	330	360	26
27	27	58	86	117	147	178	208	239	270	300	331	361	27
28	28	59	87	118	148	179	209	240	271	301	332	362	28
29	29	*a*	88	119	149	180	210	241	272	302	333	363	29
30	30		89	120	150	181	211	242	273	303	334	364	30
31	31		90		151		212	243		304		365	31

[a]For leap years the number of the day after February 28 is 1 greater than the number given in the table. Leap years: 2012, 2016, 2020, 2024, etc.

TABLE 5 s-Value, $s = (1 + r)^n$ Where the Compound Amount for a CV = \$1.00

	Annual Interest Rate, r				
n (years)	1%	$1\frac{1}{2}$%	2%	$2\frac{1}{2}$%	3%
1	1.0100 00	1.0150 00	1.0200 00	1.0250 00	1.0300 00
2	1.0202 00	1.0302 25	1.0404 00	1.0506 25	1.0609 00
3	1.0303 01	1.0456 78	1.0612 08	1.0768 91	1.0927 27
4	1.0406 04	1.0613 64	1.0824 32	1.1038 13	1.1255 09
5	1.0510 10	1.0772 84	1.1040 81	1.1314 08	1.1592 74
6	1.0615 20	1.0934 43	1.1261 62	1.1596 93	1.1940 52
7	1.0721 35	1.1098 45	1.1486 86	1.1886 86	1.2298 74
8	1.0828 57	1.1264 93	1.1716 59	1.2184 03	1.2667 70
9	1.0936 85	1.1433 90	1.1950 93	1.2488 63	1.3047 73
10	1.1046 22	1.1605 41	1.2189 94	1.2800 85	1.3439 16
11	1.1156 68	1.1779 49	1.2433 74	1.3120 87	1.3842 34
12	1.1268 25	1.1956 18	1.2682 42	1.3448 89	1.4257 61
13	1.1380 93	1.2135 52	1.2936 07	1.3785 11	1.4685 34
14	1.1494 74	1.2317 56	1.3194 79	1.4129 74	1.5125 90
15	1.1609 69	1.2502 32	1.3458 68	1.4482 98	1.5579 67
16	1.1725 79	1.2689 86	1.3727 86	1.4845 06	1.6047 06
17	1.1843 04	1.2880 20	1.4002 41	1.5216 18	1.6528 48
18	1.1961 47	1.3073 41	1.4282 46	1.5596 59	1.7024 33
19	1.2081 09	1.3269 51	1.4568 11	1.5986 50	1.7535 06
20	1.2201 90	1.3468 55	1.4859 47	1.6386 16	1.8061 11
21	1.2323 92	1.3670 58	1.5156 66	1.6795 82	1.8602 95
22	1.2447 16	1.3875 64	1.5459 80	1.7215 71	1.9161 03
23	1.2571 63	1.4083 77	1.5768 99	1.7646 11	1.9735 87
24	1.2697 35	1.4295 03	1.6084 37	1.8087 26	2.0327 94
25	1.2824 32	1.4509 45	1.6406 06	1.8539 44	2.0937 78
26	1.2952 56	1.4727 10	1.6734 18	1.9002 93	2.1565 91
27	1.3082 09	1.4948 00	1.7068 86	1.9478 00	2.2212 89
28	1.3212 91	1.5172 22	1.7410 24	1.9964 95	2.2879 28
29	1.3345 04	1.5399 81	1.7758 45	2.0464 07	2.3565 66
30	1.1478 49	1.5630 80	1.8113 62	2.0975 68	2.4272 62

TABLE 5   (*Continued*)

n (years)	Annual Interest Rate, r				
	$3\frac{1}{2}\%$	4%	$4\frac{1}{2}\%$	5%	$5\frac{1}{2}\%$
1	1.0350 00	1.0400 00	1.0450 00	1.0500 00	1.0550 00
2	1.0712 25	1.0816 00	1.0920 25	1.1025 00	1.1130 25
3	1.1087 18	1.1248 64	1.1411 66	1.1576 25	1.1742 41
4	1.1475 23	1.1698 59	1.1925 19	1.2155 06	1.2388 25
5	1.1876 86	1.2166 53	1.2461 82	1.2762 82	1.3069 60
6	1.2292 55	1.2653 19	1.3022 60	1.3400 96	1.3788 43
7	1.2722 79	1.3159 32	1.3608 62	1.4071 00	1.4546 79
8	1.3168 09	1.3685 69	1.4221 01	1.4774 55	1.5346 87
9	1.3628 97	1.4233 12	1.4860 95	1.5513 28	1.6190 94
10	1.4105 99	1.4802 44	1.5529 69	1.6288 95	1.7081 44
11	1.4599 70	1.5394 54	1.6228 53	1.7103 39	1.8020 92
12	1.5110 69	1.6010 32	1.6958 81	1.7958 56	1.9012 07
13	1.5639 56	1.6650 74	1.7721 96	1.8856 49	2.0057 74
14	1.6186 95	1.7316 76	1.8519 45	1.9799 32	2.1160 91
15	1.6753 49	1.8009 44	1.9352 82	2.0789 28	2.2324 76
16	1.7339 86	1.8729 81	2.0223 70	2.1828 75	2.3552 63
17	1.7946 76	1.9479 01	2.1133 77	2.2920 18	2.4848 02
18	1.8574 89	2.0258 17	2.2084 79	2.4066 19	2.6214 66
19	1.9225 01	2.1068 49	2.3078 60	2.5269 50	2.7656 47
20	1.9897 89	2.1911 23	2.4117 14	2.6532 98	2.9177 57
21	2.0594 31	2.2787 68	2.5202 41	2.7859 63	3.0782 34
22	2.1315 12	2.3699 19	2.6336 52	2.9252 61	3.2475 37
23	2.2061 14	2.4647 16	2.7521 66	3.0715 24	3.4261 52
24	2.2833 28	2.5633 04	2.8760 14	3.2251 00	3.6145 90
25	2.3632 45	2.6658 36	3.0054 34	3.3863 55	3.8133 92
26	2.4459 59	2.7724 70	3.1406 79	3.5556 73	4.0231 29
27	2.5315 67	2.8833 69	3.2820 10	3.7334 56	4.2444 01
28	2.6201 72	2.9987 03	3.4297 00	3.9201 29	4.4778 43
29	2.7118 78	3.1186 51	3.5840 36	4.1161 36	4.7241 24
30	2.8067 94	3.2433 98	3.7453 18	4.3219 42	4.9839 51

TABLE 5 (*Continued*)

n (years)	6%	$6\frac{1}{2}\%$	7%	$7\frac{1}{2}\%$	8%
			Annual Interest Rate, r		
1	1.0600 00	1.0650 00	1.0700 00	1.0750 00	1.0800 00
2	1.1236 00	1.1342 25	1.1449 00	1.1556 25	1.1664 00
3	1.1910 16	1.2079 50	1.2250 43	1.2422 97	1.2597 12
4	1.2624 77	1.2864 66	1.3107 96	1.3354 69	1.3604 89
5	1.3382 26	1.3700 87	1.4025 52	1.4356 29	1.4693 28
6	1.4185 19	1.4591 42	1.5007 30	1.5433 02	1.5868 74
7	1.5036 30	1.5539 87	1.6057 81	1.6590 49	1.7138 24
8	1.5938 48	1.6549 96	1.7181 86	1.7834 78	1.8509 30
9	1.6894 79	1.7625 70	1.8384 59	1.9172 39	1.9990 05
10	1.7908 48	1.8771 37	1.9671 51	2.0610 32	2.1589 25
11	1.8982 99	1.9991 51	2.1048 52	2.2156 09	2.3316 39
12	2.0121 96	2.1290 96	2.2521 92	2.3817 80	2.5181 70
13	2.1329 28	2.2674 88	2.4098 45	2.5604 13	2.7196 24
14	2.2609 04	2.4148 74	5.5785 34	2.7524 44	2.9371 94
15	2.3965 58	2.5718 41	2.7590 31	2.9588 77	3.1721 69
16	2.5403 52	2.7390 11	2.9521 64	3.1807 93	3.4259 43
17	2.6927 73	2.9170 46	3.1588 15	3.4193 53	3.7000 18
18	2.8543 39	3.1066 54	3.3799 32	3.6758 04	3.9960 20
19	3.0256 00	3.3085 87	3.6165 28	3.9514 89	4.3157 01
20	3.2071 35	3.5236 45	3.8696 84	4.2478 51	4.6609 57
21	3.3995 64	3.7526 82	4.1405 62	4.5664 40	5.0338 34
22	3.6035 37	3.9966 06	4.4304 02	4.9089 23	5.4365 40
23	3.8197 50	4.2563 86	4.7405 30	5.2770 92	5.8714 64
24	4.0489 35	4.5330 51	5.0723 67	5.6728 74	6.3411 81
25	4.2918 71	4.8276 99	5.4274 33	6.0983 40	6.8484 75
26	4.5493 83	5.1415 00	5.8073 53	6.5557 15	7.3963 53
27	4.8223 46	5.4756 97	6.2138 68	7.0473 94	7.9880 61
28	5.1116 87	5.8316 17	6.6488 38	7.5759 48	8.6271 06
29	5.4183 88	6.2106 72	7.1142 57	8.1441 44	9.3172 75
30	5.7434 91	6.6143 66	7.6122 55	8.7549 55	10.0626 57

TABLE 5   *(Continued)*

n (years)	Annual Interest Rate, r			
	$8\frac{1}{2}\%$	9%	$9\frac{1}{2}\%$	10%
1	1.0850 00	1.0900 00	1.0950 00	1.1000 00
2	1.1772 25	1.1881 00	1.1990 25	1.2100 00
3	1.2772 89	1.2950 29	1.3129 32	1.3310 00
4	1.3858 59	1.4115 82	1.4376 61	1.4641 00
5	1.5036 57	1.5386 24	1.5742 39	1.6105 10
6	1.6314 68	2.6771 00	1.7237 91	1.7715 61
7	1.7701 42	1.8280 39	1.8875 52	1.9487 17
8	1.9206 04	1.9925 63	2.0668 69	2.1435 89
9	2.0838 56	2.1718 93	2.2632 21	2.3579 48
10	2.2609 83	2.3673 64	2.4782 28	2.5937 42
11	2.4531 67	2.5804 26	2.7136 59	2.8531 17
12	2.6616 86	2.8126 65	2.9714 57	3.1384 28
13	2.8879 30	3.0658 05	3.2537 45	3.4522 71
15	3.1334 04	3.3417 27	3.5628 51	3.7974 98
15	3.3997 43	3.6424 82	3.9013 22	4.1772 48
16	3.6887 21	3.9703 06	4.2719 48	4.5949 73
17	4.0022 62	4.3276 33	4.6777 83	5.0544 70
18	4.3424 55	4.7171 20	5.1221 72	5.5599 17
19	4.7115 63	5.1416 61	5.6087 78	6.1159 09
20	5.1120 46	5.6044 11	6.1416 12	6.7275 00
21	5.5465 70	6.1088 08	6.7250 65	7.4002 50
22	6.0180 29	6.6586 00	7.3639 46	8.1402 75
23	6.5295 61	7.2578 74	8.0635 21	8.9543 02
24	7.0845 74	7.9110 83	8.8295 56	9.8497 33
25	7.6867 62	8.6230 81	9.6683 64	10.8347 06
26	8.3401 37	9.3991 58	10.5868 58	11.9181 77
27	9.0490 49	10.2450 82	11.5926 10	13.1099 94
28	9.8182 18	11.1671 40	12.6939 08	14.4209 94
29	10.6527 66	12.1721 82	13.8998 29	15.8630 93
30	11.5582 52	13.2676 78	15.2203 13	17.4494 02

TABLE 6 Value of $v^n(1 + r)^{-n}$ = PVIF for a Compound Amount of $1.00

	Interest Rate, r				
n (years)	1%	$1\frac{1}{2}\%$	2%	$2\frac{1}{2}\%$	3%
1	0.9900 99	0.9852 22	0.9803 92	0.9756 10	0.9708 74
2	0.9802 96	0.9706 62	0.9611 69	0.9518 14	0.9425 96
3	0.9705 90	0.9563 17	0.9423 22	0.9285 99	0.9151 42
4	0.9609 80	0.9421 84	0.9238 45	0.9059 51	0.8884 87
5	0.9514 66	0.9282 60	0.9057 31	0.8838 54	0.8626 09
6	0.9420 45	0.9145 42	0.8879 71	0.8622 97	0.8374 84
7	0.9327 18	0.9010 27	0.8705 60	0.8412 65	0.8130 92
8	0.9234 83	0.8877 11	0.8534 90	0.8207 47	0.7894 09
9	0.9143 40	0.8745 92	0.8367 55	0.8007 28	0.7664 17
10	0.9052 87	0.8616 67	0.8203 48	0.7811 98	0.7440 94
11	0.8963 24	0.8489 33	0.8042 63	0.7621 45	0.7224 21
12	0.8874 49	0.8363 87	0.7884 93	0.7435 56	0.7013 80
13	0.8786 63	0.8240 27	0.7730 32	0.7254 20	0.6809 51
14	0.8699 63	0.8118 49	0.7578 75	0.7077 27	0.6611 18
15	0.8613 49	0.7998 51	0.7430 15	0.6904 66	0.6418 62
16	0.8528 21	0.7880 31	0.7284 46	0.6736 25	0.6231 67
17	0.8443 77	0.7763 85	0.7141 63	0.6571 95	0.6050 16
18	0.8360 17	0.7649 12	0.7001 59	0.6411 66	0.5873 95
19	0.8277 40	0.7536 07	0.6864 31	0.6255 28	0.5702 86
20	0.8195 44	0.7424 70	0.6729 71	0.6102.71	0.5536 76
21	0.8114 30	0.7314 98	0.6597 76	0.5953 86	0.5375 49
22	0.8033 96	0.7206 88	0.6468 39	0.5808 65	0.5218 93
23	0.7954 42	0.7100 37	0.6341 56	0.5666 97	0.5066 92
24	0.7875 66	0.6995 44	0.6271 21	0.5528 75	0.4919 34
25	0.7797 68	0.6892 06	0.6095 31	0.5393 91	0.4776 06
26	0.7720 48	0.6790 21	0.5975 79	0.5262 35	0.4636 95
27	0.7644 04	0.6689 86	0.5858 62	0.5134 00	0.4501 89
28	0.7568 36	0.6590 99	0.5743 75	0.5008 78	0.4370 77
29	0.7493 42	0.6493 59	0.5631 12	0.4886 61	0.4243 46
30	0.7419 23	0.6397 62	0.5520 71	0.4767 43	0.4119 87
31	0.7345 77	0.6303 08	0.5412 46	0.4651 15	0.3999 87
32	0.7273 04	0.6209 93	0.5306 33	0.4537 71	0.3883 37
33	0.7201 03	0.6118 16	0.5202 29	0.4427 03	0.3770 26
34	0.7129 73	0.6027 74	0.5100 28	0.4319 05	0.3660 45
35	0.7059 14	0.5938 66	0.5000 28	0.4213 71	0.3553 83
36	0.6989 25	0.5850 90	0.4902 23	0.4110 94	0.3450 32
37	0.6920 05	0.5764 43	0.4806 11	0.4010 67	0.3349 83
38	0.6851 53	0.5679 24	0.4711 87	0.3912 85	0.3252 26
39	0.6783 70	0.5595 31	0.4619 48	0.3817 41	0.3157 54
40	0.6716 53	0.5512 62	0.4528 90	0.3724 31	0.3065 57

TABLE 6   (*Continued*)

n (years)	Interest Rate, r				
	$3\frac{1}{2}\%$	4%	$4\frac{1}{2}\%$	5%	$5\frac{1}{2}\%$
1	0.9661 84	0.9615 38	0.9569 38	0.9523 81	0.9478 67
2	0.9335 11	0.9245 56	0.9157 30	0.9070 29	0.8984 52
3	0.9019 43	0.8889 96	0.8762 97	0.8638 38	0.8516 14
4	0.8714 42	0.8548 04	0.8385 61	0.8227 02	0.8072 17
5	0.8419 73	0.8219 27	0.8024 51	0.7835 26	0.7651 34
6	0.8135 01	0.7903 15	0.7678 96	0.7462 15	0.7252 46
7	0.7859 91	0.7599 18	0.7348 28	0.7106 81	0.6874 37
8	0.7594 12	0.7306 90	0.7031 85	0.6768 39	0.6515 99
9	0.7337 31	0.7025 87	0.6729 04	0.6446 09	0.6176 29
10	0.7089 19	0.6755 64	0.6439 28	0.6139 13	0.5854 31
11	0.6849 46	0.6495 81	0.6161 99	0.5846 79	0.5549 11
12	0.6617 83	0.6245 97	0.5896 64	0.5568 37	0.5259 82
13	0.6394 04	0.6005 74	0.5642 72	0.5303 21	0.4985 61
14	0.6177 82	0.5774 75	0.5399 73	0.5050 68	0.4725 69
15	0.5968 91	0.5552 65	0.5167 20	0.4810 17	0.4479 33
16	0.5767 06	0.5339 08	0.4944 69	0.4581 12	0.4245 81
17	0.5572 04	0.5133 73	0.4731 76	0.4362 97	0.4024 47
18	0.5383 61	0.4936 28	0.4528 00	0.4155 21	0.3814 66
19	0.5201 56	0.4746 42	0.4333 02	0.3957 34	0.3615 79
20	0.5025 66	0.4563 87	0.4146 43	0.3768 89	0.3427 29
21	0.4855 71	0.4388 34	0.3967 87	0.3589 42	0.3248 62
22	0.4691 51	0.4219 55	0.3797 01	0.3418 50	0.3079 26
23	0.4532 86	0.4057 26	0.3633 50	0.3255 71	0.2918 73
24	0.4379 57	0.3901 21	0.3477 03	0.3100 68	0.2766 57
25	0.4231 47	0.3751 17	0.3327 31	0.2953 03	0.2622 34
26	0.4088 38	0.3606 89	0.3184 02	0.2812 41	0.2485 63
27	0.3950 12	0.3468 17	0.3046 91	0.2678 48	0.2356 05
28	0.3816 54	0.3334 77	0.2915 71	0.2550 94	0.2233 22
29	0.3687 48	0.3206 51	0.2790 15	0.2429 46	0.2116 79
30	0.3562 78	0.3083 19	0.2670 00	0.2313 77	0.2006 44
31	0.3442 30	0.2964 60	0.2555 02	0.2203 59	0.1901 84
32	0.3325 90	0.2850 58	0.2445 00	0.2098 66	0.1802 69
33	0.3213 43	0.2740 94	0.2339 71	0.1998 73	0.1708 71
34	0.3104 76	0.2635 52	0.2238 96	0.1903 55	0.1619 63
35	0.2999 77	0.2534015	0.2142 54	0.1812 90	0.1535 20
36	0.2898 33	0.2436 69	0.2050 28	0.1726 57	0.1455 16
37	0.2800 32	0.2342 97	0.1961 99	0.1644 36	0.1379 30
38	0.2705 62	0.2252 85	0.1877 50	0.1566 05	0.1307 39
39	0.2614 13	0.2166 21	0.1796 65	0.1494 48	0.1239 24
40	0.2525 72	0.2082 89	0.1719 29	0.1420 46	0.1174 63

TABLE 6 (*Continued*)

			Interest Rate, r		
n (years)	6%	$6\frac{1}{2}\%$	7%	$7\frac{1}{2}\%$	8%
1	0.9433 96	0.9389 67	0.9345 79	0.9302 33	0.9259 26
2	0.8899 96	0.8816 59	0.8734 39	0.8653 33	0.8573 39
3	0.8396 19	0.8278 49	0.8162 98	0.8049 61	0.7938 32
4	0.7920 94	0.7773 23	0.7628 95	0.7488 01	0.7350 30
5	0.7472 58	0.7298 81	0.7129 86	0.6965 59	0.6805 83
6	0.7049 61	0.6853 34	0.6663 42	0.6479 62	0.6301 70
7	0.6650 57	0.6435 06	0.6227 50	0.6027 55	0.5834 90
8	0.6274 12	0.6042 31	0.5820 09	0.5607 02	0.5402 69
9	0.5918 98	0.5673 53	0.5439 34	0.5215 83	0.5002 49
10	0.5583 95	0.5327 26	0.5083 49	0.4851 94	0.4631 93
11	0.5267 88	0.5002 12	0.4750 93	0.4513 43	0.4288 83
12	0.4969 69	0.4696 83	0.4440 12	0.4198 54	0.3971 14
13	0.4688 39	0.4410 17	0.4149 64	0.3905 62	0.3676 98
14	0.4423 01	0.4141 00	0.3878 17	0.3633 13	0.3404 61
15	0.4172 65	0.3888 27	0.3624 46	0.3379 66	0.3152 42
16	0.3936 46	0.3650 95	0.3387 35	0.3143 87	0.2918 90
17	0.3713 64	0.3428 13	0.3165 74	0.2924 53	0.2702 69
18	0.3503 44	0.3218 90	0.2958 64	0.2720 49	0.2502 49
19	0.3305 13	0.3022 44	0.2765 08	0.2530 69	0.2317 12
20	0.3118 05	0.2837 97	0.2584 19	0.2354 13	0.2145 48
21	0.2941 55	0.2664 76	0.2415 13	0.2189 89	0.1986 56
22	0.2775 05	0.2502 12	0.2257 13	0.2037 11	0.1839 41
23	0.2617 97	0.2349 41	0.2109 47	0.1894 98	0.1703 15
24	0.2469 79	0.2206 02	0.1971 47	0.1762 77	0.1576 99
25	0.2329 99	0.2071 38	0.1842 49	0.1639 79	0.1460 18
26	0.2198 10	0.1944 96	0.1721 95	0.1525 39	0.1352 02
27	0.2073 68	0.1826 25	0.1609 30	0.1418 96	0.1251 87
28	0.1956 30	0.1714 79	0.1504 02	0.1319 97	0.1159 14
29	0.1845 57	0.1610 13	0.1405 63	0.1227 88	0.1073 28
30	0.1741 10	0.1511 86	0.1313 67	0.1142 21	0.0993 77
31	0.1642 55	0.1419 59	0.1227 73	0.1062 52	0.0920 16
32	0.1549 57	0.1332 95	0.1147 41	0.0988 39	0.0852 00
33	0.1461 86	0.1251 59	0.1072 35	0.0919 43	0.0788 89
34	0.1379 12	0.1175 20	0.1002 19	0.0855 29	0.0730 45
35	0.1301 06	0.1103 48	0.0936 63	0.0795 62	0.0676 35
36	0.1227 41	0.1036 13	0.0875 35	0.0740 11	0.0626 25
37	0.1157 93	0.0972 89	0.0818 09	0.0688 47	0.0579 86
38	0.1092 39	0.0913 51	0.0764 57	0.0640 44	0.0536 90
39	0.1030 56	0.0857 76	0.0714 55	0.0595 76	0.0497 13
40	0.0972 22	0.0805 41	0.0667 80	0.0554 19	0.0460 31

TABLE 6   (*Continued*)

n (years)	Interest Rate, r			
	$8\frac{1}{2}\%$	9%	$9\frac{1}{2}\%$	10%
1	0.9216 59	0.9174 31	0.9132 42	0.9090 91
2	0.8494 55	0.8416 80	0.8340 11	0.8264 46
3	0.7829 08	0.7721 83	0.7616 54	0.7513 15
4	0.7215 74	0.7084 25	0.6955 74	0.6830 13
5	0.6650 45	0.6499 31	0.6352 28	0.6209 21
6	0.6129 45	0.5962 67	0.5801 17	0.5644 74
7	0.5649 26	0.5470 34	0.5297 87	0.5131 58
8	0.5206 69	0.5018 66	0.4838 24	0.4665 07
9	0.4798 80	0.4604 28	0.4418 48	0.4240 98
10	0.4422 85	0.4224 11	0.4035 14	0.3855 43
11	0.4076 36	0.3875 33	0.3685 06	0.3504 94
12	0.3757 02	0.3555 35	0.3365 35	0.3186 31
13	0.3462 69	0.3261 79	0.3073 38	0.2896 64
13	0.3191 42	0.2992 46	0.2806 74	0.2633 31
15	0.2941 40	0.2745 38	0.2563 23	0.2393 92
16	0.2710 97	0.2518 70	0.2340 85	0.2176 29
17	0.2498 59	0.2310 73	0.2137 77	0.1978 45
18	0.2302 85	0.2119 94	0.1952 30	0.1798 59
19	0.2122 44	0.1944 90	0.1782 92	0.1635 08
20	0.1956 16	0.1784 31	0.1628 24	0.1486 44
21	0.1802 92	0.1636 98	0.1486 97	0.1351 31
22	0.1661 67	0.1501 82	0.1357 97	0.1228 46
23	0.1531 50	0.1377 81	0.1240 15	0.1116 78
24	0.1411 52	0.1264 05	0.1132 56	0.1015 26
25	0.1300 94	0.1159 68	0.1034 30	0.0922 96
26	0.1199 02	0.1063 93	0.0944 57	0.0839 05
27	0.1105 09	0.0976 08	0.0862 62	0.0762 78
28	0.1018 51	0.0895 48	0.0787 78	0.0693 43
29	0.0938 72	0.0821 54	0.0719 43	0.0630 39
30	0.0865 18	0.0753 71	0.0657 02	0.0573 09
31	0.0797 40	0.0691 48	0.0600 02	0.0520 99
32	0.0734 93	0.0634 38	0.0547 96	0.0473 62
33	0.0677 36	0.0582 00	0.0500 42	0.0430 57
34	0.0624 29	0.0533 95	0.0457 00	0.0391 43
35	0.0575 39	0.0489 86	0.0417 36	0.0355 84
36	0.0530 31	0.0449 41	0.0381 15	0.0323 49
37	0.0488 76	0.0412 31	0.0348 08	0.0294 08
38	0.0450 47	0.0378 26	0.0317 88	0.0267 35
39	0.0415 18	0.0347 03	0.0290 30	0.0243 04
40	0.0382 66	0.0318 38	0.0265 12	0.0220 95

TABLE 7   Value of $s_{\overline{n}|\,r} = \frac{(1+r)^n - 1}{r}$ for a Periodic Payment of \$1.00

	Interest Rate, r				
n (years)	1%	$1\frac{1}{2}$%	2%	$2\frac{1}{2}$%	3%
1	1.0000 00	1.0000 00	1.0000 00	1.0000 00	1.0000 00
2	2.0100 00	2.0150 00	2.0200 00	2.0250 00	2.0300 00
3	3.0301 00	3.0452 25	3.0604 00	3.0756 25	3.0909 00
4	4.0604 01	4.0909 03	4.1216 08	4.1525 16	4.1836 27
5	5.1010 05	5.1522 67	5.2040 40	5.2563 29	5.3091 36
6	6.1520 15	6.2295 51	6.3081 21	6.3877 37	6.4684 10
7	7.2135 35	7.3229 94	7.4342 83	7.5474 30	7.6624 62
8	8.2856 71	8.4328 39	8.5829 69	8.7361 16	8.8923 36
9	9.3685 27	9.5593 32	9.7546 28	9.9545 19	10.1591 06
10	10.4622 13	10.7027 22	10.9497 21	11.2033 82	11.4638 79
11	11.5668 35	11.8632 62	12.1687 15	12.4834 66	12.8077 96
12	12.6825 03	13.0412 11	13.4120 90	13.7955 53	14.1920 30
13	13.8093 28	14.2368 30	14.6803 32	15.1404 42	15.6177 90
14	14.9474 21	15.4503 82	15.9739 38	16.5189 53	17.0863 24
15	16.0968 96	16.6821 38	17.2934 17	17.9319 27	18.5989 14
16	17.2578 64	17.9323 70	18.6392 85	19.3802 25	20.1568 81
17	18.4304 43	19.2013 55	20.0120 71	20.8647 30	21.7615 88
18	19.6147 48	20.4893 76	21.4123 12	22.3863 49	23.4144 35
19	20.8108 95	21.7967 16	22.8405 59	23.9460 07	25.1168 68
20	22.0190 04	23.1236 67	24.2973 70	25.5446 58	26.8703 74
21	23.2391 94	24.4705 22	25.7833 17	27.1832 74	28.6764 86
22	24.4715 86	25.8375 80	27.2989 84	28.8628 56	30.5367 80
23	25.7163 02	27.2251 44	28.8449 63	30.5844 27	32.4528 84
24	26.9734 65	28.6335 21	30.4218 62	32.3479 38	34.4264 70
25	28.2432 00	30.0630 24	32.0303 00	34.1577 64	36.4592 64
26	29.5256 32	31.5139 69	33.6709 06	36.0117 08	38.5530 42
27	30.8208 88	32.9866 79	35.3443 24	37.9120 01	40.7096 34
28	32.1290 97	34.4814 79	37.0512 10	39.8598 01	42.9309 23
29	33.4503 88	35.9987 01	38.7922 35	41.8562 96	45.2188 50
30	34.7848 92	37.5386 81	40.5680 79	43.9027 03	47.5754 16
31	36.1327 40	39.1017 62	42.3794 41	46.0002 71	50.0026 78
32	37.4940 68	40.6882 88	44.2270 30	48.1502 78	52.5027 59
33	38.8690 09	42.2986 12	46.1115 70	50.3540 34	55.0778 41
34	40.2576 99	43.9330 92	48.0338 02	52.6128 85	57.7301 77
35	41.6602 76	45.5920 88	49.9944 78	54.9282 07	60.4620 82
36	43.0768 78	47.2759 69	51.9943 67	57.3014 13	63.2759 44
37	44.5076 47	48.9851 09	54.0342 55	59.7339 48	66.1742 23
38	45.9527 24	50.7198 85	56.1149 40	62.2272 97	69.1594 49
39	47.4122 51	52.4806 84	58.2372 38	64.7829 79	72.2342 33
40	48.8863 73	54.2678 94	60.4019 83	67.4025 54	75.4012 60

TABLE 7 (*Continued*)

n years)	$3\frac{1}{2}\%$	4%	$4\frac{1}{2}\%$	5%	$5\frac{1}{2}\%$
			Interest Rate, r		
1	1.0000 00	1.0000 00	1.0000 00	1.0000 00	1.0000 00
2	2.0350 00	2.0400 00	2.0450 00	2.0500 00	2.0550 00
3	3.1062 25	3.1216 00	3.1370 25	3.1525 00	3.1680 25
4	4.2149 43	4.2464 64	4.2781 91	4.3101 25	4.3422 66
5	5.3624 66	5.4163 23	5.4707 10	5.5256 31	5.5810 91
6	6.5501 52	6.6329 75	6.7168 92	6.8019 13	6.8880 51
7	7.7794 08	7.8982 94	8.0191 52	8.1420 08	8.2668 94
8	9.0516 87	9.2142 26	9.3800 14	9.5491 09	9.7215 73
9	10.3684 96	10.5827 95	10.8021 14	11.0265 64	11.2562 60
10	11.7313 93	12.0061 07	12.2882 09	12.5778 93	12.8753 54
11	13.1419 92	13.4863 51	13.8411 79	14.2067 87	14.5834 98
12	14.6019 62	15.0258 05	15.4640 32	15.9171 27	16.3855 91
13	16.1130 30	16.6268 38	17.1599 13	17.7129 83	18.2867 98
14	17.6769 86	18.2919 11	18.9321 09	19.5986 32	20.2925 72
15	19.2956 81	20.0235 88	20.7840 54	21.5785 64	22.4086 64
16	20.9710 30	21.8245 31	22.7193 37	23.6574 92	24.6411 40
17	22.7050 16	23.6975 12	24.7417 07	25.8403 66	26.9964 03
18	24.4996 91	25.6454 13	26.8550 84	28.1323 85	29.4812 05
19	26.3571 81	27.6712 29	29.0635 62	30.5390 04	32.1026 71
20	28.2796 82	29.7780 79	31.3714 23	33.0659 54	34.8683 18
21	30.2694 71	31.9692 02	33.7831 37	35.7192 52	37.7860 76
22	32.3289 02	34.2479 70	36.3033 78	38.5052 14	40.8643 10
23	34.4604 14	36.6178 89	38.9370 30	41.4304 75	44.1118 47
24	36.6665 28	39.0826 04	41.6891 96	44.5019 99	47.5379 98
25	38.9498 57	41.6459 08	44.5652 10	47.7270 99	51.1525 88
26	41.3131 02	44.3117 45	47.5706 45	51.1134 54	54.9659 81
27	43.7590 60	47.0842 14	50.7113 24	54.6691 26	58.9891 09
28	46.2906 27	49.9675 83	53.9933 33	58.4025 83	63.2335 10
29	48.9107 99	52.9662 86	57.4230 33	62.3227 12	67.7113 54
30	51.6226 77	56.0849 38	61.0070 70	66.4388 48	72.4354 78
31	54.4294 71	59.3283 35	64.7523 88	70.7607 90	77.4194 29
32	57.3345 02	62.7014 69	68.6662 45	75.2988 29	82.6774 98
33	60.3412 10	66.2095 27	72.7562 26	80.0637 71	88.2247 60
34	63.4531 52	69.8579 09	77.0302 56	85.0669 59	94.0771 22
35	66.6740 13	73.6522 25	81.4966 18	90.3203 07	100.2513 64
36	70.0076 03	77.5983 13	86.1639 66	95.8363 23	106.7651 89
37	73.4578 69	81.7022 46	91.0413 44	101.6281 39	113.6372 74
38	77.0288 95	85.9703 36	96.1382 05	107.7095 46	120.8873 24
39	80.7249 06	90.4091 50	101.4644 24	114.0950 23	128.5361 27
40	84.5502 78	95.0255 16	107.0303 23	120.7997 74	136.6056 14

TABLE 7 (*Continued*)

n (years)	6%	6½%	7%	7½%	8%
			Interest Rate, r		
1	1.0000 00	1.0000 00	1.0000 00	1.0000 00	1.0000 00
2	2.0600 00	2.0650 00	2.0700 00	2.0750 00	2.0800 00
3	3.1836 00	3.1992 25	3.2149 00	3.2306 25	3.2464 00
4	4.3746 16	4.4071 75	4.4399 43	4.4729 22	4.5061 12
5	5.6370 93	5.6936 41	5.7507 39	5.8083 91	5.8666 01
6	6.9753 19	7.0637 28	7.1532 91	7.2440 20	7.3359 29
7	8.3938 38	8.5228 70	8.6540 21	8.7873 22	8.9228 03
8	9.8974 68	10.0768 56	10.2598 03	10.4463 71	10.6366 28
9	11.4913 16	11.7318 52	11.9779 89	12.2298 49	12.4875 58
10	13.1807 95	13.4944 23	13.8164 48	14.1470 88	14.4865 62
11	14.9716 43	15.3715 60	15.7835 99	16.2081 19	16.6454 87
12	16.8699 41	17.3707 11	17.8884 51	18.4237 28	18.9771 26
13	18.8821 38	19.4998 08	20.1406 43	20.8055 08	21.4952 97
14	21.0150 66	21.7672 95	22.5504 88	23.3659 21	24.2149 20
15	23.2759 70	24.1821 69	25.1290 22	26.1183 65	27.1521 14
16	25.6725 28	26.7540 10	27.8880 54	29.0772 42	30.3242 83
17	28.2128 80	29.4930 21	30.8402 17	32.2580 35	33.7502 26
18	30.9056 53	32.4100 67	33.9990 33	35.6773 88	37.4502 44
19	33.7599 92	35.5167 22	37.3789 65	39.3531 92	41.4462 63
20	36.7855 91	38.8253 09	40.9954 92	43.3046 81	45.7619 64
21	39.9927 27	42.3489 54	44.8651 77	47.5525 32	50.4229 21
22	43.3922 90	46.1016 36	49.0057 39	52.1189 72	55.4567 55
23	46.9958 28	50.0982 42	53.4361 41	57.0278 95	60.8932 96
24	50.8155 77	54.3546 28	58.1766 71	62.3049 87	66.7647 59
25	54.8645 12	58.8876 79	63.2490 38	67.9778 62	73.1059 40
26	59.1563 83	63.7153 78	68.6764 70	74.0762 01	79.9544 15
27	63.7057 66	68.8568 77	74.4838 23	80.6319 16	87.3507 68
28	68.5281 12	74.3325 74	80.6976 91	87.6793 10	95.3388 30
29	73.6397 98	80.1641 92	87.3465 29	95.2552 58	103.9659 36
30	79.0581 86	86.3748 64	94.4607 86	103.3994 03	113.2832 11
31	84.8016 77	92.9892 30	102.0730 41	112.1543 58	123.3458 68
32	90.8897 78	100.0335 30	110.2181 54	121.5659 35	134.2135 37
33	97.3431 65	107.5357 10	118.9334 25	131.6833 80	145.9506 20
34	104.1837 55	115.5255 31	128.2587 65	142.5596 33	158.6266 70
35	111.4347 80	124.0346 90	138.2368 78	154.2516 06	172.3168 04
36	119.1208 67	133.0969 45	148.9134 60	166.8204 76	187.1021 48
37	127.2681 19	142.7482 47	160.3374 02	180.3320 12	203.0703 20
38	135.9042 06	153.0268 83	172.5610 20	194.8569 13	220.3159 45
39	145.0584 58	163.9736 30	185.6402 92	210.4711 81	238.9412 21
40	154.7619 66	175.6319 16	199.6351 12	227.2565 20	259.0565 19

TABLE 7   (*Continued*)

n (years)	8½%	9%	9½%	10%
	\multicolumn{4}{c}{Interest Rate, r}			
1	1.0000 00	1.0000 00	1.0000 00	1.0000 00
2	2.0850 00	2.0900 00	2.0950 00	2.1000 00
3	3.2622 25	3.2781 00	3.2940 25	3.3100 00
4	4.5395 14	4.5731 29	4.6069 57	4.6410 00
5	5.9253 73	5.9847 11	6.0446 18	6.1051 00
6	7.4290 30	7.5233 35	7.6188 57	7.7156 10
7	9.0604 97	9.2004 35	9.3426 48	9.4871 71
8	10.8306 39	11.0284 74	11.2302 00	11.4358 88
9	12.7512 44	13.0210 37	13.2970 69	13.5794 77
10	14.8350 99	15.1929 30	15.5602 91	15.9374 25
11	17.0960 83	17.5602 93	18.0385 18	18.5311 67
12	19.5492 50	20.1407 20	20.7521 78	21.3842 84
13	22.2109 36	22.9533 85	23.7236 34	24.5227 12
13	25.0988 66	26.0191 89	26.9773 80	27.9749 83
15	28.2322 69	29.3609 16	30.5402 31	31.7724 82
16	31.6320 12	33.0033 99	34.4415 53	35.9497 30
17	35.3207 33	36.9737 05	38.7135 00	40.5447 03
18	39.3229 95	41.3013 38	43.3912 83	45.5991 73
19	43.6654 50	46.0184 58	48.5134 55	51.1590 90
20	48.3770 13	51.1601 20	54.1222 33	57.2749 99
21	53.4890 59	56.7645 30	60.2638 45	64.0024 99
22	59.0356 29	62.8733 38	66.9889 10	71.4027 49
23	65.0536 58	69.5319 39	74.3528 56	79.5430 24
24	71.5832 19	76.7898 13	82.4163 78	88.4973 27
25	78.6677 92	84.7008 96	91.2459 34	98.3470 59
26	86.3545 55	93.3239 77	100.9142 97	109.1817 65
27	94.6946 92	102.7231 35	111.5011 56	121.0999 42
28	103.7437 41	112.9682 17	123.0937 66	134.2099 36
29	113.5619 59	124.1353 56	135.7876 73	148.6309 30
30	124.2147 25	136.3075 39	149.6875 02	164.4940 23
31	135.7729 77	149.5752 17	164.9078 15	181.9434 25
32	148.3136 80	164.0369 87	181.5740 57	210.1377 67
33	161.9203 43	179.8003 15	199.8235 93	222.2515 44
34	176.6835 72	196.9823 44	219.8068 34	245.4766 99
35	192.7016 75	215.7107 55	241.6884 83	271.0243 68
36	210.0813 18	236.1247 23	265.6488 89	299.1268 05
37	228.9382 30	258.3759 48	291.8855 34	330.0394 86
38	249.3979 79	282.6297 83	320.6146 59	364.0434 34
39	271.5968 08	309.0664 63	352.0730 52	401.4477 78
40	295.6825 36	337.8824 45	386.5199 92	442.5925 56

TABLE 8 Value of $a_{\overline{n}|r} = \frac{a-(1+r)^{-n}}{r} = \text{PVIFA}_{r,n}$ for a Periodic Payment of \$1.00

	Interest Rate, r				
n (years)	1%	$1\frac{1}{2}\%$	2%	$2\frac{1}{2}\%$	3%
1	0.9900 99	0.9852 22	0.9803 92	0.9756 10	0.9708 74
2	1.9703 95	1.9558 83	1.9415 61	1.9274 24	1.9134 70
3	2.9409 85	2.9122 00	2.8838 83	2.8560 24	2.8286 11
4	3.9019 66	3.8543 85	3.8077 29	3.7619 74	3.7170 98
5	4.8534 31	4.7826 45	4.7134 60	4.6458 29	4.5797 07
6	5.7954 76	5.6971 87	5.6014 31	5.5081 25	5.4171 91
7	6.7281 95	6.5982 14	6.4719 91	6.3493 91	6.2302 83
8	7.6516 78	7.4859 25	7.3254 81	7.1701 37	7.0196 92
9	8.5660 18	8.3605 17	8.1622 37	7.9708 66	7.7861 09
10	9.4713 05	9.2221 85	8.9825 85	8.7520 64	8.5302 03
11	10.3676 28	10.0711 18	9.7868 48	9.5142 09	9.2526 24
12	11.2550 77	10.9075 05	10.5753 41	10.2577 65	9.9540 04
13	12.1337 40	11.7315 32	11.3483 74	10.9831 85	10.6349 55
14	13.0037 03	12.5433 82	12.1062 49	11.6909 12	11.2960 73
15	13.8650 52	13.3432 33	12.8492 64	12.3813 78	11.9379 35
16	14.7178 74	14.1312 64	13.5777 09	13.0550 03	12.5611 02
17	15.5622 51	14.9076 49	14.2918 72	13.7121 98	13.1661 18
18	16.3982 69	15.6725 61	14.9920 31	14.3533 64	13.7535 13
19	17.2260 09	16.4261 68	15.6784 62	14.9788 91	14.3237 99
20	18.0455 53	17.1686 39	16.3514 33	15.5891 62	14.8774 75
21	18.8569 83	17.9001 37	17.0112 09	16.1845 49	15.4150 24
22	19.6603 79	18.6208 24	17.6580 48	16.7654 13	15.9369 17
23	20.4558 21	19.3308 61	18.2922 04	17.3321 10	16.4436 08
24	21.2433 87	20.0304 05	18.9039 26	17.8849 86	16.9355 42
25	22.0231 56	20.7196 11	19.5234 56	18.4243 76	17.4131 48
26	22.7952 04	21.3986 32	20.1210 44	18.9506 11	17.8768 42
27	23.5596 08	22.0676 17	20.7068 98	19.4640 11	18.3270 31
28	24.3164 43	22.7267 17	21.2812 72	19.9648 89	18.7641 08
29	25.0657 85	23.3760 76	21.8443 85	20.4535 50	19.1884 55
30	25.8077 08	24.0158 38	22.3964 56	20.9302 93	19.6004 41
31	26.5422 85	24.6461 46	22.9377 02	21.3954 07	20.0004 28
32	27.2695 89	25.2671 39	23.4683 35	21.8491 78	20.3887 66
33	27.9896 93	25.8789 54	23.9885 64	22.2918 81	20.7657 92
34	28.7026 66	26.4817 28	24.4985 92	22.7237 86	21.1318 37
35	29.4085 80	27.0755 95	24.9986 19	23.1451 57	21.4872 20
36	30.1075 05	27.6606 84	25.4888 42	23.5562 51	21.8322 53
37	30.7995 10	28.2371 27	25.9694 53	23.9573 18	22.1672.35
38	31.4846 63	28.8050 52	26.4406 41	24.3486 03	22.4924 62
39	32.1630 33	29.3645 83	26.9025 89	24.7303 44	22.8082 15
40	32.8346 86	29.9158 45	27.3554 79	25.1027 75	23.1147 72
41	33.4996 89	30.4589 61	27.7994 89	25.4661 22	23.4124 00
42	34.1581 08	30.9940 50	28.2347 94	25.8206 07	23.7013 59
43	34.8100 08	31.5212 32	28.6615 62	26.1664 46	23.9819 02
44	35.4554 53	32.0406 22	29.0799 63	26.5038 49	24.2542 74
45	36.0945 08	32.5523 37	29.4901 60	26.8330 24	25.5187 13
46	36.7272 36	33.0564 90	29.8923 14	27.1541 70	24.7754 49
47	37.3536 99	33.5531 92	30.2865 82	27.4674 83	25.0247 08
48	37.9739 59	34.0425 54	30.6731 20	27.7731 54	25.2667 07
49	38.5880 79	34.5246 83	31.0520 78	28.0713 69	25.5016 57
50	39.1961 18	34.9996 88	31.4236 06	28.3623 12	25.7297 64

TABLE 8   (*Continued*)

n (years)	Interest Rate, r				
	$3\frac{1}{2}\%$	4%	$4\frac{1}{2}\%$	5%	$5\frac{1}{2}\%$
1	0.9661 84	0.9615 38	0.9569 38	0.9523 81	0.9478 67
2	1.8996 94	1.8860 95	1.8726 68	1.8594 10	1.8463 20
3	2.8016 37	2.7750 91	2.7489 64	2.7232 48	2.6979 33
4	3.6730 79	3.6298 95	3.5875 26	3.5459 51	3.5051 50
5	4.5150 52	4.4518 22	4.3899 77	4.3294 77	4.2702 84
6	5.3285 53	5.2421 37	5.1578 72	5.0756 92	4.9955 30
7	6.1145 44	6.0020 55	5.8927 01	5.7863 73	5.6829 67
8	6.8739 56	6.7327 45	6.5958 86	6.4632 13	6.3345 66
9	7.6076 87	7.4353 32	7.2687 91	7.1078 22	6.9521 95
10	8.3166 05	8.1108 96	7.9127 18	7.7217 35	7.5376 26
11	9.0015 51	8.7604 77	8.5289 17	8.3604 14	8.0925 36
12	9.6633 34	9.3850 74	9.1185 81	8.8632 52	8.6185 18
13	10.3027 38	9.9856 48	9.6828 52	9.3935 73	9.1170 79
14	10.9205 20	10.5631 23	10.2228 25	9.8986 41	9.5896 48
15	11.5274 11	11.1183 87	10.7395 46	10.3796 58	10.0375 81
16	12.0941 17	11.6522 96	11.2340 15	10.8377 70	10.4621 62
17	12.6513 21	12.1656 69	11.7071 91	11.2740 66	10.8646 09
18	13.1896 82	12.6592 97	12.1599 92	11.6895 87	11.2460 74
19	13.7098 37	13.1339 39	12.5932 94	12.0853 21	11.6076 54
20	14.2124 03	13.5903 26	13.0079 36	12.4622 10	11.9503 82
21	14.6979 74	14.0291 60	13.4047 24	12.8211 53	12.2752 44
22	15.1671 25	14.4511 15	13.7844 25	13.1630 03	12.5831 70
23	15.6204 10	14.8568 42	14.1477 75	13.4885 74	12.8750 42
24	16.0583 68	15.2469 63	14.4954 78	13.7986 42	13.1516 99
25	16.4815 15	15.6220 80	14.8282 09	14.0939 45	13.4139 33
26	16.8903 52	15.9827 69	15.1466 11	14.3751 85	13.6624 95
27	17.2853 65	16.3295 86	15.4513 03	14.6430 34	13.8981 00
28	17.6670 19	16.6630 63	15.7428 74	14.8981 27	14.1214 22
29	18.0357 67	16.9837 15	16.0218 89	15.1410 74	14.3331 01
30	18.3920 45	17.2920 33	16.2888 89	15.3724 51	14.5337 45
31	18.7362 76	17.5884 94	16.5443 91	15.5928 11	14.7239 29
32	19.0688 65	17.8735 52	16.7888 91	15.8026 77	14.9041 98
33	19.3902 08	18.1476 46	17.0228 62	16.0025 49	15.0750 69
34	19.7006 84	18.4111 98	17.2467 58	16.1929 04	15.2370 33
35	20.0006 61	18.6646 13	17.4610 12	16.3741 94	15.3905 52
36	20.2904 94	18.9082 82	17.6660 41	16.5468 52	15.5360 68
37	20.5705 25	19.1425 79	17.8622 40	16.7112 87	15.6739 98
38	20.8410 87	19.3678 64	18.0499 90	16.8678 93	15.8047 38
39	21.1025 00	19.5844 85	18.2296 56	17.0170 41	15.9286 62
40	21.3550 72	19.7927 74	18.4015 84	17.1590 86	16.0461 25
41	21.5991 04	19.9930 52	18.5661 09	17.2943 68	16.1574 64
42	21.8348 83	20.1856 27	18.7235 50	17.4232 08	16.2629 99
43	22.0626 89	20.3707 95	18.8742 10	17.5459 12	16.3630 32
44	22.2827 91	20.5488 41	19.0183 83	17.6627 73	16.4578 51
45	22.4954 50	20.7200 40	19.1563 47	17.7740 70	16.5477 26
46	22.7009 18	20.8846 54	19.2883 71	17.8800 67	16.6329 15
47	22.8994 38	21.0429 36	19.4147 09	17.9810 16	16.7136 64
48	23.0912 44	21.1951 31	19.5356 07	18.0771 58	16.7902 03
49	23.2765 65	21.3414 72	19.6512 98	18.1687 22	16.8627 51
50	23.4556 18	21.4821 85	19.7620 08	18.2559 25	16.9315 18

TABLE 8 (*Continued*)

n (years)	Interest Rate, r				
	6%	$6\frac{1}{2}\%$	7%	$7\frac{1}{2}\%$	8%
1	0.9433 96	0.9389 67	0.9345 79	0.9302 33	0.9259 26
2	1.8333 93	1.8206 26	1.8080 18	1.7955 65	1.7832 65
3	2.6730 12	2.6484 76	2.6243 16	2.6005 26	2.5770 97
4	3.4651 06	3.4257 99	3.3872 11	3.3493 26	3.3121 27
5	4.2123 64	4.1556 79	4.1001 97	4.0458 85	3.9927 10
6	4.9173 24	4.8410 14	4.7665 40	4.6938 46	4.6228 80
7	5.5823 81	5.4845 20	5.3892 89	5.2966 01	5.2063 70
8	6.2097 94	6.0887 51	5.9712 99	5.8573 04	5.7466 39
9	6.8016 92	6.6561 04	6.5152 32	6.3788 87	6.2468 88
10	7.3600 87	7.1888 30	7.0235 82	6.8640 81	6.7100 81
11	7.8868 75	7.6890 42	7.4986 74	7.3154 24	7.1389 64
12	8.3838 44	8.1587 25	7.9426 86	7.7352 78	7.5360 78
13	8.8526 83	8.5997 42	8.3576 51	8.1258 40	7.9037 76
14	9.2949 84	9.0138 42	8.7454 68	8.4891 54	8.2442 37
15	9.7122 49	9.4026 69	9.1079 14	8.8271 20	8.5594 79
16	10.1058 95	9.7677 64	9.4466 49	9.1415 07	8.8513 69
17	10.4772 60	10.1105 77	9.7632 23	9.4339 60	9.1216 38
18	10.8276 03	10.4324 66	10.0590 87	9.7060 09	9.3718 87
19	11.1581 16	10.7347 10	10.3355 95	9.9590 78	9.6035 99
20	11.4699 21	11.0185 07	10.5940 14	10.1944 91	9.8181 47
21	11.7640 77	11.2849 83	10.8355 27	10.4134 80	10.0168 03
22	12.0415 82	11.5351 96	11.0612 41	10.6171 91	10.2007 44
23	12.3033 79	11.7701 37	11.2721 87	10.8066 89	10.3710 59
24	12.5503 58	11.9907 39	11.4693 34	10.9829 67	10.5287 58
25	12.7833 56	12.1978 77	11.6535 83	11.1469 46	10.6747 76
26	13.0031 66	12.3923 73	11.8257 79	11.2994 85	10.8099 78
27	13.2105 34	12.5749 98	11.9867 09	11.4413 81	10.9351 65
28	13.4061 64	12.7464 77	12.1371 11	11.5733 78	11.0510 78
29	13.5907 21	12.9074 90	12.2776 74	11.6961 65	11.1584 06
30	13.7648 31	13.0586 76	12.4090 41	11.8103 86	11.2577 83
31	13.9290 86	13.2006 35	12.5318 14	11.9166 38	11.3497 99
32	14.0840 43	13.3339 29	12.6465 55	12.0154 78	11.4349 99
33	14.2302 30	13.4590 89	12.7537 90	12.1074 21	11.5138 88
34	14.3681 41	13.5766 09	12.8540 09	12.1929 50	11.5869 34
35	14.4982 46	13.6869 57	12.9476 72	12.2725 11	11.6545 68
36	14.6209 87	13.7905 70	13.0352 08	12.3465 22	11.7171 93
37	14.7367 80	13.8878 59	13.1170 17	12.4153 70	11.7751 79
38	14.8460 19	13.9792 10	13.1934 73	12.4794 14	11.8288 69
39	14.9490 75	14.0649 86	13.2649 28	12.5389 89	11.8785 82
40	15.0462 97	14.1455 27	13.3317 09	12.5944 09	11.9246 13
41	15.1380 16	14.2211 52	13.3941 20	12.6459 62	11.9672 35
42	15.2245 43	14.2921 61	13.4524 49	12.6939 18	12.0066 99
43	15.3061 73	13.3588 37	13.5069 62	12.7385 28	12.0432 40
44	15.3831 82	14.4214 43	13.5579 08	12.7800 26	12.0700 74
45	15.4558 32	14.4802 28	13.6055 22	12.8186 29	12.1084 02
46	15.5243 70	14.5354 26	13.6500 20	12.8545 39	12.1374 09
47	15.5890 28	14.5872 54	13.6916 08	12.8879 43	12.1642 67
48	15.6500 27	14.6359 19	13.7304 74	12.9190 17	12.1891 36
49	15.7075 72	14.6816 15	13.7667 98	12.9479 22	12.2121 63
50	15.7618 61	14.7245 21	13.8007 46	12.9748 12	12.2334 85

TABLE 8   (*Continued*)

n (years)	Interest Rate, r			
	$8\frac{1}{2}\%$	9%	$9\frac{1}{2}\%$	10%
1	0.9216 59	0.9174 31	0.9132 42	0.9090 91
2	1.7711 14	1.7591 11	1.7472 53	1.7355 37
3	2.5540 22	2.5312 95	2.5089 07	2.4868 52
4	3.2755 97	3.2397 20	3.2044 81	3.1698 65
5	3.9406 42	3.8896 51	3.8397 09	3.7907 87
6	4.5535 87	4.4859 19	4.4198 25	4.3552 61
7	5.1185 14	5.0329 53	4.9496 12	4.8684 19
8	5.6391 82	5.5348 19	5.4334 36	5.3349 26
9	6.1190 63	5.9952 47	5.8752 84	5.7590 24
10	6.5613 48	6.4176 58	6.2787 98	6.1445 67
11	6.9689 84	6.8051 91	6.6473 04	6.4950 61
12	7.3446 86	7.1607 25	6.9838 39	6.8136 92
13	7.6909 55	7.4869 04	7.2911 78	7.1033 56
13	8.0100 97	7.7861 50	7.5718 52	7.3666 87
15	8.3042 37	8.0606 88	7.8281 75	7.6060 80
16	8.5753 33	8.3125 58	8.0622 60	7.8237 09
17	8.8251 92	8.5436 31	8.2760 37	8.0215 53
18	9.0554 76	8.7556 25	8.4712 66	8.2014 12
19	9.2677 20	8.9501 15	8.6495 58	8.3649 20
20	9.4633 37	9.1285 46	8.8123 82	8.5135 64
21	9.6436 28	9.2922 44	8.9610 80	8.6486 94
22	9.8097 96	9.4424 25	9.0968 76	8.7715 40
23	9.9629 45	9.5802 07	9.2208 92	8.8832 18
24	10.1040 97	9.7066 12	9.3341 48	8.9847 44
25	10.2341 91	9.8225 80	9.4375 78	9.0770 40
26	10.3540 93	9.9289 72	9.5320 34	9.1609 45
27	10.4646 02	10.0265 80	9.6182 96	9.2372 23
28	10.5664 53	10.1161 28	9.6970 74	9.3065 67
29	10.6603 26	10.1982 83	9.7690 18	9.3696 06
30	10.7468 44	10.2736 54	9.8347 19	9.4269 14
31	10.8265 84	10.3428 02	9.8947 21	9.4790 13
32	10.9000 78	10.4062 40	9.9495 17	9.5263 76
33	10.9678 13	10.4644 41	9.9995 59	9.5694 32
34	11.0302 43	10.5178 35	10.0452 59	9.6085 75
35	11.0877 91	10.5668 21	10.0869 95	9.6441 59
36	11.1408 12	10.6117 63	10.1251 09	9.6765 08
37	11.1896 89	10.6529 93	10.1599 17	9.7059 17
38	12.2347 36	10.6908 20	10.1917 05	9.7326 51
39	11.2762 55	10.7255 23	10.2207 35	9.7569 56
40	11.3145 20	10.7573 60	10.2472 47	9.7790 51
41	11.3479 88	10.7865 69	10.2714 58	9.7991 37
42	11.3822 93	10.8133 66	10.2935 69	9.8173 97
43	11.4122 52	10.8379 51	10.3137 62	9.8339 97
44	11.4398 64	10.8605 05	10.3322 03	9.8490 89
45	11.4653 12	10.8811 97	10.3490 43	9.8628 08
46	11.4887 67	10.9001 81	10.3644 23	9.8752 80
47	11.5103 84	10.9175 97	10.3784 69	9.8866 18
48	11.5303 08	10.9335 75	10.3912 96	9.8969 25
49	11.5486 71	10.9482 34	10.4030 10	9.9062 96
50	11.5655 95	10.9616 82	10.4137 07	9.9148 14

TABLE 9 Call Option Values and Percentage of Share Price

	MP/CV(SP)								
$\sigma\sqrt{T}$	.40	.45	.50	.55	.60	.65	.70	.75	.80
.05	.0	.0	.0	.0	.0	.0	.0	.0	.0
.10	.0	.0	.0	.0	.0	.0	.0	.0	.0
.15	.0	.0	.0	.0	.0	.0	.1	.2	.5
.20	.0	.0	.0	.0	.0	.1	.4	.8	1.5
.25	.0	.0	.0	.1	.2	.5	1.0	1.8	2.8
.30	.0	.1	.1	.3	.7	1.2	2.0	3.1	4.4
.35	.1	.2	.4	.8	1.4	2.3	3.3	4.6	6.2
.40	.2	.5	.9	1.6	2.4	3.5	4.8	6.3	8.0
.45	.5	1.0	1.7	2.6	3.7	5.0	6.5	8.1	9.9
.50	1.0	1.7	2.6	3.7	5.1	6.6	8.2	10.0	11.8
.55	1.7	2.6	3.8	5.1	6.6	8.3	10.0	11.9	13.8
.60	2.5	3.7	5.1	6.6	8.3	10.1	11.9	13.8	15.8
.65	3.6	4.9	6.5	8.2	10.0	11.9	13.8	15.8	17.8
.70	4.7	6.3	8.1	9.9	11.9	13.8	15.8	17.8	19.8
.75	6.1	7.9	9.8	11.7	13.7	15.8	17.8	19.8	21.8
.80	7.5	9.5	11.5	13.6	15.7	17.7	19.8	21.8	23.7
.85	9.1	11.2	13.3	15.5	17.6	19.7	21.8	23.8	25.7
.90	10.7	13.0	15.2	17.4	19.6	21.7	23.8	25.8	27.7
.95	12.5	14.8	17.1	19.4	21.6	23.7	25.7	27.7	29.6
1.00	14.3	16.7	19.1	21.4	23.6	25.7	27.7	29.7	31.6
1.05	16.1	18.6	21.0	23.3	25.6	27.7	29.7	31.6	33.5
1.10	18.0	20.6	23.0	25.3	27.5	29.6	31.6	33.5	35.4
1.15	20.0	22.5	25.0	27.3	29.5	31.6	33.6	35.4	37.2
1.20	21.9	24.5	27.0	29.3	31.5	33.6	35.5	37.3	39.1
1.25	23.9	26.5	29.0	31.3	33.5	35.5	37.4	39.2	40.9
1.30	25.9	28.5	31.0	33.3	35.4	37.4	39.3	41.0	42.7
1.35	27.9	30.5	33.0	35.2	37.3	39.3	41.1	42.8	44.4
1.40	29.9	32.5	34.9	37.1	39.2	41.1	42.9	44.6	46.2
1.45	31.9	34.5	36.9	39.1	41.1	43.0	44.7	46.4	47.9
1.50	33.8	36.4	38.8	40.9	42.9	44.8	46.5	48.1	49.6
1.55	35.8	38.4	40.7	42.8	44.8	46.6	48.2	49.8	51.2
1.60	37.8	40.3	42.6	44.6	46.5	48.3	49.9	51.4	52.8
1.65	39.7	42.2	44.4	46.4	48.3	50.0	51.6	53.1	54.4
1.70	41.6	44.0	46.2	48.2	50.0	51.7	53.2	54.7	56.0
1.75	43.5	45.9	48.0	50.0	51.7	53.4	54.8	56.2	57.5
2.00	52.5	54.6	56.5	58.2	59.7	61.1	62.4	63.6	64.6
2.25	60.7	62.5	64.1	65.6	66.8	68.0	69.1	70.0	70.9
2.50	67.9	69.4	70.8	72.0	73.1	74.0	74.9	75.7	76.4
2.75	74.2	75.4	76.6	77.5	78.4	79.2	79.9	80.5	81.1
3.00	79.5	80.5	81.4	82.2	82.9	83.5	84.1	84.6	85.1
3.50	87.6	88.3	88.8	89.3	89.7	90.1	90.5	90.8	91.1
4.00	92.9	93.3	93.6	93.9	94.2	94.4	94.6	94.8	94.9
4.50	96.2	96.4	96.6	96.7	96.9	97.0	97.1	97.2	97.3
5.00	98.1	98.2	98.3	98.3	98.4	98.5	98.5	98.6	98.6

Source: R. A. Brealey and C. M. Stewart (1991). *Principles of Corporate Finance*. McGraw-Hill, New York.

TABLE 9   (*Continued*)

$\sigma\sqrt{T}$	MP/CV(SP)									
	.82	.84	.86	.88	.90	.92	.94	.96	.98	1.00
.05	.0	.0	.0	.0	.0	.1	.3	.6	1.2	2.0
.10	.1	.2	.3	.5	.8	1.2	1.7	2.3	3.1	4.0
.15	.7	1.0	1.3	1.7	2.2	2.8	3.5	4.2	5.1	6.0
.20	1.9	2.3	2.8	3.4	4.0	4.7	5.4	6.2	7.1	8.0
.25	3.3	3.9	4.5	5.2	5.9	6.6	7.4	8.2	9.1	9.9
.30	5.0	5.7	6.3	7.0	7.8	8.6	9.4	10.2	11.1	11.9
.35	6.8	7.5	8.2	9.0	9.8	10.6	11.4	12.2	13.0	13.9
.40	8.7	9.4	10.2	11.0	11.7	12.5	13.4	14.2	15.0	15.9
.45	10.6	11.4	12.2	12.9	13.7	14.5	15.3	16.2	17.0	17.8
.50	12.6	13.4	14.2	14.9	15.7	16.5	17.3	18.1	18.9	19.7
.55	14.6	15.4	16.1	16.9	17.7	18.5	19.3	20.1	20.9	21.7
.60	16.6	17.4	18.1	18.9	19.7	20.5	21.3	22.0	22.8	23.6
.65	18.6	19.3	20.1	20.9	21.7	22.5	23.2	24.0	24.7	25.5
.70	20.6	21.3	22.1	22.9	23.6	24.4	25.2	25.9	26.6	27.4
.75	22.5	23.3	24.1	24.8	25.6	26.3	27.1	27.8	28.5	29.2
.80	24.5	25.3	26.0	26.8	27.5	28.3	29.0	29.7	30.4	31.1
.85	26.5	27.2	28.0	28.7	29.4	30.2	30.9	31.6	32.2	32.9
.90	28.4	29.2	29.9	30.6	31.3	32.0	32.7	33.4	34.1	34.7
.95	30.4	31.1	31.8	32.5	33.2	33.9	34.6	35.2	35.9	36.5
1.00	32.3	33.0	33.7	34.4	35.1	35.7	36.4	37.0	37.7	38.3
1.05	34.2	34.9	35.6	36.2	36.9	37.6	38.2	38.8	39.4	40.0
1.10	36.1	36.7	37.4	38.1	38.7	39.3	40.0	40.6	41.2	41.8
1.15	37.9	38.6	39.2	39.9	40.5	41.1	41.7	42.3	42.9	43.5
1.20	39.7	40.4	41.0	41.7	42.3	42.9	43.5	44.0	44.6	45.1
1.25	41.5	42.2	42.8	43.4	44.0	44.6	45.2	45.7	46.3	46.8
1.30	43.3	43.9	44.5	45.1	45.7	46.3	46.8	47.4	47.9	48.4
1.35	45.1	45.7	46.3	46.8	47.4	47.9	48.5	49.0	49.5	50.0
1.40	46.8	47.4	47.9	48.5	49.0	49.6	50.1	50.6	51.1	51.6
1.45	48.5	49.0	49.6	50.1	50.7	51.2	51.7	52.2	52.7	53.2
1.50	50.1	50.7	51.2	51.8	52.3	52.8	53.3	53.7	54.2	54.7
1.55	51.8	52.3	52.8	53.3	53.8	54.3	54.8	55.3	55.7	56.2
1.60	53.4	53.9	54.4	54.9	55.4	55.9	56.3	56.8	57.2	57.6
1.65	54.9	55.4	55.9	56.4	56.9	57.3	57.8	58.2	58.6	59.1
1.70	56.5	57.0	57.5	57.9	58.4	58.8	59.2	59.7	60.1	60.5
1.75	58.0	58.5	58.9	59.4	59.8	60.2	60.7	61.1	61.5	61.8
2.00	65.0	65.4	65.8	66.2	66.6	66.9	67.3	67.6	67.9	68.3
2.25	71.3	71.6	71.9	72.2	72.5	72.8	73.1	73.4	73.7	73.9
2.50	76.7	77.0	77.2	77.5	77.7	78.0	78.2	78.4	78.7	78.9
2.75	81.4	81.6	81.8	82.0	82.2	82.4	82.6	82.7	82.9	83.1
3.00	85.3	85.4	85.6	85.8	85.9	86.1	86.2	86.4	86.5	86.6
3.50	91.2	91.3	91.4	91.5	91.6	91.6	91.7	91.8	91.9	92.0
4.00	95.0	95.0	95.1	95.2	95.2	95.3	95.3	95.4	95.4	95.4
4.50	97.3	97.3	97.4	97.4	97.4	97.5	97.5	97.5	97.5	97.6
5.00	98.6	98.6	98.7	98.7	98.7	98.7	98.7	98.7	98.7	98.8

TABLE 9   (*Continued*)

$\sigma\sqrt{T}$	1.02	1.04	1.06	1.08	1.10	1.12	1.14	1.16	1.18
.05	3.1	4.5	6.0	7.5	9.1	10.7	12.3	13.8	15.3
.10	5.0	6.1	7.3	8.6	10.0	11.3	12.7	14.1	15.4
.15	7.0	8.0	9.1	10.2	11.4	12.6	13.8	15.0	16.2
.20	8.9	9.9	10.9	11.9	13.0	14.1	15.2	16.3	17.4
.25	10.9	11.8	12.8	13.7	14.7	15.7	16.7	17.7	18.7
.30	12.8	13.7	14.6	15.6	16.5	17.4	18.4	19.3	20.3
.35	14.8	15.6	16.5	17.4	18.3	19.2	20.1	21.0	21.9
.40	16.7	17.5	18.4	19.2	20.1	20.9	21.8	22.6	23.5
.45	18.6	19.4	20.3	21.1	21.9	22.7	23.5	24.3	25.1
.50	20.5	21.3	22.1	22.9	23.7	24.5	25.3	26.1	26.8
.55	22.4	23.2	24.0	24.8	25.5	26.3	27.0	27.8	28.5
.60	24.3	25.1	25.8	26.6	27.3	28.1	28.8	29.5	30.2
.65	26.2	27.0	27.7	28.4	29.1	29.8	30.5	31.2	31.9
.70	28.1	28.8	29.5	30.2	30.9	31.6	32.3	32.9	33.6
.75	29.9	30.6	31.3	32.0	32.7	33.3	34.0	34.6	35.3
.80	31.8	32.4	33.1	33.8	34.4	35.1	35.7	36.3	36.9
.85	33.6	34.2	34.9	35.5	36.2	36.8	37.4	38.0	38.6
.90	35.4	36.0	36.6	37.3	37.9	38.5	39.1	39.6	40.2
.95	37.2	37.8	38.4	39.0	39.6	40.1	40.7	41.3	41.8
1.00	38.9	39.5	40.1	40.7	41.2	41.8	42.4	42.9	43.4
1.05	40.6	41.2	41.8	42.4	42.9	43.5	44.0	44.5	45.0
1.10	42.3	42.9	43.5	44.0	44.5	45.1	45.6	46.1	46.6
1.15	44.0	44.6	45.1	45.6	46.2	46.7	47.2	47.7	48.2
1.20	45.7	46.2	46.7	47.3	47.8	48.3	48.7	49.2	49.7
1.25	47.3	47.8	48.4	48.8	49.3	49.8	50.3	50.7	51.2
1.30	48.9	49.4	49.9	50.4	50.9	51.3	51.8	52.2	52.7
1.35	50.5	51.0	51.5	52.0	52.4	52.9	53.3	53.7	54.1
1.40	52.1	52.6	53.0	53.5	53.9	54.3	54.8	55.2	55.6
1.45	53.6	54.1	54.5	55.0	55.4	55.8	56.2	56.6	57.0
1.50	55.1	55.6	56.0	56.4	56.8	57.2	57.6	58.0	58.4
1.55	56.6	57.0	57.4	57.8	58.2	58.6	59.0	59.4	59.7
1.60	58.0	58.5	58.9	59.2	59.6	60.0	60.4	60.7	61.1
1.65	59.5	59.9	60.2	60.6	61.0	61.4	61.7	62.1	62.4
1.70	60.9	61.2	61.6	62.0	62.3	62.7	63.0	63.4	63.7
1.75	62.2	62.6	62.9	63.3	63.6	64.0	64.3	64.6	64.9
2.00	68.6	68.9	69.2	69.5	69.8	70.0	70.3	70.6	70.8
2.25	74.2	74.4	74.7	74.9	75.2	75.4	75.6	75.8	76.0
2.50	79.1	79.3	79.5	79.7	79.9	80.0	80.2	80.4	80.6
2.75	83.3	83.4	83.6	83.7	83.9	84.0	84.2	84.3	84.4
3.00	86.8	86.9	87.0	87.1	87.3	87.4	87.5	87.6	97.7
3.50	92.1	92.1	92.2	92.3	92.4	92.4	92.5	92.6	92.6
4.00	95.5	95.5	95.6	95.6	95.7	95.7	95.7	95.8	95.8
4.50	97.6	97.6	97.6	97.6	97.7	97.7	97.7	97.7	97.8
5.00	98.8	98.8	98.8	98.8	98.8	98.8	98.8	98.8	98.9

MP/CV(SP)

TABLE 9 (*Continued*)

$\sigma\sqrt{T}$	MP/CV(SP)									
	1.20	1.25	1.30	1.35	1.40	1.45	1.50	1.75	2.00	2.50
.05	16.7	20.0	23.1	25.9	28.6	31.0	33.3	42.9	50.0	60.0
.10	16.8	20.0	23.1	25.9	28.6	31.0	33.3	42.9	50.0	60.0
.15	17.4	20.4	23.3	26.0	28.6	31.1	33.3	42.9	50.0	60.0
.20	18.5	21.2	23.9	26.4	28.9	31.2	33.5	42.9	50.0	60.0
.25	19.8	22.3	24.7	27.1	29.4	31.7	33.8	42.9	50.0	60.0
.30	21.2	23.5	25.8	28.1	30.2	32.3	34.3	43.1	50.1	60.0
.35	22.7	24.9	27.1	29.2	31.2	33.2	35.1	43.5	50.2	60.0
.40	24.3	26.4	28.4	30.4	32.3	34.2	36.0	44.0	50.5	60.1
.45	25.9	27.9	29.8	31.7	33.5	35.3	37.0	44.6	50.8	60.2
.50	27.6	29.5	31.3	33.1	34.8	36.4	38.1	45.3	51.3	60.4
.55	29.2	31.0	32.8	34.5	36.1	37.7	39.2	46.1	51.9	60.7
.60	30.9	32.6	34.3	35.9	37.5	39.0	40.4	47.0	52.5	61.0
.65	32.6	34.2	35.8	37.4	38.9	40.3	41.7	48.0	53.3	61.4
.70	34.2	35.8	37.3	38.8	40.3	41.6	43.0	49.0	54.0	61.9
.75	35.9	37.4	38.9	40.3	41.7	43.0	44.3	50.0	54.9	62.4
.80	37.5	39.0	40.4	41.8	43.1	44.4	45.6	51.1	55.8	63.0
.85	39.2	40.6	41.9	43.3	44.5	45.8	46.9	52.2	56.7	63.6
.90	40.8	42.1	43.5	44.7	46.0	47.1	48.3	53.3	57.6	64.3
.95	42.4	43.7	45.0	46.2	47.4	48.5	49.6	54.5	58.6	65.0
1.00	44.0	45.2	46.5	47.6	48.8	49.9	50.9	55.6	59.5	65.7
1.05	45.5	46.8	48.0	49.1	50.2	51.2	52.2	56.7	60.5	66.5
1.10	47.1	48.3	49.4	50.5	51.6	52.6	53.5	57.9	61.5	67.2
1.15	48.6	49.8	50.9	51.9	52.9	53.9	54.9	59.0	62.5	68.0
1.20	50.1	51.3	52.3	53.3	54.3	55.2	56.1	60.2	63.5	68.8
1.25	51.6	52.7	53.7	54.7	55.7	56.6	57.4	61.3	64.5	69.6
1.30	53.1	54.1	55.1	56.1	57.0	57.9	58.7	62.4	65.5	70.4
1.35	54.6	55.6	56.5	57.4	58.3	59.1	59.9	63.5	66.5	71.1
1.40	56.0	56.9	57.9	58.7	59.6	60.4	61.2	64.8	67.5	71.9
1.45	57.4	58.3	59.2	60.0	60.9	61.6	62.4	65.7	68.4	72.7
1.50	58.8	59.7	60.5	61.3	62.1	62.9	63.6	66.8	69.4	73.5
1.55	60.1	61.0	61.8	62.6	63.3	64.1	64.7	67.8	70.3	74.3
1.60	61.4	62.3	63.1	63.8	64.5	65.2	65.9	68.8	71.3	75.1
1.65	62.7	63.5	64.3	65.0	65.7	66.4	67.0	69.9	72.2	75.9
1.70	64.0	64.8	65.5	66.2	66.9	67.5	68.2	70.9	73.1	76.6
1.75	65.3	66.0	66.7	67.4	68.0	68.7	69.2	71.9	74.0	77.4
2.00	71.1	71.7	72.3	72.9	73.4	73.9	74.4	76.5	78.3	81.0
2.25	76.3	76.8	77.2	77.7	78.1	78.5	78.9	80.6	82.1	84.3
2.50	80.7	81.1	81.5	81.9	82.2	82.6	82.9	84.3	85.4	87.2
2.75	84.6	84.9	85.2	85.5	85.8	86.0	86.3	87.4	88.3	89.7
3.00	87.8	88.1	88.3	88.5	88.8	89.0	89.2	90.0	90.7	91.8
3.50	92.7	92.8	93.0	93.1	93.3	93.4	93.5	94.0	94.4	95.1
4.00	95.8	95.9	96.0	96.1	96.2	96.2	96.3	96.6	96.8	97.2
4.50	97.8	97.8	97.9	97.9	97.9	98.0	98.0	98.2	98.3	98.5
5.00	98.9	98.9	98.9	98.9	99.0	99.0	99.0	99.1	99.1	99.2

TABLE 10   Commutation Table (Interest of 5%)

x	l_x	d_x	q_x	D_x	N_x	C_x	M_x	L.E.
0	100,000	1,260	0.012600	100,000,000	1,992,208.86	1,200.00	5,132.91	74.4
1	98,740	90	0.000932	94,038.10	1,892,208.86	83.45	3,932.91	74.3
2	98,648	64	0.000649	89,476.64	1,798,170.76	55.29	3,849.46	73.4
3	98,584	49	0.000497	85,160.57	1,708,694.12	40.31	3,794.18	72.4
4	98,535	40	0.000406	81,064.99	1,623,533.55	31.34	3,753.87	71.5
5	98,495	36	0.000366	77,173.41	1,542,468.56	26.86	3,722.53	70.5
6	98,459	33	0.000335	73,471.62	1,465,295.15	23.45	3,695.66	69.5
7	98,426	30	0.000305	69,949.52	1,391,823.53	20.31	3,672.21	68.5
8	98,396	26	0.000264	66,598.29	1,321,874.01	16.76	3,652.90	67.6
9	98,370	23	0.000234	63,410.18	1,255,275.73	14.12	3,635.14	66.6
10	98,347	19	0.000193	60,376.53	1,191,865.55	11.11	3,621.02	65.6
11	98,328	19	0.000193	57,490.35	1,131,489.02	10.58	3,609.92	64.6
12	98,309	24	0.000244	54,742.13	1,073,998.67	12.73	3,599.34	63.6
13	98,285	37	0.000376	52,122.63	1,019,256.54	18.69	3,586.61	62.6
14	98,248	52	0.000529	49,621.92	967,133.91	25.01	3,567.92	61.7
15	98,196	67	0.000682	47,233.95	917,511.99	30.69	3,542.91	60.7
16	98,129	82	0.000836	44,954.03	870,278.04	35.78	3,512.21	59.7
17	98,047	94	0.000959	42,777.58	825,324.01	39.06	3,476.44	58.8
18	97,953	102	0.001041	40,701.49	782,546.43	40.36	3,437.38	57.8
19	97,851	110	0.001124	38,722.96	741,844.94	41.46	3,397.01	56.9
20	97,741	118	0.001207	36,837.55	703,121.97	42.36	3,355.56	56.0
21	97,623	124	0.001270	35,041.03	666,284.42	42.39	3,313.20	55.0
22	97,499	129	0.001323	33,330.02	631,243.39	42.00	3,270.81	54.1
23	97,370	130	0.001335	31,700.88	597,913.37	40.31	3,228.81	53.2
24	97,240	130	0.001337	30,151.00	566,212.49	38.39	3,188.50	52.2
25	97,110	128	0.001318	28,676.85	536,061.49	36.00	3,150.11	51.3
26	96,982	126	0.001299	27,275.29	507,384.63	33.75	3,114.12	50.4
27	96,856	126	0.001301	25,942.72	480,109.35	32.14	3,080.37	49.4
28	96,730	126	0.001303	24,675.21	454,166.63	30.61	3,048.23	48.5
29	96,604	127	0.001315	23,469.59	429,491.42	29.38	3,017.61	47.6
30	96,477	127	0.001316	22,322.60	406,021.84	27.99	2,988.23	46.6
31	96,350	130	0.001349	21,231.64	383,699.23	27.28	2,960.24	45.7
32	96,220	132	0.001372	20,193.32	362,467.60	26.38	2,932.96	44.7
33	96,088	137	0.001326	19,205.35	342,274.28	26.08	2,906.58	43.8
34	95,951	143	0.001490	18,264.73	323,068.92	25.92	2,880.50	42.9
35	95,808	153	0.001597	17,369.06	304,804.19	26.42	2,854.57	41.9
36	95,655	163	0.001704	16,515.54	287,435.13	26.80	2,828.16	41.0
37	95,492	175	0.001833	15,702.29	270,919.58	27.41	2,801.35	40.1
38	95,317	188	0.001972	14,927.15	255,217.30	28.04	2,773.95	39.1
39	95,129	203	0.002134	14,188.30	240,290.14	28.84	2,745.91	38.2
40	94,926	220	0.002318	13,483.83	226,101.85	29.76	2,717.07	37.3
41	94,706	241	0.002545	12,811.98	212,618.02	31.05	2,687.31	36.4
42	94,465	264	0.002795	12,170.83	199,806.04	32.39	2,656.26	35.5
43	94,201	288	0.003057	11,558.88	187,635.20	33.66	2,623.87	34.6
44	93,913	314	0.003344	10,974.80	176,076.33	34.95	2,590.21	33.7
45	93,599	343	0.003665	10,417.24	165,101.53	36.36	2,555.26	32.8
46	93,256	374	0.004010	9,884.83	154,684.29	37.76	2,518.91	31.9
47	92,882	410	0.004414	9,376.36	144,799.46	39.42	2,481.15	31.0
48	92,472	451	0.004877	8,890.45	135,423.10	41.30	2,441.73	30.1
49	92,021	495	0.005379	8,425.80	126,532.64	43.17	2,400.44	29.3
50	91,526	540	0.005900	7,981.41	118,106.84	44.85	2,357.27	28.4

TABLE 10   (*Continued*)

x	l_x	d_x	q_x	D_x	N_x	C_x	M_x	L.E.
51	90,986	584	0.006419	7,556.49	110,125.43	46.19	2,312.43	27.6
52	90,402	631	0.006980	7,150.47	102,568.94	47.53	2,266.23	26.8
53	89,771	684	0.007619	6,762.44	95,418.47	49.07	2,218.70	26.0
54	89,087	739	0.008295	6,391.34	88,656.03	50.49	2,169.63	24.1
55	88,348	797	0.009021	6,036.50	82,264.69	51.86	2,119.13	24.4
56	87,551	856	0.009777	5,697.19	76,228.19	53.05	2,067.27	23.6
57	86,695	919	0.010600	5,372.84	70,531.00	54.24	2,014.22	22.8
58	85,776	987	0.011507	5,062.75	65,158.16	55.48	1,959.98	22.0
59	84,789	1,063	0.012537	4,766.18	60,095.41	56.91	1,904.50	21.3
60	83,726	1,143	0.013676	4,482.32	55,329.23	58.38	1,847.58	20.5
61	82,581	1,233	0.014931	4,210.49	50,846.91	59.87	1,789.21	19.8
62	81,348	1,324	0.016276	3,950.12	46,636.42	61.23	1,729.34	19.1
63	80,024	1,415	0.017682	3,700.79	42,686.30	62.32	1,668.11	18.4
64	78,609	1,502	0.019107	3,462.24	38,985.51	63.00	1,605.79	17.7
65	77,107	1,587	0.020582	3,234.37	35,523.27	63.40	1,542.78	17.0
66	75,520	1,674	0.022166	3,016.95	32,288.90	63.69	1,479.38	16.3
67	73,846	1,764	0.023888	2,809.60	29,271.95	63.92	1,415.69	15.7
68	72,082	1,864	0.025859	2,611.89	26,462.35	64.33	1,351.78	15.1
69	70,218	1,970	0.028055	2,423.19	23,850.47	64.75	1,287.45	14.4
70	68,248	2,083	0.030521	2,243.05	21,427.28	65.20	1,222.70	13.8
71	66,165	2,193	0.033144	2,071.04	19,184.23	65.37	1,157.50	13.2
72	63,972	2,299	0.035938	1,907.04	17,113.19	65.27	1,092.13	12.6
73	61,673	2,394	0.038818	1,750.96	15,206.15	64.73	1,026.86	12.1
74	59,279	2,480	0.041836	1,602.85	13,455.19	63.86	962.13	11.5
75	56,799	2,560	0.045071	1,462.66	11,852.34	62.78	898.26	11.0
76	54,239	2,640	0.048673	1,330.22	10,389.68	61.66	835.48	10.5
77	51,599	2,721	0.052734	1,205.22	9,059.46	60.53	773.81	9.9
78	48,878	2,807	0.057429	1,087.30	7,854.24	59.47	713.29	9.4
79	46,071	2,891	0.062751	976.05	6,766.94	58.33	653.82	8.9
80	43,180	2,972	0.068828	871.24	5,790.89	57.11	595.49	8.5
81	40,208	3,036	0.075507	772.64	4,919.65	55.56	538.37	8.0
82	37,172	3,077	0.082777	680.29	4,147.00	53.63	482.81	7.6
83	34,095	3,083	0.090424	594.26	3,466.72	51.18	429.18	7.2
84	31,012	3,052	0.098414	514.79	2,872.45	48.25	378.00	6.8
85	27,960	2,999	0.107260	442.02	2,357.66	45.15	329.76	6.5
86	24,961	2,923	0.117103	375.82	1,915.64	41.91	284.60	6.1
87	22,038	2,803	0.127189	316.01	1,539.82	38.28	242.69	5.8
88	19,235	2,637	0.137094	262.68	1,223.81	34.30	204.41	5.5
89	16,598	2,444	0.147247	215.88	961.12	30.27	170.11	5.2
90	14,154	2,246	0.158683	175.32	745.24	26.50	139.84	4.9
91	11,908	2,045	0.171733	140.48	569.92	22.98	113.34	4.7
92	9,863	1,831	0.185643	110.81	429.44	19.59	90.36	4.4
93	8,032	1,608	0.200199	85.94	318.63	16.39	70.77	4.2
94	6,424	1,381	0.214975	65.47	232.68	13.40	54.39	4.0
95	5,043	1,159	0.229824	48.94	167.22	10.71	40.98	3.8
96	3,884	945	0.243306	35.90	118.27	8.32	30.27	3.7
97	2,939	754	0.256550	25.87	82.37	6.32	21.95	3.5
98	2,185	587	0.268650	18.32	56.60	4.69	15.63	3.4
99	1,598	448	0.280350	12.76	38.18	3.41	10.94	3.3
100	1,150	335	0.291304	8.75	25.42	2.43	7.53	3.2

Source: Based on R. Muksian (2003). *Mathematics of Interest Rates, Insurance, Social Security, and Pensions.* Prentice Hall, Upper Saddle River, NJ.

INDEX

Mathematical Finance, First Edition. M. J. Alhabeeb.
© 2012 John Wiley & Sons, Inc. Published 2012 by John Wiley & Sons, Inc.